FOSSIL MEN

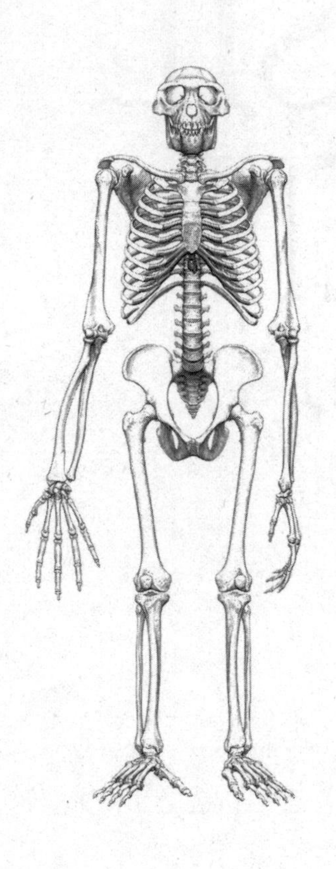

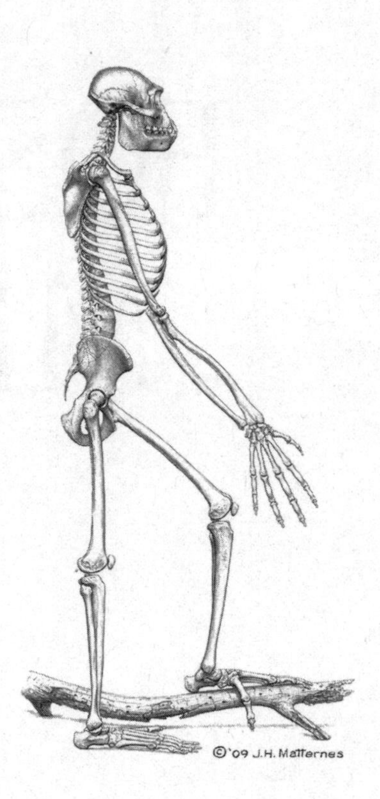
©'09 J.H. Matternes

FOSSIL MEN

The Quest for the Oldest Skeleton and the Origins of Humankind

KERMIT PATTISON

wm

WILLIAM MORROW

An Imprint of HarperCollins*Publishers*

HarperCollins books may be purchased for educational, business, or sales promotional use. For information, please e-mail the Special Markets Department at SPsales@harpercollins.com.

FIRST EDITION

Design by Elina Cohen
Frontispiece image by © 2009 J. H. Matternes

Library of Congress Cataloging-in-Publication Data

Names: Pattison, Kermit, author.
Title: Fossil men : the quest for the oldest skeleton and the origins of humankind / Kermit Pattison.
Description: First edition. | New York, NY : William Morrow, [2020] | Includes bibliographical references and index. | Summary: "Fossil Men is the riveting science-adventure story of the brilliant team who discovered the "Ardi" skeleton, a human more than a million years older than the famous Lucy, and their 20-year quest to redefine our understanding of human evolution"—Provided by publisher.
Identifiers: LCCN 2019056406 (print) | LCCN 2019056407 (ebook) | ISBN 9780062410283 (hardcover) | ISBN 9780062410290 (trade paperback) | ISBN 9780062410306 (ebook)
Subjects: LCSH: Fossil hominids. | Human evolution.
Classification: LCC GN282 .P38 2020 (print) | LCC GN282 (ebook) | DDC 569.9—dc23
LC record available at https://lccn.loc.gov/2019056406
LC ebook record available at https://lccn.loc.gov/2019056407

ISBN 978-0-06-241028-3

20 21 22 23 24 LSC 10 9 8 7 6 5 4 3 2 1

For Maja

A man has come; a quarrel will come.

—ETHIOPIAN PROVERB

Contents

Introduction
T. REX

This book is a history of science and a detective story about the most fundamental mystery of all: where did we come from? Like any good mystery, it begins with a body.

My journey started when I became intrigued by an ancient cold case—the oldest known skeleton of a member of the human family. In 2012, I flew out to the University of California at Berkeley to meet one of the world's most successful fossil hunters, and talk about his most recent major discovery—a 4.4-million-year-old skeleton of the species *Ardipithecus ramidus*, nicknamed Ardi. Initially, I did not intend to write much about Ardi, which I envisioned as just a bit of background to the more interesting drama later in human evolution. The more I learned, however, the more unsettled I became because Ardi seemed to refute so many prevailing theories about evolution.

Ardi was an inconvenient woman, one who disturbed scholars of human origins more than many cared to admit. Her skeleton challenged core beliefs about how we became human, how our ancestors split from the other apes, how we came to stand upright, how we evolved our nimble hands, and whether the savannas were truly the crucibles of humanity as depicted in countless museum dioramas and textbooks. Most importantly, it showed that these early human ancestors looked surprisingly unlike the modern chimpanzees often touted as models of the human past. In some aspects of anatomy, humans turned out to be more primitive than living African apes, a finding that reversed forty years of conventional wisdom. "It is so rife with anatomical surprises," the discovery team reported of the skeleton, "that no one could have imagined it without direct fossil evidence."

Fossils are often called bones of contention. But the odd thing about this skeleton was not controversy but the *lack* of it. Something very curious happened after Ardi was revealed to the world in 2009. A bombshell dropped and then . . . silence. The very people who *should* have been most excited by this discovery seemed to shrug off the findings. As I later discovered, there were myriad reasons: some peers vehemently disagreed with the conclusions; others dreaded engaging in arguments likely to end unpleasantly; and some sought to consign the fossil to irrelevance by ignoring it. Ardi, and the team that discovered her, seemed to be personae non gratae. One of them was even called "He-Who-Must-Not-Be-Named."

My curiosity was aroused. Anybody who must not be named certainly must be interviewed.

TIM WHITE, THE FRONT MAN OF THE ARDI TEAM, WAS A PALEOANTHROPOLOgist, a scientist of the human fossil record, with a reputation for having a razor intellect, hair-trigger bullshit detector, short temper, long list of discoveries, and longer list of enemies. His department webpage at the University of California at Berkeley showed a picture that made him look like a warlord in the Ethiopian badlands surrounded by a security detail brandishing assault rifles.

White ignored my initial messages, except for a curt reply that he was busy in Ethiopia and would not make himself available. I persisted, and months later, he finally agreed to talk. He recently had returned from his annual fossil-hunting mission. and suffered from a fever with cyclical headaches. Ordered by his doctor to remain in bed, White cancelled all his classes for two weeks. Yet he rose and kept his appointment. Later I realized this was characteristic: White habitually sacrificed his own comfort, appearance, and health to advance his scientific mission.

To get to his office, I entered the massive Neo-Babylonian complex of the Valley Life Sciences Building on the Berkeley campus and walked past a life-size skeleton cast of a *Tyrannosaurus rex*. Like Ardi, it was an extinct bipedal creature unimaginable to modern eyes until it was actually found. I passed a small display with a replica of the famous Lucy skeleton, a human ancestor whose species, *Australopithecus afarensis*, White had co-named three decades earlier. I took an elevator to the fifth floor, knocked, and heard a groan from within. The door opened and there stood a skinny man, uncombed and rumpled, looking like he'd been pulled from the bottom of the hamper.

I held out my hand to introduce myself. He bumped it with a clenched fist.

"Ethiopian handshake," he deadpanned. "You don't want to get what I have."

A huge snake skin hung on the office wall. White had killed and eaten the puff adder in Tanzania in his younger days, back when he remained on friendly terms with at least some of the Leakeys, the famous fossil dynasty and his former employers. Nearby hung a Victorian-era caricature of Darwin parodied as an ape—a reminder that scientific achievement is not always honored by contemporaries. Another cartoon depicted a Victorian evolutionist who more closely matched White in temperament, Thomas Henry Huxley, a pugnacious anatomist known as "Darwin's bulldog." A framed photo contained a snapshot of a hunchbacked man with an assault rifle; I later learned it was a friend from the Ethiopian desert who'd been killed in tribal warfare.

White gestured for me to take a seat in an office lined with books and replicas of fossil skulls. We talked . . . and talked . . . and talked. About science, there seemed to be no end to his willingness to give time, fever be damned. He was encyclopedic, sarcastic, and uproariously impolitic—he labeled one colleague a moron, another a bottom feeder, another a bozo, and chucked many more into a bulk bin of *assholes.* White seemed engaged in perpetual struggle against somebody—celebrity scientists, academic critics, university administrators, Ethiopian antiquities officials, journal editors, incompetents of all types, and so on. The loathing flowed both ways: some colleagues refused to attend conferences if White was present. At that moment, he was engaged in litigation against his own employer, after having sued the University of California over a pair of 9,500-year-old skeletons that the university wanted to hand over to Native American tribes. (White and his co-plaintiffs complained that the university had favored ideology over science.)

His longtime Ethiopian colleague Berhane Asfaw later explained:

> Do you know why he is feared by his colleagues? If there is something wrong in the science, he will not be diplomatic. He will just tell you straightforward—*this is totally wrong.* Most people will never say that and try to dodge the issue. Even if they are going to control his resources and deny him grant money, Tim will say, "Forget it, the guy is wrong. If we don't tell him it's wrong, it is like disseminating false information." That is why most people hate him. He is not nice to bad science.

These observations were echoed by geologist Maurice Taieb:

> Tim White is *brutal.* He's a real scientist. His literature will stay forever. He doesn't care about publishing books, being on media, et cetera. He wants to do the job before popularizing. People think he wants to keep for himself. *No, no!* He wants to be *sure.*

There was another view, of course. Don Johanson, who rose to fame with his discovery of Lucy, described his estranged partner White in dark terms:

> He feels he has done so much work on these fossils himself that our colleagues who sit in their little air conditioned offices—as Tim would say—don't deserve to see these fossils. And besides, he would say very often, even if they *did* see the fossils they wouldn't even know what they were looking at. He degrades his colleagues with his ad hominem arguments, making people feel inadequate.

Meave Leakey, the reigning matriarch of the world-famous fossil-hunting dynasty, described White with a note of sympathy:

> He sees the world as being very against him. He doesn't like to go to lots of meetings. It's a shame—he is a hell of a good scientist, has really good fossils, and has lots to contribute, but makes it very difficult because he's so defensive, and then he gets nasty.

In the course of writing this book, I would spend countless hours with White, his colleagues, and rivals. I got to know him over eight years and he proved—contrary to my initial impressions and those of adversaries—enormously generous when it came to his science, on his exacting terms. White would prove to be perhaps the single most intense character I have ever encountered in my half century on this planet. He was profane, persnickety, judgmental, wildly irreverent, crazy hilarious, and a rough-edged field guy who made some of his celebrity adversaries seem like cardboard cutouts. I must confess, I found him quite likable and fabulously entertaining—even when I, too, endured his flayings or outbursts. "And you should have met him twenty years ago!" laughed his wife, Leslea Hlusko, when I later spoke with her. "He's *way* toned down from when I first met him."

That visit to the Berkeley lair opened a door into a strange world—past and present. Behind the man Ethiopians called *Teem* stood a *team,* many with backgrounds almost unimaginable to western readers. They included a man who survived imprisonment and torture during the Ethiopian Red Terror purges and became the first person in his country to earn a doctorate in physical anthropology; an Ethiopian peasant who became a U.S. government scientist with a top secret clearance and used personal vacation time to conduct expeditions in his home country; a former evangelical Christian turned-evolutionary theorist; tribal gunmen who littered the desert with skeletons of enemies and later searched that same land for fossil bones; and a Japanese polymath who became so absorbed in rebuilding the Ardi skull that he didn't go home at night.

Several Ardi investigators were veterans of the team that interpreted the Lucy skeleton in the 1970s and 1980s. In terms of name recognition, Lucy remains the best-known human ancestor. In terms of science, however, Ardi was more revelatory. Lucy represented a new species within an already-known genus (a genus is a taxonomic category that may include multiple species that are closely related and share similar adaptations); she was an older variation on an anatomical theme that gradually came to light over half a century. Ardi represented something entirely novel—not only a new species but a new genus and a hitherto-unknown hybrid of arboreal ape and terrestrial biped. Due to the secretive nature of the team and their all-at-once publishing strategy, it essentially appeared overnight. Despite the best efforts of detractors and the intellectual resistance typically provoked by such disruptions to the status quo, it likely will go down in history as one of the major discoveries of our era.

This does not mean that every one of the claims presented by the interpreters will stand forever—they rarely do. Fossil milestones tend to outlast the arguments made on their behalf. Even so, her discovery revealed an evolutionary stage never seen before and demanded a wholesale rethinking of our origins. In their quest, Tim White, Berhane Asfaw, and their team crossed paths with most of the major personalities, debates, and discoveries in anthropology over the last half century.

The events described here coincided with Ethiopia's emergence as one of the great fossil-producing nations in the world. The country is layered with geologic and human history, from mountains shrouded in clouds to deserts below sea level. It is a crossroads of ancient Christianity, Islam, and Judaism; the ruins of a once-great kingdom in the African highlands; the birthplace of Blue Nile; and—thanks to its unique geology—producer of many of the oldest fossils in the human family. This tale played out as the country endured two revolutions, multiple wars, shifting geopolitical alliances, and evolved from feudal kingdom to Marxist dictatorship to modern "developmental" state (as political economists label such governments aggressively pursuing economic growth)—all united by a longstanding tradition of authoritarianism.

The Ardi investigators spent years in the crossfire of warring nomadic tribes. They labored with dental tools and porcupine quills to rescue fragile fossils from blocks of earth, and reassembled a skeleton from fragments. At times, their mission was nearly terminated by shootings or hostile bureaucrats. At one point, the Ethiopian government revoked their permit for fieldwork, locked them out of the national museum, and prohibited them from examining the fossils they had found. The scientists fought with academic rivals who complained that they were hoarding one of the world's most valuable fossils and hadn't shared a useful detail for more than a decade.

The Ardi team operated as a largely self-contained unit, indifferent or sometimes hostile to the larger academic "profession" (they supplied the quotation marks). They strove to integrate Africans into a science that previously granted no seats at the table for indigenous people from fossil-producing nations. They labored to establish Ethiopia as an international center of science—not just an exporter of fossils for the convenience of western scholars. Over the years, the team scored a record of discoveries beyond almost any rival, much as critics loathed to admit it. White—always ready to map fault lines—saw his crew as an embattled minority in a struggle between true *scientists* and *careerists*. In turn, opponents denounced him as a self-righteous scold who practiced an antiquated form of science. Something about the pursuit of our own origins arouses great passions not seen in other disciplines. In an ideal world, the task should be left to more dispassionate investigators but, since no other species has volunteered, the job is left to us imperfect humans.

I SPENT EIGHT YEARS REPORTING THIS STORY. I INTERVIEWED PEOPLE around the world, burrowed into long-forgotten archives, read hundreds of papers, and joined two expeditions into the field. The year after the initial meeting in Berkeley, I rode into the Afar Depression of Ethiopia beside White in a caravan of safari vehicles. He warned the expedition would probe particularly dangerous areas that the team had avoided for most of the last twenty years. Locals had shot at them more than once—most recently the previous year. Crammed between us sat one of the team's best fossil scouts, an Afar man named Elema. (Both roared with laughter as they recounted their first meeting when Elema stormed into camp with two guns and tried to expel the expedition from his territory.) The landscape shimmered with heat as the caravan threaded through the brush and needle-sharp thorn trees clawed at the car with a discordant *screeeeeccch*. From that moment onward, White kept up a running tutorial: Duck when the car scraped past the bushes to avoid laceration. Watch where you step to avoid vipers, cobras, and scorpions. Take a buddy when you relieve yourself in the bush at night so the hyenas don't sneak up behind and kill you. Through it all, he shared the tricks of the trade that had produced so many discoveries by his team. Use the glint of the sun to find fossils. Always survey walking uphill, never down. Stay with the group and their armed escorts. "If you get lost or separated, the first thing that is going to kill you is thirst," White cautioned, "if the local yokels don't kill you first."

He corrected himself: actually the locals *probably* won't kill you—not anymore. The wild nomads were being modernized, the desert cultivated, and the crusty characters of the science replaced by polished professionals. But then everything became precarious again. As this book was being completed a few years later, Ethiopia entered another period of turbulence, tribal warfare forced the team to suspend fieldwork, and prospects for future discoveries became uncertain. I was lucky enough to catch a glimpse in the twilight before it all vanished, to see the T. rexes, as it were, while they were still flesh and blood. The mere description of the Ardi contrarians, and serious consideration of their ideas, is likely to annoy, even enrage, opponents. Some people refused to have anything to do with me once they realized my reporting concerned Those Who Shall Not Be Named.

The title *Fossil Men*—which may seem an odd choice for a saga about a female skeleton—refers to these central characters, the lead investigators of a team that collected truckloads of old bones and occupied a lonely branch of the science that some peers disdained as outdated. The great irony is that this team proved more forward-thinking than many contemporaries, and fossils never go obsolete—and sometimes force us to write history anew. Once upon a time, "fossil men" was also a term for human ancestors, but the title should not be read as endorsement of bygone sexist language nor a dismissal of the contributions of female scientists on the Ardi team or any other. If anything, this field needs more women.

This is a scientific odyssey that began before there was something called the Internet and which spanned the careers of six U.S. presidents. Critics groused about the long delay and excessive secrecy of the so-called Manhattan Project of paleoanthropology. In the meantime, the Internet revolutionized modern life. Scientists sequenced the human genome, then the chimpanzee genome, and eventually transcribing the code of life became commonplace. Members of the team died from violence or old age.

When Ardi finally was revealed to the world, many people found the skeleton—or at least the arguments presented with it—too outlandish to believe. This story is a voyage to the deep past to encounter ancestors, animals, environments, and even a tree of life unlike any we would recognize in our modern world.

Science is not only a quest for information. It also is a competition between paradigms, or models for understanding nature. *Fossil Men* is, in part, a tale of how scientists discover, analyze, grapple with dissonance, shed old beliefs, and arrive at new understanding—in other words, the evolution of thought. It also is an account of the nonscientific dynamics of human psychology, bias, resentment, rival camps, and tribalism. The Ardi team bet that consensus was a poor predictor of being right. That conventional wisdom wasn't any less wrong because a lot of people happened to believe it. That human intelligence still was capable of better insight than computer-generated analysis. That the old masters of anatomy, whose work had been largely forgotten and entombed in dusty library stacks, sometimes offered better insight into human evolution than the cutting-edge technologies that now commanded the attention of the field. That fossils could provide better information about human evolution than predictions based on molecular biology and modern apes. That "the profession" had ventured too far out on a limb in concocting a narrative of human origins, and they had no choice but to saw it off.

Chapter 1
THE ROOTS OF HUMANITY

Tim White had endless patience for pursuing human ancestors but limited patience for their living descendants. In October 1981, the skinny fossil hunter labored over a line of Land Rovers and trailers in the eucalyptus-shaded driveway of the American Embassy in Addis Ababa, the highland capital of Ethiopia. A crew of Ethiopians and foreign scientists were packing the vehicles with tools, jerrycans, pickaxes, shovels, sieves, and provisions to sustain an expedition for two months in the desert. White dropped to his hands and knees to check the suspension of an overloaded bush car, rose to second-guess how his companions loaded sacks of flour and sugar, and inserted himself in every detail of the mission because they had to get all the logistics right to maximize the odds of success—finding the apelike creatures who begat us.

White was an anthropology professor from the University of California at Berkeley, and at age thirty-one had already made headlines around the world. He had named the oldest species of human ancestor, reconstructed the most ancient skull, and excavated fossil footprints that showed the earliest evidence of upright walking. In this mission, he hoped to find something even older, although just what he couldn't be sure.

White was not the clubbable sort of academic. On the hunt for fossils, he became a sinewy human bullwhip—get moving, get it right, or get the hell out of the way. He swore. He railed against enemies. He took delight in mimicking the follies of adversaries and burst into madcap laughter. He knew how to run chainsaws, fix engines, skin snakes, and survive in the wilderness. The son of a California highway worker, he grew up a blue-collar mountain boy and had bulled his way into academia and the frontiers of his science. One graduate school instructor—now estranged—once remarked that young Tim carried not just a chip on his shoulder but a *log.* Shortly before he flew to Ethiopia, the Berkeley anthropology department issued a stern warning to White about "actions which would diminish collegiality in the department" after an investigation found evidence of his "sarcasm, verbal abuse and rudeness." Yet even detractors had to admit that his monastic devotion to fossils made him, in the words of one mentor, "the best in the business today."

In only a few years since graduate school, White had amassed a publication record that rivaled the giants of his profession. He could recite from memory the catalog number and the anatomical minutiae of every major specimen in the human fossil record—and lament the careless mistakes of whoever cleaned it. He began his career in Kenya as an assistant to Richard Leakey, the scion of the world-famous fossil-hunting family, but their relationship soured after White accused his boss of scientific censorship and stormed out of his office. Next, White moved to the camp of Mary Leakey, the caustic, cigar-smoking, whiskey-drinking matron of the Leakey clan. They worked side-by-side excavating the famous Laetoli footprints, which proved that human ancestors walked upright by 3.6 million years ago. But White grew critical of Mary's field techniques and her views about the human family tree and she grew tired of his hectoring. They parted on bad terms.

Then White distinguished himself as the most dogged member of the team that reconstructed the most famous human ancestor ever found—Lucy, the petite skeleton of a 3.2-million-year-old upright walker with a small brain and an apelike snout. The skeleton had been discovered in Ethiopia in 1974 by American anthropologist Don Johanson, who enlisted White as the expert he most trusted to help unlock the secrets of his newly discovered trove of bones and teeth. Colleagues watched in awe as White rummaged through unsorted crates of rubble, picked out fragments, and pieced together a fossil tooth. Born color-blind, White was acutely sensitive to bone geometry and obsessive about detail. No watchmaker was more exacting. "He'll go far beyond the nth degree to verify everything and anything—to the point where you think it's somewhat pathological," said colleague Steve Ward.

Some skeptics dismissed Lucy as an extinct ape or dead-end lineage. In the end, White's view prevailed: she represented a previously unknown species of human ancestor, or hominid, the family of creatures on our side of the split from the apes.* White and Johanson named her species *Australopithecus afarensis* and declared it the direct ancestor of *all* subsequent members of the human family. In 1981—not long before this expedition getting underway in the embassy driveway—Johanson had published the book *Lucy,* which catapulted the world's oldest skeleton into a household name and proclaimed her "the beginnings of humankind."

Except Lucy couldn't truly be the beginnings of humankind. There had to be older creatures that would reveal how our peculiar primate lineage split from the other African apes, started walking on two feet, and began an evolutionary journey unlike any other

---

* In recent years, scientists have adopted the term *hominin* to refer to all creatures in the human branch after our split from our common ancestor with chimpanzees. This book will retain *hominid* because it was the preferred term during most of the events described here. In short, *hominid* and *human ancestor* will be used interchangeably for all creatures on our side of the chimp-human split—or what scientists call the *human clade*—regardless of whether they were direct ancestors or extinct side branches. Similarly, *human family* should be understood as an informal equivalent to *human clade,* not as a formal taxonomic category.

creature in the animal kingdom. Whatever came before Lucy remained hidden in the Dark Ages—a blank spot in the human fossil record beyond 4 million years ago known as "the Gap." Lucy and *Australopithecus afarensis* seemed to appear out of nowhere. White wanted to find a window into the gap—and that meant taking this mission where few dared to go.

The destination was a little-known territory reported to be littered with bones up to 6 million years old. It lay down in the Afar Depression, the same massive valley that had produced Lucy, a place of scorching heat, wild animals, and gun-wielding nomads. The local tribes had a centuries-old reputation for killing and castrating intruders. In recent years, this remote lowland had become a battleground between Ethiopia's military government and insurgents. Troops had massacred hundreds of Afars—and one foreign anthropologist. With the exception of a brief reconnaissance to prepare for this mission, no fossil expedition had ventured there for the previous four years.

The U.S. embassy compound, perched on a mountain slope above the Ethiopian capital, was a beleaguered Cold War outpost in the midst of a hostile Marxist dictatorship. Huge portraits of Marx, Engels, and Lenin loomed over the central square of the city. The American ambassador and half of his staff had been expelled. Only a skeleton crew remained at the embassy, many of them undercover CIA employees. In the driveway, a U.S. Marine guard in camouflage fatigues leaned against the pole of a basketball hoop and shot the breeze with the scientists as they loaded their cars. An American newspaper correspondent perched in the open rear door of a vehicle and scribbled in his notebook. The resumption of research in the land of Lucy was big news back in the United States.

"With Lucy, you have a creature with a brain size a third as big as modern humans, and yet is walking fully on two legs," White told the reporter. "Is that adaptation a very long one, or does Lucy represent just the beginnings of that trend? . . . The thing that

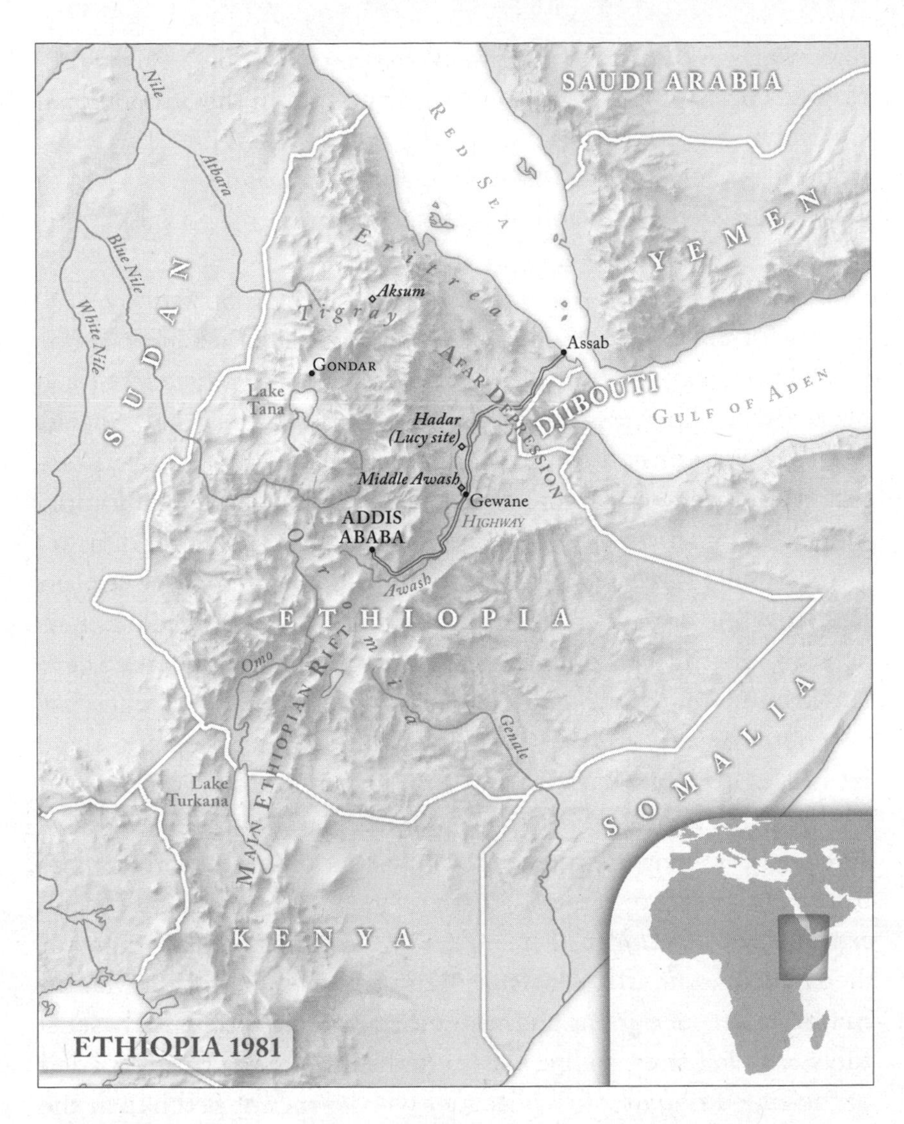

really set humans apart from the apes was this peculiar form of walking around."

Charles Darwin, the father of evolution, had theorized that humans evolved big brains, tool use, and erect walking simultaneously, as a package. But a series of discoveries had destroyed that theory, culminating with Lucy, who showed that upright walking came at least one million years before big brains or stone implements. Like many anthropologists, White suspected our odd form of locomotion was perhaps *the* original distinction that sent our ancestors on their own evolutionary path. "The question is," White said as he loaded the cars, "how far back does that adaptation go?"

The answer waited in the desert.

THE ETHIOPIAN GOVERNMENT HAD PROHIBITED AMERICAN EMBASSY PERsonnel from traveling outside the capital. The scientific team, however, had won permission to explore the Afar Depression, thanks to two of White's companions, one an old-school English archeologist, and the other an Ethiopian student who had survived imprisonment and torture.

Desmond Clark, the leader of the expedition, supervised the loading while dressed in a khaki safari suit. He seemed like a character from central casting who strode through the field with a cane and uttered *Jolly good show! God bless, old boy! Bloody awful people!* without a hint of irony. He was erudite, impeccably polite, and a bit nostalgic for the days when when the sun never set on the British Empire. He packed his bush car with a leather case of stainless steel cups; no matter how far from civilization he ventured, Clark invariably found time—and scarce water—for evening cocktails with his wife, Betty.

Clark was an archeologist at Berkeley, then one of the leading human origins programs on the globe. He specialized in early stone tools, not old bones, and he had invited White to lead the fossil side of the expedition despite colleagues who warned that the cocksure young man could be, to soften it with British understatement, *a bit difficult*. By some accounts, White had arrived in their department like a grenade tossed into a small academic pond. But Clark saw great potential in his protégé. As an empirical scientist, Clark had been perturbed by the "mass of rather airy fairy model building" that had permeated his corner of academia in recent years—and he found a like-minded skeptic in White, a discovery-hungry fossil man with zero tolerance for nonsense. Young Tim was a damn fine scientist with the grit to survive in a place like Ethiopia, where life could be bloody awful.

Clark had spent nearly half a century on the continent. Born in England, he was shipped off to boarding school at age six, and learned Latin, rugby, and rowing. After studying archeology at Cambridge, he took a position in 1938 at the Rhodes-Livingstone Museum near Victoria Falls in what was then the colony of Northern Rhodesia (now Zambia). Clark rowed on the Zambezi River, taking care to avoid truculent hippos and crocodiles, and drank with the expats in the boat club. During World War II, he served in an ambulance crew as the British army expelled the Italian fascist forces from Ethiopia and Somalia. When soldiers dug trenches, garbage pits, and latrines, Clark dropped into the holes and picked up ancient stone tools. By the end of the war, he had filled a couple dozen gasoline containers with artifacts, which became the basis for his doctorate and a major work, *The Prehistoric Cultures of the Horn of Africa*. Clark began his career as an "Africanist" at a time when most scholars dismissed the world's second-largest continent as a backwater of human evolution.

In 1961, he joined the faculty at Berkeley. His arrival coincided with the dawning of a new paradigm of human origins. Biochemical studies were demonstrating a close and jarringly recent relationship between humans, chimpanzees, and gorillas (two apes indigenous to the continent). New fossil discoveries were providing hard evidence of primitive human ancestors in Africa with apelike traits—and far older than fossils from Europe or Asia. Scholars were showing that stone tools in Africa were much more ancient than those elsewhere. Clark argued that the continent represented not only the birthplace of the human lineage but the seedbed of "just about every significant biological and cultural advance" including upright locomotion, stone tools, animal butchery, brain expansion, and more—a view that later became conventional wisdom. Without Africa, argued Clark, there would be no prehistory, no civilization, and no humanity. And no single place would better document the early chapters of the human story than the destination of the 1981 expedition—the Afar Depression.

"If we find hominids," Clark said as his crew readied the cars, "that would be marvelous."

Getting to that point had not been so marvelous. In recent years, Ethiopia had been racked by revolution, making fieldwork almost impossible. Until 1974, the country of 30 million, mostly peasant farmers, was ruled by a man whose grandiose titles trailed his name like a retinue of royal attendants—Emperor Haile Selassie, King of Kings, Elect of God, Conquering Lion of the Tribe of Judah. (Before being crowned, he was a nobleman named Ras Tafari, who inspired Jamaican Rastafarianism.) The national constitution declared him a direct descendant of the Queen of Sheba and King Solomon of Jerusalem, and few dared question the myth. In the streets, Ethiopians threw themselves onto the ground when he passed in one of his limousines.

Clark met the emperor once. Haile Selassie recognized that prehistory research brought prestige to his country in the eyes of the world, and in 1971 he had invited delegates of a scientific conference to his palace where liveried butlers served glasses of *tej*, an alcohol made from honey. ("You could drink quite a lot of it and it didn't seem to worry you at all," recalled Clark, "but then it later hit you rather suddenly.") The grand doors of the throne room opened and royal attendants ushered each delegate down a pillared hall to stand before the throne. The frail old monarch sat in robes like a soft-spoken aesthete and cradled a small dog in his lap as he exchanged a few pleasantries with the archeologist. It reminded Clark of an audience with a medieval king.

It was a last glimpse of a vanishing world. Ethiopia's ancien régime was dying. The monarchy had endured two millennia but by the 1970s observers sensed the emperor had gone senile. The bureaucracy he had built to prevent intrigue also forestalled progress; the competing security services founded to keep eyes on one another became nests of plotters. In 1972, the government tried to suppress news of famine in the northern provinces to avoid embarrassment. By the time foreign aid arrived the following year, one hundred thousand people had perished. The scandal destroyed the monarchy's last vestiges of legitimacy. Soldiers mutinied, students demonstrated, and workers went on strike. In September 1974, a group of military officers staged a coup. Haile Selassie was arrested in his palace, denied the dignity of departing in one of his many limousines, and stuffed into the backseat of a Volkswagen. The eighty-three-year-old monarch died while under arrest—maybe of natural causes, maybe smothered in his bed, depending on who you believed—and his body was discovered years later hidden beneath a toilet in the palace. On November 23, 1974, the military government executed fifty-nine senior officers and cabinet ministers from the old regime. The following day, a foreign expedition discovered Lucy down in the Afar Depression. Fossil bones were harvested in the Afar lowlands while corpses piled up in the highland capital. A secretive military committee called the Derg eventually came under the command of the most ruthless revolutionary—Mengistu Haile Mariam.

Under the emperor, Ethiopia had been a staunch Cold War ally of the United States. Under the Derg, it declared itself a socialist state and swung to the Soviet bloc. The archeologists and anthropologists who had been welcomed by the emperor suddenly fell under suspicion as potential spies. The government expelled American military personnel and closed an American listening base in the highlands of Eritrea that had monitored Soviet space missions. Soviet advisors and military hardware poured into the country. Fidel Castro sent Cuban troops to repel an invasion by Somalia. East Germans from the infamous Stasi security service advised Ethiopians on surveillance, interrogation, and hunting down "counter-revolutionaries." The capital erupted in urban warfare between the Derg and other revolutionary factions. One diplomat in the American embassy recalled those years as "bazookas, machine guns, and rifles being fired every night."

With a prototypical British stiff upper lip, Clark continued his expeditions after nearly all other U.S.-based researchers ceased fieldwork in the country. When his team flew into Addis Ababa, they were greeted by banners: "The victory of socialism is inevitable!" To get to his equipment storage shed in the U.S. embassy compound, his team passed roadblocks where they were searched by menacing militiamen with rifles and fixed bayonets—an ordeal repeated block after block within sight of the embassy gates. "We'd hear gunshots all night," recalled Steve Brandt, one of Clark's students at the time. "We'd get up in the morning and there were bodies in the street. There were murals depicting Uncle Sam and with his head cut off and rolling into a basket." The violence mostly spared foreigners—but not entirely. In 1977, a British geologist failed to show up at Clark's excavation. The scientist and three Ethiopian companions had been killed at a militia roadblock. The state news agency denounced the victims as a British spy and Ethiopian "counterrevolutionaries." They were among 360 killed that week.

Despite the Cold War tensions, the antiquities bureaucracy remained surprisingly cordial to their old friends from the west. Ethiopians revere their past, which seemed a glorious balm for the agonies of the present. In the northern highlands, stone obelisks and ruined palaces marked the remains of the ancient empire of Aksum, which rivaled Persia and Rome, and adopted Christianity in the fourth century C.E., making Ethiopia the oldest Christian nation in the world. Supposedly the Ark of the Covenant containing the tablets of the Ten Commandments sat in a church in Aksum—though no outsider was allowed to see it because it was guarded by high priests (one British scholar who got a peek inside during World War II reported the box was empty). Hidden in the highlands were ancient churches carved from solid volcanic rock. The country boasted the oldest written history of any African country except Egypt. In Gondar, a historic city near the source of the Blue Nile, stood castles of a seventeenth-century African Camelot. Lucy added yet another point of pride because Ethiopia now could claim humanity's oldest ancestors. The communist bloc, however, showed little interest in the apelike fossils because Marxist-Leninist ideologues took a dim view of the notion that human behavior had biological origins. So antiquities officials had to discreetly maintain their alliance with western scientists like Clark.

Years of delicate negotiation had been required to smooth the way for the 1981 expedition. The territory they planned to explore had fallen into Clark's hands with—as many things did in Ethiopia—a bit of intrigue. An American geologist named Jon Kalb had run expeditions into a section of the Afar Depression called the Middle Awash until he was expelled by the security ministry in 1978 after being accused of being a CIA agent—but whether those rumors actually had anything to do with his expulsion was just another one of the uncertainties often aswirl in Ethiopia. A few days afterward, Clark went to see his old friend Berhanu Abebe, a Sorbonne-educated historian who headed the antiquities agency and guided foreign researchers through the labyrinthine bureaucracy. He agreed to let Clark take over the orphaned territory. The state apparatus seemed like something designed by Kafka—a costive, bewildering ordeal of permission letters, rubber stamps, and doors that mysteriously opened or shut depending on political winds and maneuvering of factions. The antiquities administration was still dominated by veterans of the emperor's civil service, many of them educated in Europe or the United States. Even as the dictator Mengistu (Ethiopians refer to people by first names)* raged against American imperialism, in the antiquities offices there was no Cold War. (One visitor from the U.S. National Science Foundation reported that cultural officials like Berhanu wanted to keep the science alive because "Ethiopia is intensely nationalistic and takes great pride in being the homeland of the earliest humans.")

---

* This book will use two styles: first names for Ethiopians and last names for westerners.

Clark convinced the U.S. National Science Foundation to finance an expedition. "I hope this can be arranged before some eastern bloc scientists take them over," Clark urged the funding agency. In the new era of nationalism, however, the government mantra was *Ethiopia First,* and antiquities administrators pressed foreign researchers to provide funds to improve the national museum and train Ethiopian scholars. Long frustrated by the lack of local investment by foreign scientists, the antiquities administration had threatened to suspend fieldwork unless the Americans provided money for a new museum building. Clark secured another NSF grant to construct a new fossil and archeology lab at the Ethiopian National Museum and recruited a promising Ethiopian scholar named Berhane Asfaw.

BERHANE ASFAW HAD NO MEMORY OF HEARING OF LUCY'S DISCOVERY. WHEN Lucy made international news in 1974, Berhane was a student at Haile Selassie University, the training ground of the country's elite. The main campus was a former palace of the emperor where the salon was pockmarked by bullet holes from a failed coup and massacre of senior government officials and members of the royal family. By the 1970s, the emperor's namesake university had become a hotbed of dissent against the regime. "In those days," recalled one student contemporary, "not to be a Marxist was considered heretical." Student radicals made the revolution, but all their rejoicing at the fall of the monarchy proved short-lived when it became clear the military planned to replace one dictatorship with another. In 1976, the Derg unleashed a campaign of Red Terror to neutralize the opposition, and tens of thousands were killed by security forces or progovernment militias. Death squads went house to house hunting "counter-revolutionary outlaws" and dispatched "revolutionary justice" in the streets—a bullet to the head. They dumped bodies in the roads with placards denouncing traitors and prohibited families from recovering loved ones. At night, hyenas descended from the forests and scavenged the corpses. Propaganda banners screamed, "Let the Red Terror Intensify!" At the university, security forces conducted nighttime raids, herded students onto soccer fields, picked out some for beatings, and dragged away others who never returned. In the security ministry, police posted yearbook photos of students to be tracked down and liquidated.

Berhane joined an underground opposition group. At a café, he met a young woman named Frehiwot Worku, who also belonged to a resistance cell. She seemed naive, and Berhane warned her to be careful and contact him if she ran into trouble. In 1977, authorities raided her house, found incriminating papers and a gun, and arrested her parents. Berhane snuck Frehiwot out of the city in disguise and hid her in his home province of Gondar, where his father had been secretary general of the provincial government under the Haile Selassie regime. Berhane took a job as a schoolteacher in a remote village, and the couple lived together as lovers. After a few months, Berhane sensed they were being watched. He sent Frehiwot to his parents and promised to follow. The next day, he was arrested.

Hauled back to his hometown in chains, Berhane was thrown into a prison under the command of a Derg official nicknamed "the Butcher of Gondar." Torturers locked his wrists in metal U-bolts on the floor or hung him upside down, beat him, and demanded information. Many died during torture sessions, and others met their ends in firing squads. When prisoners heard their names were about to be called, they groomed themselves and tried to meet death with dignity. In Ethiopian tradition, families are responsible for feeding prisoners, so Frehiwot brought meals to Berhane every day and wondered if each visit might be the last. Miraculously, he was released after six months. Of the seven men chained together when Berhane arrived in prison, only two survived.

He returned to the university, completed a degree in geology, and renounced politics. "It was totally purposeless," he recounted with disgust. "All those things I trusted—I found them to be not trustworthy." Soon he found a new mission. During a summer job in the antiquities ministry, he was assigned to write a survey of Ethiopian archeological sites and plunged into the literature—all of it written by Americans and Europeans. "I didn't see a single Ethiopian name!" he recalled. "Why were there no Ethiopians?" The country so proud of its heritage had relied on foreigners to tell its story. Berhane's report circulated in the ministry and university and distinguished him as a potential scholar. One day in 1979, he was summoned to the university to meet a foreign archeologist. As he sat waiting, a tall *farenji*, "a foreigner," approached with a white goatee and brisk walk. "Jolly good!" sang Desmond Clark, shaking Berhane's hand. "Where can we sit and talk?"

Soon Berhane was on his way to the doctoral program at Berkeley. Clark invited Berhane and Frehiwot—by then married with a son—to his home for holiday parties where guests sat around crackling fires, drank sherry, smoked cigars, and listened to the archeologist recount past adventures such as picking up stone tools from a minefield during the liberation of Gondar during World War II. Despite all his old-world mannerisms, Clark strode ahead of his field in nurturing African scholarship. He believed Ethiopia should be represented by a properly trained scientist—and Berhane was his choice. In the Berkeley anthropology department, Berhane encountered another exotic character—Tim White. Ethiopian culture emphasized reserve and discretion. Professor White cursed and didn't give a shit about politics or offending powerful people. Berhane had never met a scientist so exhaustive, nor so exhausting. "If it was possible, he would want you to work more than twenty-four hours a day," said Berhane. "There is no limit for Tim. The difficult thing is he expects everybody else to work like him." White instilled his students with reverence: fossils were treasures—*fucking jewels!*—and impoverished Ethiopia was the richest country in the world. In the years ahead, White would teach Berhane how to identify fragments of bone, rescue fossils from blocks of rock, and reassemble broken skulls like puzzles. "Now that I can work on such things in Ethiopia," Berhane reported to his government, "I can do what no Ethiopian has done before."

Photograph © 1982 David L. Brill

Berhane Asfaw and Tim White examine a fossil skull at Berkeley.

With Clark's blessing, Berhane later would switch out of archeology to become the first Ethiopian to earn a doctorate in physical anthropology. Instead of tools, he would specialize in bones. His imprisonment and torture would not be an excuse to abandon his country, but an obligation to serve it. "After that experience, every day is a gift," Berhane explained. "And every day has to be worth living for. If I go through all this hardship, there is no reason to go anywhere else. I have to stick here and make a difference in anything that I do." In 1981, his presence helped his Berkeley mentors secure permissions for the expedition getting underway in the American embassy driveway. Clark reported to the NSF: "Our relations with the Ethiopian authorities are better than they have ever been, thanks in no small part to Berhane Asfaw."

WHEN THE CARS WERE PACKED, NINE FOREIGNERS AND SIX ETHIOPIANS squeezed into the seats and the expedition began its long journey. The caravan descended into the Afar Depression, dropping two kilometers in elevation from the cool highlands to the broiling lowlands. On the highway, they passed tanks, military trucks, and donkey-drawn farm carts. Barefoot peasants labored in fields of teff, and Soviet-built MiG jets streaked overhead and left vapor trails across the azure sky. Two days of driving took them to their target—an expanse the size of Rhode Island along both sides of the Awash River.

They erected a camp of green tents and a "laboratory" beneath a white canvas fly. In the open-air kitchen, the Ethiopian cooks roasted goats over fires, baked bread in an oven burrowed in a huge termite mound, and hung pots and pans from acacia trees. As the dry season advanced, the water hole dwindled to a concentrate of dissolved salts and minerals and stirred revolt in their intestines. "When friends wrote letters saying they were green with envy over our work in Ethiopia," recalled paleontologist Kris Krishtalka, "little did they know that we were green with dysentery." Guards hired from the local Afar tribe stood watch with old bolt-action rifles and kept an eye out for lions and raiders from the hostile Issa tribe. When the sun blazed at midday, the searchers rested beneath shade trees and dreamed of spending years harvesting these badlands to fill Darwin's tree of life with bones.

For two months, crews surveyed the territory and found it rich beyond their dreams. The searchers perched atop roof racks as bush cars jounced over the badlands from one fossil bed to another. The desert floor was covered with bones of extinct creatures—ancient monkeys, horses, hippos, giraffes, rodents, pigs, birds, turtles, snakes, crocodiles, even croc *eggs*. Fossils were so plentiful that local tribes piled them up to construct stone graves. "You can't even walk without stepping on fossils in some places," enthused White.

As an archeologist, Clark sought stone tools and butchery sites—animal fossils bearing evidence of being hacked apart by primitive *Homo* with stone blades. As an anthropologist, White lusted after fossils—particularly human ancestors from the nearly unknown period beyond 4 million years ago, the time before Lucy. In sediments more than 3 million years old, he picked up an upper thighbone from another member of Lucy's species, a first hint that the searchers had picked up the trail of more early bipeds. A teammate found pieces of skull around 4 million years old, the then-oldest braincase fragment in the human family. At that time, biochemical studies estimated that the human lineage, known as the *Hominidae,* split from the African apes around 5 million or 6 million years ago—exactly the time period preserved by the oldest fossil beds in the new project area. Further searching, Clark and White reported excitedly, might reveal "the deepest roots of the Hominidae."

"It was certainly by far the most successful field season in all my forty odd years of work in Africa," Clark gushed afterward. "Everything seemed to be there." Clark and White declared the Middle Awash "perhaps the most important study area in the world for the documentation and understanding of human origins and evolution." That valley of dry bones not only opened a window into earliest humanity; it promised to become a laboratory of the *entire* human past where they might discover how human ancestors split from the apes, began to walk upright, evolved from head to toe, used tools, and grew big brains. White predicted the Middle Awash would someday eclipse Olduvai Gorge, the locale in Tanzania made famous by the Leakeys.

Back at Berkeley a few months later, Clark and White announced their discoveries in a press conference. In one photo, White held a big-fanged chimpanzee skull and positioned the newly discovered piece of fossil braincase above it while Clark held a big-brained modern human skull, underscoring their interpretation of the new finds—an apelike ancestor on its way to becoming like us. Eventually, they would make a much bigger discovery that would rewrite this simple narrative and shake the family tree down to its roots.

Tim White and Desmond Clark show their new fossil alongside chimpanzee and human skulls.

White had passed almost within sight of the very fossils he sought. Midway through the 1981 expedition, he and three young scientists had crossed the Awash River on a tippy raft made from planks and oil drums, and explored the west side of the river near an Afar village called Aramis. The surveyors took only a quick glance that left White unimpressed with the scatter of broken fossils. "Fauna here is scrappy," he wrote in his field notes. Years later, a closer inspection would reveal something more. On the ground—in fragments so small that they could only be seen when crawling inches above the dirt—lay the broken bones of a primitive species of the human family that eventually would be named after the indigenous word for *root*. And just below the surface lay a skeleton far older and even more complete than Lucy. It would take another thirteen years to encounter it.

Chapter 2
BANNED

Humans are an egocentric species—the only animal that views its own existence as a problem to solve. Where do we come from? In the late nineteenth century, Charles Darwin and Thomas Henry Huxley pointed to Africa as the birthplace of our species based on the anatomical similarities between humans and African apes—but at that point there was no fossil evidence. In 1871, in *The Descent of Man*, Darwin wrote, "Those regions which are most likely to afford remains connecting man with some extinct ape-like creature, have not as yet been searched by geologists." Early evolutionists were slow to take Darwin's hint, and few bothered to look in what explorer Henry Morton Stanley labeled "the dark continent."

Early fossil discoveries pointed elsewhere. In 1856, the first fossil member of the human family was discovered in the Neander Valley of Germany. Its stout bones and apelike brow ridges distinguished it as a creature more primitive than modern humans and it was named *Homo neanderthalensis*, or Neanderthal Man. In 1891, a fossil skullcap and femur were found in Java—the first representative of a species now called *Homo erectus*. In England, the Piltdown man discovery of 1912 seemed to affirm the origins of humanity in Europe; decades passed before the so-called missing link was exposed as a forgery of a human skull joined to an orangutan jaw with its teeth filed down. At that point, there was no way to reliably determine dates of fossils, so their ages remained sheer guesswork.

The first African fossil from the human family was found in 1924, a partial skull recovered from a quarry in South Africa. Discoverer Raymond Dart named the species *Australopithecus africanus* and identified it as a "man-ape." But leading authorities dismissed the fossil as just an extinct ape—a response that would routinely follow discoveries of primitive forms. Many were blinded by racism and simply refused to accept a human ancestor from Africa. Over the next two decades, more fossils appeared in the karst caves of South Africa. In the 1940s and 1950s, the scientific establishment finally recognized *Australopithecus* as a member of the human family.

In 1959, Mary and Louis Leakey, a British couple who persisted in the lonely pursuit of human ancestors in Africa, found a hominid skull with robust jaws and big chewing teeth that they named *Zinjanthropus* at Olduvai Gorge in Tanzania. The discovery marked two breakthroughs: the first hominid from East Africa, and the first with a firm age produced by new advances in radiometric dating. *Zinjanthropus* was 1.75 million years old, far more ancient than any member of the human family known in Europe or Asia at that time. Soon afterward, the Leakeys found a larger-brained, small-toothed member of the human family they called *Homo habilis*, or the "Handyman," because it was found near stone tools. The quest for human origins shifted to East Africa, and the great bone rush began in the Great Rift Valley.

East Africa was not necessarily the *only* place where early human ancestors dwelled, but it contains geology where the oldest remains are best *preserved*. The Great Rift Valley is a vast crack in the earth's surface that stretches up eastern Africa from Mozambique to the Red Sea. The separation of continental plates has created a series of rift basins extremely well suited to the preservation of fossils. For millions of years, water has flowed down the slopes and deposited sediments on the valley floor, forming a geological

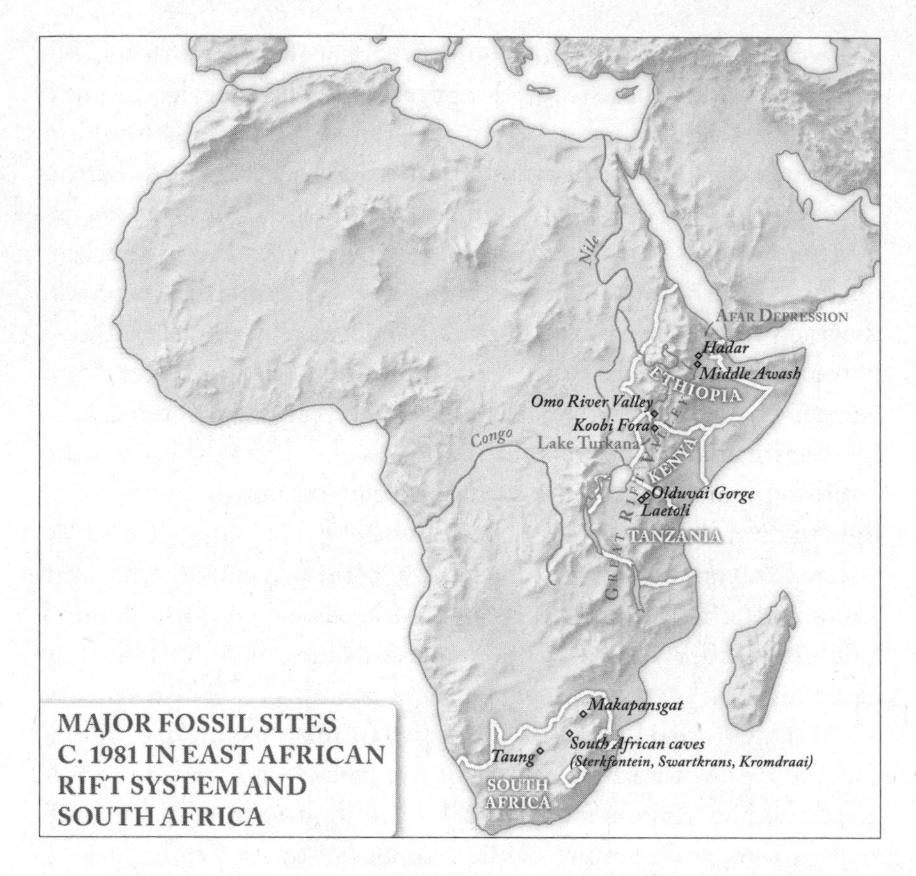

MAJOR FOSSIL SITES C. 1981 IN EAST AFRICAN RIFT SYSTEM AND SOUTH AFRICA

layer cake whose beds entombed the remains of human ancestors and other species that lived alongside them. Alkaline volcanic soils helped bones fossilize into stone. Millions of years later, fossils come to the surface by geological faulting and erosion.

In northern Ethiopia, the rift widens into the delta-shaped Afar Depression, a slowly sinking basin that someday will become another sea. (Here the rift involves three continental plates, not two as it does farther south.) The northern Afar is a parched desert whose lowest points lie below sea level and is one of the hottest places on earth. Salt flats mark where an extinct sea perished from thirst and where camel caravans still ply centuries-old routes of the salt trade. Through this desert winds the Awash River, which drains the Ethiopian highlands. It is one of the few waterways in the world that do not empty into an ocean; instead, it dead-ends into a briny lake whose waters evaporate in a scorching death valley. Along the 1,200-kilometer length of the Awash, tributaries flow down from the highlands and slice open cross sections of geological strata and expose the remains of ancient worlds—including a once-more-hospitable valley inhabited by ancient human ancestors.

By the early 1980s, the human family tree seemed a Y-shaped diagram with Lucy's species, *Australopithecus afarensis*, recently named as the oldest-known part of the trunk. She belonged to the genus *Australopithecus*, which was distinguished by big chewing teeth, small brains, and upright posture. The genus persisted 3 million years and diversified into at least two lineages. One was a now-extinct line of robust creatures with big chewing teeth of the genus *Paranthropus* (*Zinjanthropus* was later put into this group), and the other was a more gracile lineage of *Homo*, which spread across the globe and produced several species, including the last survivor, *Homo sapiens*.

But the deepest roots beneath that family tree remained a mystery. In the 1980s, the fossil record between 4 and 8 million years ago remained almost entirely blank. A few teeth, a jaw fragment, and an arm bone around 5 million years old had been found in Kenya in the 1960s, but they were too fragmentary to reveal much. The badlands of the Afar Depression offered the best hope of filling in the unknown earliest chapters. "Deposits of the right age seem to exist in northern Ethiopia," Ernst Mayr, the senior statesman of evolutionary biology, wrote in 1982, "and one can hope for crucial discoveries in the not-too-distant future."

TO CLARK AND WHITE, THAT FUTURE SEEMED WITHIN GRASP. FOLLOWING the promising finds of the first expedition, the U.S. National Science Foundation awarded a multiyear grant of nearly half a million dollars to the Berkeley team in 1982. Clark invited White to become his coleader because the young anthropologist had proven himself a "tower of strength" in the previous expedition. "He is a superb field and lab man," judged Clark. In August 1982, the team arrived in Addis Ababa to find the new American-funded lab building at the national museum nearly complete.

But the thaw in Cold War relations proved fleeting. The team spent three weeks in Addis Ababa obtaining security clearances, preparing vehicles, and stockpiling supplies—invariably a time-consuming process of pleading for permits, waiting in lines, and scrounging materials in a land of scarcity. On the eve of departure to the fossil fields, Clark was summoned to the Ministry of Culture and informed that his clearances had been revoked. Nobody could explain why. Civilian officials had little authority, and all power flowed from the military government, a secretive force answerable to no one. David Korn, the chargé d'affaires at the U.S. embassy, cabled the U.S. State Department: "The principal problem has been to try to identify which official or officials took the decision in question. . . . Everyone seems to be passing the proverbial buck."

Factions jockeyed to control Ethiopia's proudest science. The commissioner of science and technology sought to wrest control from the Ministry of Culture; meanwhile, Ethiopian academics from the national university clamored to claim a larger role in the research long dominated by foreigners. One official warned the scientists not to discuss their predicament with others in the government. Berhane Asfaw was called into the office of the Ethiopian commissioner of science and technology and interrogated. What were the Americans doing? Had they stolen fossils? Berhane discovered that the Berkeley team had been denounced by rivals—the Ethiopian students of Jon Kalb, the geologist who had been expelled. "These student associates of Kalb have falsely accused me of helping my supervisors to steal and smuggle fossils out of Ethiopia," Berhane wrote after the incident. "This is totally false and slanderous." When the questioning turned to the CIA rumors about Kalb, Berhane became terrified. He had been lucky to survive prison once and feared he might not a second time.

By early October, the mission had become a lost cause. The government suspended all anthropology and archeology expeditions until Ethiopia could rewrite its antiquities laws, which dated to the era of the emperor. "Foreign expeditions have often been perceived as 'ripping off' the Ethiopians, removing fossils and contributing little or nothing to local institutions or scholars," Clark and White reported. At the Addis Hilton, some scientists were evicted to make room for a delegation from Zimbabwe, and secret police entered their rooms to remove the listening devices. "The telephones in every room were bugged," recalled geologist Martin Williams. The scientists returned their vehicles and equipment to the storage shed beneath the eucalyptus trees at the U.S. embassy compound, flew back to California, and planned to resume research the following year. As it turned out, the ban lasted nine years.

Chapter 3
ORIGINS

For millennia, Ethiopia has occupied a mystical place in the imagination of the world. The name of the country was coined by ancient Greeks to describe a place reputed to be close to the sun (Ethiopia comes from the Greek compound word for "burned faces"). In the *Odyssey*, Homer wrote about "the distant Ethiopians, the farthest outpost of mankind." The Egyptians called it Kush—a land beyond the unnavigable, upper reaches of the Nile. The biblical book of Genesis described Ethiopia as a land watered by the river Gihon that flowed out of Eden. During the Middle Ages, European legends told of a king named Prester John who ruled over a Christian kingdom beyond the known world—and in the sixteenth century, Portuguese explorers claimed to have located that mythic land in the highlands of Ethiopia. "The Ethiopians slept near a thousand years," wrote eighteenth-century British historian Edward Gibbon in his *The History of the Decline and Fall of the Roman Empire*, "forgetful of the world, by whom they were forgotten." Isolation bred distinctiveness. Ethiopian languages are written in Ge'ez, an ancient script used nowhere else in the planet. Ethiopia remains the last country using the thirteen-month pre-Julian calendar (based on the annual flooding of the Nile), which lags about seven years behind the rest of the world.

The country is layered in time. Prehistory research began in

Ethiopia in 1906 after Emperor Menelik II invited German archeologists to excavate the ruins of the kingdom of Aksum under a secret agreement. The Germans wanted to study ancient civilization while Menelik sought scientific endorsement for the legitimacy of his dynasty and his territorial claims. It is axiomatic that science should remain free of political influence, but in Ethiopia prehistory research became interwoven with politics, national identity, and the country's wary relationship with foreigners—and remained so through a succession of regimes.

Modern paleoanthropology began in Ethiopia in the 1960s when Emperor Haile Selassie asked Louis Leakey why Ethiopia produced no magnificent fossil discoveries like the ones in Kenya or Tanzania. Leakey assured the emperor that Ethiopia surely *could* prove rich with early human fossils and stone tools, if the government made research less difficult. (A few years earlier, F. Clark Howell, then of the University of Chicago and later of Berkeley, made a reconnaissance in the Omo Valley in Southwest Ethiopia, but Ethiopian officials blocked his access to the field and then, after allowing him to continue, confiscated and lost his fossils.) The emperor promised to expedite matters. Soon afterward, Leakey was summoned to the Ethiopian embassy in Nairobi, where the ambassador read aloud a declaration from the monarch authorizing research. In 1967, teams from America, France, and Kenya began a joint expedition in the Omo Valley, where they found plenty of animal fossils but only fragmentary remains of human ancestors.

Meanwhile, an unheralded young researcher found a richer target on the other side of the country. In the 1960s, French-Tunisian geologist Maurice Taieb began surveying the Afar Depression for his doctoral dissertation. The Afar remained the Wild West of Ethiopia—a deadly valley of dangerous animals and fierce nomads with a reputation for castrating intruders. Even Ethiopians avoided it. Many early explorers left their skeletons there. "In this country every stranger is an enemy worth killing," wrote nineteenth-century Swiss adventurer Werner Munzinger. He proved his point with his life. In 1875, Munzinger led a mercenary force into Ethiopia that was wiped out by Afar clans. Similarly, in 1881, a group of Italian explorers was massacred. Three years later, another group of Italians went looking for the previous one and met the same fate. The threat of annihilation deterred other adventurers for the next half century. In 1928, mining engineer Louis Mariano Nesbitt set out to explore the Afar with two Italian adventurers and fifteen Ethiopians, trekking eight hundred miles through desert and sometimes going days without water in temperatures over 150 degrees Fahrenheit. Nesbitt—who traveled through what later became the research territory of the Berkeley team—painted a fearsome portrait of the Afar people, especially the male warriors who adorned themselves with kill trophies. One member of his party was assassinated by an Afar guide and two others disappeared, presumably murdered. "Men must think of blood," one Afar man purportedly told the explorers, "for it is better to die than to live without killing."

By the time Taieb ventured into the Afar Depression in the 1960s, a foreigner could travel with less fear of dismemberment. Lacking money for a vehicle, Taieb took donkeys, swam across rivers, and caught boat rides from crocodile hunters. Later he acquired a Land Rover and probed deeper into the wilderness with Afar guides, following bulldozer tracks left by a German company constructing a highway to the Red Sea. Speeding across the desert, he ran down ostriches to provide fresh meat for his Afar friends. Jon Kalb, his assistant, described Taieb as "impetuous, excitable, often impatient, chain-smoking, and just slightly crazy." Taieb expected to find the Awash Valley covered by recent volcanic soils, which are generally devoid of fossils. To his surprise, he found sedimentary rocks rich in petrified bone.

Late one afternoon in 1969, Taieb reached the top of a hill above the Awash River and spied a magnificent geological exposure below. His Afar guide feared that warriors from the enemy Issa tribe might be lurking in the canyons, so he remained atop the hill. Taieb descended by himself and found the ground scattered with bones of rhinos, bovids, and elephants—the type of fauna typically found alongside early human ancestors. The Afar called the place *ahdi d'ar,* or "treaty stream." Foreigners shortened the name to Hadar—a place that would become famous in the annals of paleoanthropology.

But when Taieb tried to recruit the big names in the field, all were busy elsewhere or skeptical of the unproven youngster and his unknown territory. Later in Paris, Taieb encountered American anthropologist Don Johanson, a student on the Omo expedition. "I met this effervescent guy," recalled Johanson, "with constant Gauloises hanging from his lip, thick glasses, and bubbling, 'If you think you have fossils at the Omo, you should come with me to the Afar.'"

Don Johanson did not fit the stereotype of a field scientist. He was an urbane dandy with tastes for designer clothes, fine wines, and French cuisine. He wore bushy sideburns, and his dark eyebrows leapt up and down as he talked excitedly. The son of Swedish immigrants, Johanson lost his father when he was only two years old and was raised by a single working mother. A scholarly neighbor introduced the boy to anthropology, and Johanson became captivated by the pursuit of human origins. As a graduate student at the University of Chicago, Johanson had gone to Africa in 1970 with F. Clark Howell's Omo expedition. Like many newcomers, he initially wilted in the heat and couldn't identify fossils, but he eventually became adept on the outcrops and determined to make a name for himself. Some mentors doubted he would last long in the field. They underestimated him.

In 1972, Taieb organized a trip to Hadar with Johanson, Kalb, and French paleontologist Yves Coppens. They stepped to the edge of a plateau and beheld kilometer after kilometer of canyons where fossils spilled from eroding sediments. The four explorers formed the International Afar Research Expedition, and the Ethiopian government issued a permit for a huge territory roughly the size of Djibouti. "Being strong willed as I was in those days," Johanson recalled, "I was absolutely certain we would find fossils of human ancestors."

In 1973, the Hadar expedition scored the first major discovery. One day, Johanson searched some deposits from the Pliocene epoch more than 3 million years old and came upon what he mistook as a hippo rib. He kicked it and realized it was the shinbone of a small primate. A monkey, he figured. He picked it up and spotted a lower thighbone a few yards away.

He matched the two bones. They fit together perfectly.

But something didn't make sense. The bones joined at an angle with the thighbone canted slightly off the vertical centerline of the shin. No monkey or ape did that. All other primates have thighbones aligned with shinbones. By contrast, human thighbones angle inward from hip to knee, which helps us balance on one foot when we walk or run. The human femur and tibia form a so-called bicondylar angle—and so did Johanson's fossil knee. That posed a conundrum. At the time, some leading authorities insisted that our ancestors began walking upright only half a million years ago. Others recognized clues that it might be far older, but proof remained scarce. This knee was more than 3 million years old. Could it really belong to a human ancestor? Lacking comparative samples, Johanson did something that haunted him for years afterward: he robbed a grave.

The hilltops of the Afar desert were dotted with circular stone crypts where the Afar clans entombed the dead. Johanson and his graduate student, Tom Gray, snuck out of camp and stole a femur from a nearby burial mound. Back in the tent that night, Johanson compared the purloined bone with his fossil knee and saw that both shared the anatomy of bipeds. After returning to the USA, Johanson hurried to consult an authority on human locomotion—Owen Lovejoy of Kent State University, who affirmed the fossil belonged to an upright walker, one far older than any other biped known at that time. "They won't believe you," Lovejoy told Johanson. "You'd better go back and find a whole one."

The following year Johanson returned to Hadar and found the skeleton that would make him famous. On November 24, 1974, Johanson and Gray spent the morning surveying some barren hills. By midday, the desert broiled with heat and the two men headed back toward their car. Johanson spotted an arm bone and pointed it out to Gray. As their eyes adjusted to a new search image, the bare moonscape suddenly came alive with fossils—pieces of skull, vertebrae, pelvis, and ribs. "It was almost like bones started popping out of the ground," recalled Gray. They drove back to camp tooting the horn, and all hands went to search the area. Three weeks of collecting and sieving produced several hundred pieces.

The team reconstructed the skeleton on a table in camp as a battery-powered cassette player blared the Beatles song "Lucy in the Sky with Diamonds." The skeleton took on the name Lucy. She was the oldest skeleton ever found in the human family. In fact, she was the oldest by nearly 3 million years (a 75,000-year-old Neanderthal was distant second). The team recovered about 40 percent of her bones; lab technicians later reproduced missing parts by mirror imaging and reassembled about 70 percent. In camp, Johanson pored over each piece. Maurice Taieb watched his companion become terse and preoccupied. "Every night," recalled Taieb, "he puts everything back in the box and takes it into his tent."

At the end of the season, the expedition returned to a capital unsettled by revolution. The discovery already had made international news and the Ethiopian minister of culture reproached the scientists for giving the skeleton a foreign name. He suggested an Ethiopian one—Dinknesh, which means "she is wonderful" in the Amharic language. But the skeleton already had been introduced to the world press as Lucy, and so she remained. The ministry had other things to worry about: government officials were being lined up and shot. "You could not predict what would happen the next day," recalled Aklilu Habte, a former minister of culture. "People were being killed right and left." Johanson and Taieb hurried to write a scientific report and secure permissions to whisk their fossils out of the country. Johanson described a thrilling realization as he flew home: "I was no longer an unknown anthropology graduate, but a promising young field worker with fossils dazzling enough to match those of paleoanthropology's certified supernova, Richard Leakey."

Don Johanson (*left*) and Maurice Taieb (*right*) with the Lucy skeleton in the Hadar camp.

Lucy was the biggest discovery in a generation. Only one year later, however, the Hadar team made yet another find just as extraordinary. In 1975, medical student Mike Bush spotted two teeth in an eroding hillside, prompting an intense collection. When a *National Geographic* photographer complained about the light, Johanson suspended the operation until the next day to ensure good photos. The following morning, a member of the French film crew sought shade beneath an acacia tree and almost sat on a fossil heel and thighbone. Other people began shouting about finding fossils up and down the hillside. Over the remainder of the field season, the Hadar team found 216 hominid fossils from at least thirteen individuals. The Lucy skeleton gained a comparative population, which spurred a research boom in comparative anatomy, paleo demography, and analysis of intraspecies variation. Taieb and Johanson initially surmised the concentration of bones belonged to a troop killed by a flash flood. (Years later they abandoned this claim after other experts suggested the remains might have been preyed upon by carnivores and collected by flowing water.) The Hadar team dubbed them "the First Family."

The expedition managed only one more field season before Ethiopia became too dangerous due to purges in the cities and guerrilla insurgency and tribal warfare in the Afar Depression. The U.S. embassy urged Johanson to cancel his 1976 expedition. Unable to dissuade him, the embassy requested a list of personnel and their next of kin. At the end of that expedition, the Hadar team returned to a capital in the grip of the terror. With the bureaucracy nearly frozen by fear, Johanson worried about getting his fossils safely out of the country. To his relief, his export permissions were signed by a friendly young minister. When the man went home that evening, he was assassinated.

BUT WHAT WERE ALL THESE NEW FOSSILS? AFTER THE 1975 FIELD SEASON, Johanson stopped in Nairobi to show his First Family fossils to the royal family of human origins, the Leakeys. Patriarch Louis Leakey had died three years earlier and his widow, Mary, continued her long-standing research mission at Olduvai Gorge in Tanzania. Their son and heir apparent, Richard Leakey, was making headlines with his own discoveries in Kenya. Johanson courted the family, but he aspired to eclipse his more famous competitor, Richard Leakey. Johanson later recalled: "Our conversation always had a dimension of competition . . . we were looking for chinks in each other's armor."

At the National Museum of Kenya, Johanson laid out his Hadar fossils for inspection by Richard Leakey and his team. Some of the First Family individuals were much larger than Lucy. Initially, Johanson and the Leakeys agreed the Hadar fossils represented two separate groups: one *Australopithecus* (in Leakey orthodoxy, these were "near men" doomed to extinction) and the other *Homo* (deemed ancestral to modern humans). Nearby lurked a junior scientist whom Johanson did not recognize—a young American graduate student with big glasses, lank hair, and white lab coat who intently examined the fossils. Johanson mistook his silence for timidity.

It was Tim White, then an apprentice on the Leakey teams. His problem wasn't shyness; he just wasn't buying the two-lineage interpretation from the big shots around the table. White had been studying hominid jaws for his dissertation, including a batch of 3.75-million-year-old fossils discovered by Mary's team at Laetoli, Tanzania. He believed Mary's Tanzanian fossils and Johanson's Ethiopian fossils belonged to the same species—one new to science. When White spoke up, Johanson was taken aback. The two populations were separated by a thousand miles and at least half a million years. Johanson recounted the exchange:

"One species?" asked Johanson.

"One species," insisted White.

"What about Lucy?"

"Lucy too."

Johanson wouldn't buy it. "No way," he said.

Today Ethiopia and other countries require antiquities to be stored in their national museums and allow export only in rare circumstances. In the 1970s, however, the country still lacked a modern museum, casting lab, and trained technicians (the grant later secured by the Berkeley team to construct a new lab sought to correct this deficiency). As a consequence, the government allowed foreign scientists to borrow specimens for study. Johanson returned home to the Cleveland Museum of Natural History, where he served as curator of physical anthropology, with crate-loads of fossils. Only a few years earlier, the fossil record for early

Don Johanson showed his new fossils from Ethiopia to the Leakey team at the National Museums of Kenya. During the visit, he met Tim White, then a graduate student on the Leakey team, who made a provocative suggestion—all the new fossils from Ethiopia and Tanzania belonged to a single species. (*From left to right:*) Tim White, Richard Leakey, Bernard Wood, and Don Johanson.

humanity had been virtually empty. Suddenly the period between 3 and 4 million years ago had become crowded—and most of the crowd sat in the bunker-like basement near the southern shore of Lake Erie.

A few months after the meeting in Kenya, Johanson traveled to a conference in France, and again ran into Tim White, who was steaming after a fallout with his boss, Richard Leakey. "I finally decided I could not proceed without him," Johanson remembered. "The risk of getting mixed up on the fossils was greater than the risk of annoying the Leakeys by associating with Tim."

Their partnership would rewrite the human past. The Cleveland museum became the Camelot of human origins. A large team studied and reconstructed the Hadar fossils, but assigning them a place in the family tree largely fell to Johanson and White. They had three sets of fossils to consider: the Lucy skeleton; the First Family and other fossils from Hadar (all believed to be around 3 million years old); and casts of a third group of fossils collected by Mary Leakey's team in Tanzania (dated 3.6 to 3.8 million years old). Some aspects of the fossils seemed humanlike while others seemed apelike. What were they?

White's approach to collaboration was sometimes adversarial, and the lab became a courtroom where hypotheses stood trial. "Instead of having somebody else pick on us in print after publication," White said at the time, "we decided to pick on each other before publication." Johanson adhered to the Leakey orthodoxy: some of the new fossils belonged to the genus *Homo* and were ancestral to humans, and others belonged to a dead-end *Australopithecus* lineage. White insisted the fossils all represented one species. The two men and their colleagues spent months arguing, often shouting at each other deep into the night. "Tim kept me in line," said Johanson. White won the argument his usual way: by breaking down the matter to a series of fundamental questions, exhaustively piling up evidence, and relentless advocacy. He laid out a series of the Ethiopian and Tanzanian fossil jaws in order of size on the lab table. When placed in sequence, the stark differences between the largest and smallest specimens vanished into a continuum of normal variation typical of any higher primate. For comparison, he also lined up skulls and teeth of modern chimps and gorillas from the Cleveland museum collection. Each ape species showed a huge range of variation. Pick two individuals from ends of the spectrum and the differences seemed dramatic. Put them in context of a population and they appeared to be just two samples from a sequence.

Eventually, Johanson could not argue anymore. By December 1977, he agreed with White: it was all one species. The anthropologists ascribed the size differences to sexual dimorphism (males were larger than females) and invoked the anatomical phenomenon of allometry (disproportionate changes in certain parts). Next question: what to name it? The brains were too small to be *Homo*, so Johanson and White named the new species *Australopithecus afarensis*, after the Afar region. (Under zoological convention, the genus name is capitalized and the species name is lowercase.) By the rules of taxonomy, a species must be defined by a single type specimen. In a controversial move, the anthropologists selected an older jaw from Tanzania (collected by Mary Leakey's team) as the exemplar of a species named for Ethiopia. Fossil species typically are defined by skulls and teeth, and White argued that the Tanzanian fossil best illustrated its characteristics. The choice also offered tactical advantages because it extended *afarensis** deeper into the past, secured the title of oldest African hominid, and deterred others from placing the older fossils into a new species of *Homo*.

The two Americans felt pressured to move quickly. Both the Hadar and Laetoli specimens had been described in the scientific literature and thus were fair game for anybody to assign to a new or existing species. Johanson and White feared others would name a new taxon before they did—particularly Yves Coppens, the French paleoanthropologist whom Johanson came to regard as a rival. In May 1978, Johanson announced the new species at the Nobel Symposium hosted by the Royal Swedish Academy of Sciences. His move infuriated Mary Leakey because Johanson discussed her Laetoli fossils before her scheduled talk and thereby scooped the grand dame of human origins on her own discoveries.

The young Americans made a disruptive claim—Lucy and her *afarensis* lineage represented the ancestors of *all* later members of

---

* This book will occasionally refer to taxa only by the single species name. The author begs forgiveness for this violation of the conventions of binomial nomenclature (which demands use of both genus and species names) for the sake of brevity. Therefore, "*ramidus*" should be read as shorthand for *Ardipithecus ramidus*, "*afarensis*" for *Australopithecus afarensis*, and so on. At times, the genus names will be abbreviated (e.g., *Ar. ramidus* or *Au. afarensis*).

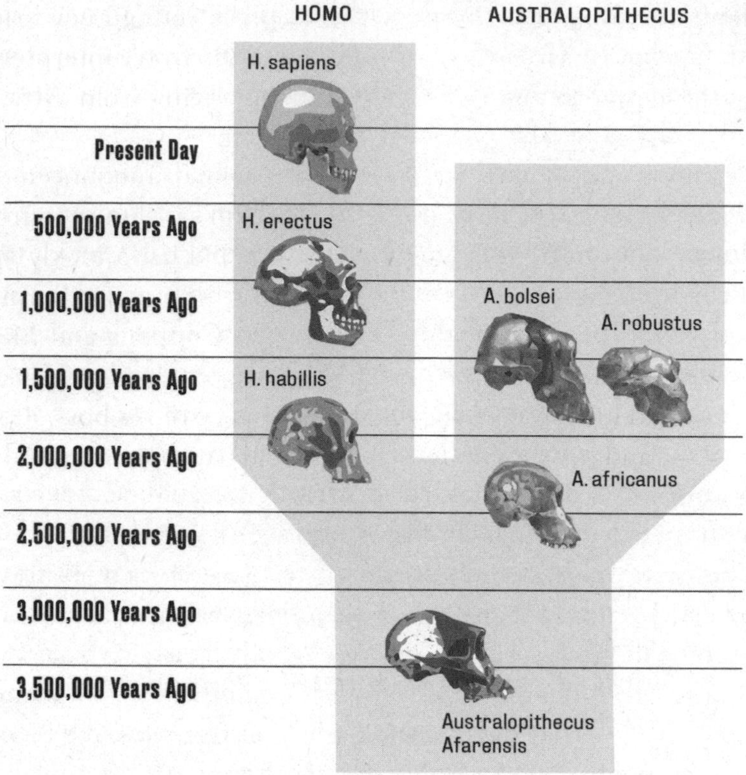

The Y-shaped family tree proposed by Don Johanson and Tim White in 1979.

the human family. The Leakeys had long believed that *Homo,* the human genus, went back tens of million of years. Problem was, nobody could find humanlike, big-brained ancestors anywhere near that age. Instead, searchers found only fossils that looked like primitive *Australopithecus*—which White and Johanson now asserted were the true human ancestors between 3 and 4 million years ago.

Such claims aroused huge objections. Some insisted that *Australopithecus afarensis* was too apelike to be a human ancestor. But molecular biologists had shown that humans had shared a common ancestor with chimps and gorillas as recently as 5 million years ago, and this revelation strongly influenced the young Lucy team. In fact, *afarensis* marked the first major fossil discovery interpreted under the new paradigm of recent human origins from African apes.

Johanson and White drafted a more formal announcement of the new species for the Cleveland museum's in-house journal *Kirtlandia*, a strategy that allowed them to publish quickly and avoid getting tangled up with peer reviewers at a major journal. As a professional courtesy, the pair invited Coppens and Mary Leakey to be coauthors. In June 1978, White returned to Nairobi and presented his theory to a skeptical audience of his boss, Richard Leakey, and his staff scientists. A journalist took notes of the discussion:

ALAN WALKER (anthropologist): If the degree of sexual dimorphism is outside the modern range, then you must justify your reasoning.

TIM WHITE: It's simplest to have only one species in the family collection.

WALKER: Numerical simplicity is not necessarily the truth.

RICHARD LEAKEY: Have you done sufficient study and measurements to convince us?

WHITE: Our scheme elucidates . . .

ANDREW HILL (paleontologist): Obscures!

WHITE: We recognize a significant . . .

HILL: How do you know it's significant if you haven't quantified?

WHITE: From my experience!

WALKER: You need to be just a little more precise . . .

LEAKEY: It's my feeling that you are guilty of imposing what you think is right upon the fossils.

Meanwhile, another debate raged: did Lucy and her species walk on two legs like modern humans? The Cleveland team argued that *Australopithecus afarensis* was fully bipedal. The fossils bore primitive traits such as longer forelimbs and long, curved fingers and toes, but the Lucy team's locomotion expert, Owen Lovejoy, dismissed those as leftover evolutionary baggage from an arboreal ancestor. He argued that Lucy's pelvis and lower limbs proved that her species already had made the transformation from climber to biped. Some scholars remained incredulous that such a primitive creature could walk upright and insisted that *afarensis* remained a tree climber, or perhaps waddled upright like a circus ape. Then came evidence beyond the dreams of any anthropologist—bipedal footprints from the time of Lucy's species.

The footprints represented one of the most amazing discoveries in the annals of paleoanthropology, and led the Leakey dynasty into another conflict with a brash young American who threw himself into the excavation—Tim White.

WHITE ARRIVED IN MARY LEAKEY'S LAETOLI CAMP ON JULY 4, 1978—American Independence Day. It was one week after his debate with Richard Leakey's team at the museum in Nairobi. The Laetoli scientists were not sure what to make of the headstrong Yank in his bush hat banded by rattlesnake skin. With blond hair and big glasses, he looked like folk singer John Denver, a country boy pumped full of piss and vinegar. The Laetoli camp was set in a hollow shaded by acacia trees infested by puff adders, an extremely dangerous snake. Undeterred, White killed the snakes, skinned them, and barbecued them on skewers. To supplement the meager camp fare, White and his pal Peter Jones, an archeologist, challenged each other to catch guinea fowl. They tried to grab the speedy birds on foot, hit them with rocks, or run them down in Land Rovers. When all these attempts failed, they strapped White to the car and he tried to bat birds out of the air while his friend raced over the savanna.

White had earned a reputation as an irritant in the camp of Richard Leakey, but that conflict initially caused no concern in the camp of the family matriarch because Mary often did not get along with her son in those days, either. Mary spoke with caustic wit and openly picked "stinkers" and favorites. "I never could understand how she could start the day as fresh as a daisy, despite having put down half a bottle of whiskey in the evening," recalled Richard. "Without exception, she would do it until the day she died. She had a liver you wouldn't have believed." Initially, Mary took a liking to White, invited him to join her for Christmas, and even nicknamed one newly discovered fossil as "Timothy." She wrote a recommendation for his job application to Berkeley and praised him as a "very able, clear-thinking young man."

At that point, the Laetoli team had been searching three years for ancient footprints. The quest had begun in 1975 when some young scientists playfully pelted each other with dried elephant dung. Andrew Hill ducked into a gully and scanned the ground for more ammo. Then something caught his eye—animal footprints. Hill dropped to his knees and examined a hard layer of fossilized ash with tiny indentations of raindrops that reminded him of Pompeii. The dung war came to a cease-fire as the scientists gawked at tracks of ancient elephants, antelopes, buffalo, giraffes, and even birds. Could the Laetoli ash contain footprints of human ancestors, too?

Shortly after White arrived, geologist Paul Abell finally found the first trace of the long-sought hominid footprints, but the impressions were ambiguous enough that some doubt remained. Already there had been one false alarm. The previous year, a strange set of tracks were mistakenly identified as a human ancestor. Mary held a press conference at the National Geographic Society offices in Washington, D.C., and announced the tracks belonged to hominids who were fleeing an erupting volcano. Her scientific team later determined those footprints belonged to a bear. This time, she remained skeptical and assigned a single African workman to conduct a preliminary excavation. He uncovered tracks that undoubtedly belonged to a humanlike biped. "At this stage," recalled Mary, "Tim White, whom I could never convince of the experience and ability as excavators of my Kenyan staff, decided that he ought to take charge of what would clearly need to be a careful and extremely precise excavation."

The Laetoli footprints were preserved by a rare combination of circumstances. About 3.6 million years ago, a volcanic eruption blanketed the landscape with ash like new-fallen snow. Rain transformed the ash into muck like wet cement. Into this scene ambled two or three human ancestors who left behind a set of tracks as vivid as footprints on a beach. The ash hardened and became buried by another eruption whose ash bore a slightly different chemical composition.

White sometimes spent three days cleaning a single footprint. He and anthropologist Ron Clarke carefully darkened the ash infill with paint thinner and excavated with dental picks. "I remember Mary getting impatient with the amount of time Tim was taking," said archeologist Peter Jones. "In the interest of speeding things up, Mary jumped in and didn't do the greatest job." With a chisel and failing eyesight, Mary gouged some of the footprints. To White, fossils were sacred objects, and the footprints among the most precious ever found. By his account, he felt compelled to protect the world's most valuable archeological discovery from his boss, herself a living legend in African archeology. "I would literally run a kilometer to get to the outcrop to maximize the time we had to excavate before Mary got there," said White. "She was just not with it."

White and Mary also argued about the footprints. The tracks clearly showed humanlike biomechanics of heel strike, rolling over an arched midfoot, and pushing off the toes. No prints suggested divergent opposable toes. No handprints suggested the creatures ambled on all fours. White believed his soon-to-be-named species

Photograph by and © Peter Jones and Tim D. White 1978

Tim White at Laetoli.

Australopithecus afarensis made the tracks. Mary thought the new species was too primitive to have made such humanlike footprints. In fact, the footprints looked so modern that some scholars attributed them to some still-undiscovered ancestor.

By that summer, White and Johanson already had drafted their paper lumping Mary's Laetoli fossils and Johanson's Hadar fossils into the new species *Australopithecus afarensis*. Mary agreed that most of the fossils could be one species—but not that dreadfully named genus *Australopithecus*. Mary despised Johanson because of the episode at the Nobel Symposium in Sweden. Increasingly, her former darling White also talked his way onto her stinkers list. "I had to listen in the workroom at my own camp at Laetoli to a long harangue from Tim White in an attempt to get me to change my mind," Mary recalled. Finally, White snapped that if Mary was so resistant she should remove her name from the upcoming scientific paper announcing the new species. Until that point, Mary claimed she did not realize her name would be on any paper. (Johanson and White insist they kept Mary fully apprised and sent her a draft.) The next morning, she rose early, drove to the nearest town and sent a curt telegram to Johanson in Cleveland: PLEASE OMIT MY NAME FROM PAPER ON NEW SPECIES REGARDS MARY.

By the end of the summer of 1978, the Laetoli excavators had uncovered thirty-four prints in two parallel trails. When they reached a geological fault, the excavators stopped for the season and planned to resume the following year. But the personal fault lines became impassable. With misgivings, Mary invited White to return to Laetoli for the next field season. "In order to avoid clashes in the field," Mary wrote White, "I must ask you to appreciate that the direction of this project is in my hands and my decision must be final, even if you do not agree." White responded that he would say whatever he thought was right. As the date approached, Mary grew anxious and hatched a plan to dispatch White to excavate a remote area where she knew no footprints would be found. She openly cackled about sending her annoying assistant on a wild goose chase. Jones met White at the airport and warned him. White abandoned his plans to rejoin the Laetoli camp, detoured to Nairobi to spend the summer in the Kenyan museum, and eventually relaunched his career in one part of the Rift Valley where the Leakeys did not exert influence: Ethiopia.

In the fall of 1978, *Kirtlandia* carried an announcement of the new species (after interrupting the press run to remove Mary's name). The journal had small circulation, so the worldwide announcement effectively occurred in January 1979, when Johanson and White published a paper in *Science*. A front-page headline in the *New York Times* trumpeted: NEW-FOUND SPECIES CHALLENGES VIEWS ON EVOLUTION OF HUMANS.

> CLEVELAND. Jan. 18—A previously unknown human ancestor that lived in Africa three million to four million years ago and had an unexpected combination of a small-brained, apelike head and a fully erect body has been discovered by two American anthropologists.
>
> The discovery, the first species of human ancestor to be named in 15 years, deals a major blow to the old but still widely held belief that erect posture, which would theoretically free the hands for tool-making, evolved in tandem with an enlarged brain.
>
> The discovery reopens the question of why a creature with a head almost resembling that of a chimpanzee, and apparently incapable of making tools, would have begun to walk on two legs.

The paper quickly drew objections. One concerned the decision to combine fossils from Ethiopia with a type specimen from Tanzania: if it became necessary to split the taxon into two species, the name would stay with the type specimen from Tanzania—thus leaving no *afarensis* in the Afar. In his *Lucy* memoir, Johanson said his new species name was endorsed by the eminent biologist and taxonomic guru Ernst Mayr. But Mayr had reviewed a draft manuscript that left him unsure of the locations of the two sites, and afterward he chided Johanson: "Your description of Australopithecus afarensis was somewhat of a nomenclatural mess."

Others contended that both Hadar and Laetoli fossils were just older representatives of the South African *Australopithecus africanus*—the species discovered by Raymond Dart in 1924. (South African anthropologist Phillip Tobias of the University of the Witwatersrand planned to declare the Hadar and Laetoli fossils as a subspecies of *africanus* at the Nobel conference, but Johanson got to the podium first and beat him to the punch.) Richard Leakey also cast doubt. "I think Don was right the first time," said Leakey. "They're sampling two different populations, *Homo* and *Australopithecus*." Johanson and White suspected the Leakeys had a personal grudge because Lucy and *afarensis* undermined the dynasty's monopoly. Johanson and Richard Leakey grew testy toward each other—partially because of scientific differences and partially because Johanson aspired to be the next Richard Leakey. The final rift came at a joint appearance at the American Museum of Natural History in New York in 1981. Richard expected to deliver a lecture; Johanson came to debate and forced Richard into a fumbling defense of his version of the family tree—all filmed for television. Richard stormed out. The Leakeys were further incensed by Johanson's best-selling *Lucy* book and his eagerness to challenge their views. Thanks to his knack for publicity, Johanson became what one author described as "the new king of anthropology in the eyes of the U.S. print and television media." In 1983, Johanson wrote Richard Leakey requesting permission to look at some Kenyan fossils. Richard told him to forget it: "I consider you a scoundrel."

In 1981, Johanson left Cleveland and founded the Institute for Human Origins, an independent center by the Berkeley campus. He selected the location to facilitate collaboration with the university's geologists and human origins scholars, especially White. In 1983, Mary Leakey retired, left her camp in Olduvai Gorge, and to her horror the world-famous site was reclaimed by her two American rivals and their Tanzanian colleagues. In 1985 and 1986, Johanson and White led expeditions to Olduvai and found a scrappy 1.8-million-year-old skeleton attributed to *Homo habilis*. White gloated: "The Leakeys took thirty years to find a hominid at Olduvai Gorge. We got one in three days." Johanson retained another ghostwriter, wrote a second memoir titled *Lucy's Child*, and developed his popular paleo franchise.

But Olduvai was just a sideshow during Ethiopia's moratorium. The real priority was resuming work in the Afar Depression to answer all the questions provoked by Lucy and *afarensis:* What came before? How did upright locomotion begin? How did the ancestors of humans split from the ancestors of African apes? In the end of his first *Lucy* book, Johanson dreamed of partnering with White again to search the place most likely to produce fossils older than Lucy—the badlands of the Middle Awash. He predicted: "What we find in them could well blow the roof off of everything." As it turned out, what blew up was their friendship.

Chapter 4
THE FALSIFIER

White once observed that scientists were people who never outgrew their childlike curiosity, and his own boyhood was a case in point. He grew up in the San Bernardino Mountains of Southern California back when bears and mountain lions still roamed the forests. The family lived alongside a mountain highway called Rim of the World for its panoramic views. On clear days, they could see the citrus groves in the valley below and the Pacific Ocean on the horizon. "As Southern California filled with people and automobiles," he reminisced in a memoir essay, "there were fewer and fewer clear days."

White and his younger brother, Scott, roamed the woods in their Davy Crockett caps and toy rifles. The brothers captured lizards, birds, turtles, and a raccoon for pets. White also became fascinated by dinosaurs and begged his parents to take him into the desert to search for fossils. They returned home with no dinosaurs—and learned that finding old bones wasn't so easy. As a child, he visited the Los Angeles Museum of Natural History, pressed his nose against the display cases, became transfixed by rows of animal teeth, and was possessed by a desire to find such things himself. His father brought home artifacts such as a Native American stone bowl that had been unearthed in a road excavation. The relics captured the boy's imagination: *who were these people?*

When White was a teenager, his grandmother gave the family a series of Time-Life books. One volume, *Evolution*, explained life in Darwinian terms. Another volume, *Early Man*, introduced him to apelike human ancestors and captivated him so much that the pages fell out from repeated turning (the author was paleoanthropologist F. Clark Howell, one of his early heroes and later a mentor). The book contained a fold-out illustration known as the *March of Progress* from ape to human, which became the iconic rendering of a now-defunct linear view of evolution. White read *National Geographic* accounts of Louis and Mary Leakey hunting for fossils in Africa and dreamed of following in their footsteps.

After the family moved down to a nearby town beside Lake Arrowhead, White led other boys on explorations of the wilderness. They taught themselves to use topographical maps and practiced spotting small artifacts on the ground. As a teenager, he and his friend Wes Reeder surveyed archeological sites, and their detailed reports became part of the county museum collection. They found traces of cultures that had occupied the mountains over the centuries—Native Americans, Spaniards, Mormons, gold rush miners, loggers, and settlers—and picked up atlatl points, arrowheads, pottery, and a cavalryman button. "He would sit and make meticulous notes about everything we encountered during our surveys," said Reeder. "He would draw every flake and artifact."

The boys earned an invitation to join an archeological excavation at Calico Hills in the Mojave Desert headed by archeologist Ruth "Dee" Simpson and Louis Leakey. In the final years of his life, Leakey championed a theory that humans occupied California more than fifty thousand years ago—tens of thousands of years before they were believed to have crossed the Bering land bridge into the Americas. Other scholars, even Leakey's wife, Mary, dismissed the notion as nonsense and a sorry consequence of the old man's thirst for public adulation.

White's father, a veteran of World War II, was a no-nonsense man who commanded road crews. When the boys were young, they visited construction sites and rode in the cabs of heavy machinery with their dad. Down in the Los Angeles Valley, the sixties exploded in drugs, sex, rock music, the antiwar movement, and civil unrest. Bob White, who also served on the local school board, made sure his boys took no part in that nonsense. He taught hard work and planning—and had little tolerance for screwing up. "My dad was a real disciplinarian," recalled White. "There was no messing around."

While most mountain boys took labor jobs or went to the Vietnam War, White attended the University of California at Riverside, a new campus in the state's vaunted university system. White began college as a C student—he considered it a success because he'd been warned that most mountain kids flunked out—until something took hold and he began to apply himself with an intensity that never subsided. He entertained vague aspirations of becoming a marine biologist but found himself drawn to the solitude of deserts with wide-open spaces, earth laid bare, and layers of geologic time stacked atop one other. For a college assignment, he spent days observing the denning habits of rattlesnakes. Being color-blind, White developed an acute perception of shape and was undeceived by camouflage. "Snake recognition," he later quipped, "would become an exceptionally valuable career asset."

The Riverside campus was modeled after a small elite college (its provost called it "the Swarthmore of the West") and its faculty encouraged critical thinking—something that came naturally to the big-grinning, sharp-tongued student. In White's first field archeology class, the instructors quickly realized that his knowledge surpassed all other students and sent him off into the wilderness on an independent field survey. "He was always pushing the envelope and trying to do more—and do more well," recalled Jim O'Connell, the instructor.

Midway through college, White was riveted by a 1970 television documentary called *The Man Hunters*. The film showed Clark Howell's expedition to the Omo Valley and gave White his first glimpse of the fossil-rich country where he would spend his career.

"In southern Ethiopia, these barren hills may conceal the earliest evidence of man's existence," intoned the narrator. "A scientific expedition seeks a creature which was neither man nor animal but something in between."

In his third year of college, White began taking anthropology courses. He stalked out of one class after concluding the young teacher spouted textbook dogma but didn't know anything about how the science really worked in the field. In another class, White grumbled after getting an A-minus on an exam. His instructor, Alan Fix, told him to study harder next time. A few weeks later, the professor read White's final exam and recalled gasping, "God, it's like the textbook!" White later wrote the best thesis his advisor had ever seen. "He was supremely confident," said Fix. "Call it competence or arrogance, Tim had it as an undergraduate."

White aspired to only one destination for graduate school—the University of California at Berkeley, then one of the premier anthropology programs in the world and home to luminaries such as Desmond Clark and Clark Howell. The department rejected him. "He was dumbfounded," recalled Fix.

White had applied to only one backup, the University of Michigan, which was such an afterthought that he had filled out the application in pencil. The school held little appeal because none of the anthropologists specialized in fieldwork. With no alternative, White headed to Ann Arbor. The Michigan physical anthropologists, C. Loring Brace and Milford Wolpoff, belonged to a minority camp called the "single species school," which held that all human ancestors belonged to one lineage. "Tim was extremely smart, very careful, and sarcastic," recalled Wolpoff. "As long as he was a student, we got along really well. We stopped talking to each other a couple of years past graduation. I think a lot of hostility must have built up in Tim that he just sat on until he was free of me."

Nobody worked harder. "He lived physical anthropology, 24-7," remembered Bill Jungers, another graduate student at the time. "He didn't have much of a social life. He loved argument. He and

Milford engaged each other on a daily basis on one argument after another. More often than not, Tim knew more about what he was talking about than Milford."

C. Loring Brace, another mentor at Michigan, foresaw that White was destined for a brilliant career as a "super producer" of his science. "His dedication to his work is almost monastic," Brace wrote in a recommendation, "but it reflects real enthusiasm and not the grim form of careerism that one sometimes encounters." White scrutinized both friend and foe, eschewed academic politics, and avoided the temptations of publicity. "He is a cautious scholar and will resist making interpretations of any kind until evidence is compelling," Brace continued. "Sometimes this has the effect of making him seem a little narrow, but it means that he will never be guilty of facile or shallow generalizations." White was too contrarian to be corralled into any school against his will and independent to the point of obstinance. He didn't have the highest grades but certainly displayed the most zealous work ethic. And he knew how to make sacrifices: to avoid being drafted and sent to Vietnam, White deprived himself of food so his weight fell below the minimum requirement for the military. "If he has any weakness," Brace added, "it is manifested in a certain rigidity of mind—he simply refuses to consider certain theoretical possibilities in the absence of overwhelming evidence."

White was so devoted to his science that he literally moved into the graduate student research laboratories, which were located in an old maternity hospital. He slept in a secret lair hidden behind a partition and saved money for travel to Africa. In 1974, he obtained a dream job—an invitation to join the expedition of Richard Leakey, who was making headlines with discoveries in Kenya. Like the Afar Depression of Ethiopia, the Turkana Basin of Kenya is one of the richest fossil fields on earth. The apprenticeship placed White at the epicenter of paleoanthropology during the fossil gold rush of East Africa—an early example of his lifelong habit of pushing his way to the frontiers of his science.

Richard Leakey cut a dashing figure. He piloted bush planes, rode camels, hobnobbed with European royal families, issued pithy quotes, and claimed several major fossil discoveries. He had supplanted his father, Louis, as the power broker of African paleoanthropology and took control of the Kenyan National Museum at age twenty-three in what his biographer called "political maneuvering worthy of Machiavelli." Leakey aspired to turn the Kenyan museum into the African equivalent of the Smithsonian Institution.

Leakey also had been running a field expedition in Koobi Fora on the east side of Lake Turkana. Before launching his own operation, Leakey had spent one season in the Omo Valley of Ethiopia with Clark Howell. Leakey had not attended university, and Howell urged him to obtain a proper academic degree. Leakey sensed condescension from the professional scientists and resented it. "I was constantly being reminded that I was sort of a 'tent boy,'" he complained. After he became famous, Leakey quipped that he went to universities only to lecture, not study. In 1968, he moved downriver and started his own operation on the east side of Lake Turkana in Kenya, Ethiopia's neighbor and historic rival for fossil glory. At Koobi Fora, he left field surveying work to his team of African fossil hunters, known as the Hominid Gang, and issued standing instructions that important discoveries should be left in place so the field team could call Nairobi on the radiophone and Richard could fly up and formally collect the specimen—often with photographers or VIP guests in tow. A reporter from the *New York Times* described one such visit in 1974:

> Once the plane touches down on the grassy landing strip at his camp, Richard goes for a quick swim, wraps a colorful African print sarong about himself and pads barefooted to the head of the dining table under a grass roof. There he presides like a firm but benevolent chieftain over the team of loyal and hardworking scientists delighted to work at as rich a fossil ground as Richard Leakey commands.

That year, one of the assistants who stepped off a plane on the dirt airstrip at Koobi Fora was Tim White. White would spend three years working at Leakey's field camp, but the relationship grew testy. In his second year, Leakey angrily charged that White damaged a fossil during recovery; White claimed the fossil was broken by one of the African workers and his real offense was "spoiling a photo-op." Afterward, Leakey sent the group to another locality—where one of the Hominid Gang found a 1.6-million-year-old *Homo erectus* cranium that falsified the single-species theory of White's graduate school advisors. White absorbed the wisdom of the Kenyan fossil hunters about how to dig water holes, how to recognize sand riverbeds that might trap a vehicle, how to duck thorns when driving through the bush, and how to spot bones. He had arrived in Africa with dreams of fossil fields ready for harvest but found a different reality on the ground—they were being stripped bare. International researchers competed to find new sites and score quick discoveries. They scoured the landscape for human ancestors but ignored other data that could be invaluable for scientific research. Over the years, White developed a different vision: sites must be managed carefully for long-term sustainability to extract as much data as possible.

At the time, Leakey was embroiled in a dispute about his prize discovery—a skull that he claimed to be 2.6 million years old and the oldest representative of *Homo*. Critics questioned the skull's age and taxonomic assignment—they thought the fossil was only 2 million years old and actually *Australopithecus*. The strongest opposition came from Berkeley geologists who specialized in dating volcanic rocks and Clark Howell, the leading field worker of his era, who had spent years collecting fossils upstream on the Omo River and recognized that Leakey's dates did not match the fauna typically found in strata of that age. Pigs are useful "biomarkers" because certain species are associated with specific time periods. Find a pig and it can tell you the time. When Leakey's pig expert also questioned the age of the skull, Leakey fired him and replaced him with Tim White. Leakey assigned White and John Harris (who also happened to be Leakey's brother-in-law) to conduct an independent analysis, but they concluded the critics were right. Even worse, the problem went well beyond pigs—the *entire* stratigraphy at Koobi Fora was screwed up. They drafted a paper that affirmed the younger dates and described the implications for Leakey's prize skull. Not ready to concede, Leakey insisted the paper stick to pigs and omit any discussion of human ancestors. "Tim got very angry," recalled Leakey. "He said, 'You're denying us academic freedom!' I said, 'It's not academic freedom. It's protocol.'" Leakey pointed out that *he* was the boss, raised the money, and could fire employees for insubordination. He had requested a report on pigs—not hominids. "Tim absolutely refused and slammed the door," said Leakey. "He never forgave me for that incident." Afterward, White confided in his journal: "I lost my temper. I was really shaken up—actually crying because of the insult." White denied slamming any doors and said Leakey expressed his displeasure through intermediaries, not directly. Nonetheless, he concluded that "the Firm"—as the Leakey team was known by insiders—demanded blind loyalty and used coercive tactics to stifle dissent. Shortly after the blowup, White joined forces with Don Johanson and filled his ears with all the shortcomings of the Leakeys. "Tim is nowhere near the P.R. guy that Don is, but make no mistake he was, particularly back then, a very ambitious guy," recalled Tom Gray, a fellow member of the Lucy team. "I'm sure Tim recognized that Don was a ticket for him." To them, Leakey became an antihero—a model to outperform.

IN 1977, WHITE WAS HIRED TO FILL A TEMPORARY VACANCY IN THE BERKELEY anthropology department—the same one that had rejected him as a graduate student. When a tenure-track position opened up, White applied but did not make the short list.

Faculty supporters intervened to reinstate White to the group of finalists. His advocates included the two Clarks—Desmond Clark and Clark Howell, the leading archeologist and anthropologist working in Africa. The disputes from Kenya seeped into department politics, and the hiring process became heated. One of Richard Leakey's closest academic collaborators, archeologist Glynn Isaac, sat on the Berkeley faculty and had an antagonistic relationship with White. Another professor complained of "irreconcilable differences in opinion between Professor White and others on this faculty concerning issues of Human evolution." Eventually, White won the job and proved himself a popular teacher with a flair for theater. In a demonstration of stone tools, he mimicked a popular skit on the TV show *Saturday Night Live* by hacking apart a putty figurine of Mr. Bill beneath the overhead projector. He entertained audiences with irreverent takedowns of Leakey orthodoxy. Former student Susan Antón remembered Introduction to Physical Anthropology as "an oversized class of some 800 students taught to rock concert perfection by Tim White. I'd waited a year

Tim White navigating with air photos in Ethiopia, 1981.

to take his class because a friend had declared that 'Tim White is God.'" Yet some faculty disdained the young professor as abrasive and unimpressive.

Among his critics was Sherwood Washburn, one of the most influential anthropologists of the twentieth century. In 1951, Washburn had written a manifesto for "new physical anthropology" that called upon the profession to abandon its archaic fixation with classification and measurement of bones and instead pursue the biological mechanisms of evolution—an agenda that basically gave the profession its marching orders for the second half of the century. Similarly, Washburn was one of the first anthropologists to recognize the revolutionary implications of molecular biology—namely that humans recently shared common ancestors with chimps and gorillas—and one of the first Americans to seriously promote studies of wild primates. He had lost faith in old bones and complained that trying to reconstruct human history from anatomy was "the most fundamental mistake in the history of evolution." Likewise, he called comparative anatomy a "primitive, nineteenth-century science." Instead, Washburn argued that evolutionists should be guided by molecular biology and primatology, and use fossils only as a third source of information to confirm the narrative told by the first two. By the time White arrived at Berkeley, Washburn had ceased original research in order to teach and theorize. The University of California appointed him a University Professor, an honor reserved for a select faculty, many of whom were Nobel Prize winners. Washburn was a Boston Brahmin, a *Mayflower* descendant, the holder of three Harvard degrees, and pugnacious. He did not like Tim White.

White heard scuttlebutt about a campaign to block his appointment, which confirmed his long-standing suspicion that power brokers conspired against him. "He always felt persecuted," recalled William Simmons, department chair at the time. "He was very opinionated. I don't know if he had strong likes, but he had strong dislikes." The tension only grew when the department

Sherwood Washburn.

assigned Washburn's protégés to serve as teaching assistants for White. "There was a Washburn-inspired graduate student instructor revolt," White recounted of his assistants. "In the meantime, my enrollments were going up and his were going down."

The big blowup came in 1981, when Washburn scheduled a seminar to critique a paper recently published by one of White's friends. White showed up, publicly jousted with Washburn, and infuriated his senior colleague. "Tim White came to the seminar and prevented any normal discussion," Washburn complained in a departmental memo. "He repeated himself time after time and seemed unable to follow other people's comments. He talked so much I found the only way of being heard was to interrupt." Washburn filed a grievance—his first, he noted, in forty years in academia. "White's behavior," he raged, "was the worst I have ever encountered."

White insisted that he spoke up only to challenge inaccurate statements by his elder colleague and portrayed the complaint as a ruse to sabotage his chance of tenure. The department already was investigating a prior complaint against White by Washburn and a group of his students. A department memo itemized their allegations:

Professor White is authoritarian.

Professor White is verbally abusive, usually in criticism of an absent person or third party.

Professor White ignores students who come to see him in his office.

Professor White violates canons of confidentiality.

Professor White violates canons of collegial behavior in his sarcastic criticism of others.

Professor White has a violent temper, sometimes physically expressed by kicking and throwing objects.

A department committee found substantial evidence to support the first and last accusations (but made no determination on the others). Although there was no question that White could be bossy and volatile, his behavior was deemed insufficient to justify disciplinary action and he was merely warned about "his style of criticism." By good fortune, a student had recorded the seminar confrontation. The committee listened to the tape and determined that Washburn's complaints were unfounded. The investigators concluded that White might have been confrontational, but no more so than his antagonists—including Washburn. Ultimately, the committee cleared White of any misconduct and attributed the clash to lingering bad blood over his hiring, department politics, and "the unyielding nature of the personalities involved."

A senior faculty member urged White to visit Washburn and make peace. When they met, the elder statesman of American physical anthropology delivered a blunt assessment of his junior colleague. White recalled: "I wrote down what he told me at that meeting: your appointment was the biggest mistake our department has ever made." The university did not agree. The following year, White got tenure.

The same qualities that made White prickly also made him a formidable scientist. He didn't care about who his work delighted or enraged. When people asked what he hoped to find in Ethiopia, White shrugged, *nothing in particular*. He just wanted to know the past, whatever it might be. He demanded data—cold, hard facts, and not academic bullshit or theoretical arm waving. Washburn suggested that anthropologists should temper their conclusions with odds because, after all, everybody was bound to be wrong about *something*. (Indeed, Washburn believed that scientific impasse often arose from the human foible of false certainty.) White responded that he had zero interest in playing games of probabilities because he wanted to know the past *for sure*. He knew he wasn't good at math or theory, but he was punctilious, relentless, and he excelled in the fossil fields. Intellectuals and theorists found him as dull and hardheaded as a battering ram—and just as dangerous when he arrived at their gate.

Shortly after White returned from his first trip to Ethiopia, another field team reported finding an ancestor older than Lucy. In the Sahabi Desert of Libya, an expedition led by Noel Boaz of New York University picked up fossils attributed to a primitive ape near the root of the human family. The prime specimen was described as a clavicle, or collarbone, bearing similarities to bipeds, robust muscle attachments like arboreal apes, and anatomy "reminiscent of a pygmy chimpanzee, the living hominoid that probably most closely resembles the common ancestor of apes and humans." Then the bone came under the cold scrutiny of Tim White.

In 1983, White stood up at a conference and announced that

the ostensible collarbone was not a human ancestor at all—it was a dolphin rib. *Flipperpithecus,* he called it. He had spent more than a year studying the bone with his graduate students and laid out a devastating indictment that his colleagues had mistaken a marine mammal for a primate. The great White shark tore apart his victim and flashed a slide of a bottlenose from a *National Geographic* cover with the title "The Trouble with Dolphins." After making a big splash in scientific journals and the media, *Flipperpithecus* sank into obscurity.

White was an apex predator of his discipline—a man who took to heart the idea that science was an endeavor distinguished by falsification, or putting theories to empirical test. Enemies not only resented him; they *fucking hated* him. He cared not one bit. "Self-criticism will enhance your science," White later wrote, "self-esteem will not."

As far as White was concerned, there was a single version of truth. And the greatest obstacle to finding such unambiguous answers was the profession's oldest bad habit—forming preconceptions before finding fossils. "The difference between science and science fiction," White declared, "is evidence."

His mission was to find it.

Chapter 5
THE FARTHEST OUTPOST OF HUMANKIND

Year after year, the fossil hunters pleaded to resume research in Ethiopia. Throughout the 1980s, Ethiopian officials assured them, *maybe next year,* and extended the moratorium another season. At one point, White, Clark, and Johanson returned to Addis Ababa to receive an award for their scientific contributions to Africa presented by the dictator Comrade Mengistu himself. The permanent secretary to the Ministry of Culture assured the scientists they would have no trouble getting into the field the following year. He was wrong: the rewriting of antiquities laws progressed at a glacial pace. Cold War suspicions continued to chill the science. In December 1983, security forces in Addis Ababa raided a clandestine meeting and caught an American CIA agent conspiring with Ethiopian dissidents. The *Ethiopian Herald* headline announced, COUNTER-REVOLUTIONARY ELEMENTS CAUGHT RED-HANDED HERE. Such intrigue did not help the plight of the foreign scientists asking to be turned loose to explore the countryside.

But the moratorium only affected research in the field—not work in the museum. The new lab building that had been constructed with American taxpayer funds sat waiting for scientific expertise. For years, the fossils and artifacts collected from Hadar, the Omo Valley, and elsewhere had sat in disorganized heaps in the museum or old storage sheds. Specimens were being lost, broken, and scattered. Insects had eaten identification labels. A beautiful skull of the ancient elephant *Deinotherium* had been smashed when somebody dropped a statue on it. Many fossils had been exported to Europe, Kenya, or the United States. The museum staff remained untrained in proper conservation techniques. In his first visit to Ethiopia in 1981, White had been horrified to find the fossils that he and his colleagues had cleaned so carefully in Cleveland rolling loose inside cardboard boxes. Lucy herself lay in an old paint container. Bones gathered by his predecessors in the Middle Awash sat in a disorganized mess. "It was in the most horrendous condition," White recalled with disgust, "covered with dust, cobwebs, rat shit, dead rat carcasses."

The new museum building was supposed to change all that. It had been financed with an $86,000 grant from the U.S. National Science Foundation to Desmond Clark and his Berkeley colleague Clark Howell, despite reservations from the U.S. State Department about investing in such a project in a hostile, turbulent dictatorship. The Berkeley scholars had argued that the lab was both an urgent scientific priority and a "moral obligation." The building would safeguard the valuable collections, serve as a show of good faith, and allow American scientists to retain a toehold in Marxist Ethiopia. It also represented an investment in the country's development: keeping antiquities in Ethiopia ensured that locals would be trained as curators and scholars.

The Berkeley scientists embarked on a decades-long effort to develop the museum lab. They drew up floor plans, ordered cabinets, and imported tools for making fossil casts, microscopes, and airscribes (fossil-cleaning devices with vibrating needles like mini jackhammers). They trained the Ethiopian staff to use equipment and curate collections. "We started opening up old boxes and cleaning bones," recalled Yonas Beyene, a museum employee. In 1984, the Berkeley team returned and found the new lab building was being used by the government to produce propaganda for the tenth anniversary of the revolution. The lab designed for making casts of fossils had been commandeered to make statues of Mengistu. Berhane went to see to the minister of culture and convinced his government to restore the building to its original purpose as a center for science.

Photograph © 1984 David L. Brill

Berhane Asfaw and Tim White examine a fossil elephant tusk in the National Museum of Ethiopia. During the moratorium on field research in Ethiopia, the Berkeley team helped outfit the new lab, organized fossil collections, and trained a core group of local personnel. In the background are Gen Suwa, Don Johanson, Woldesenbet Abomssa (the museum storekeeper), and Desmond Clark.

Even Lucy was conscripted into the revolution. In the national museum, Dinknesh joined an exhibit showing Ethiopia progressing from *Australopithecus afarensis* to the ancient Kingdom of Aksum to the socialist state. Meanwhile, the Marxist government staged a show to impress the world. In the central city plaza at Meskel Square, the military held a Soviet-style military parade. Soldiers, workers, and peasants filed past a reviewing stand where

Mengistu stood beside East German leader Erich Honecker and Soviet Politburo member Grigory Romanov. But no display could mask the unrest in the countryside. The ancient scourge of famine ravaged the northern provinces, and hundreds of thousands of peasants abandoned their homes and began walking toward wherever there was rumor of food. In relief camps, aid agencies sometimes reported thousands of deaths per day. Images of emaciated peasants became familiar to television viewers around the world and inspired the Live Aid concerts and the 1984 single "Do They Know It's Christmas?"

Berhane returned home every year during his graduate studies at Berkeley. While other Ethiopians flew home with western goods and gifts for their families, Berhane brought luggage stuffed with material for the lab. "One year I had thirteen suitcases—only one was my personal one," he recalled. He imported books, computers, and equipment. He repatriated fossils that had been exported by American scientists and dreamed of turning the national museum into a scientific center worthy of his country's treasures. White also devoted himself to the museum; he was a relentless systemizer who wanted every specimen on a properly organized shelf. Year after year, he and his students invested sweat equity and waited for the day when the government would allow them to resume their quest for early humanity. Yet the ban on foreign research did have at least one positive outcome. During those years of scientific isolation, Ethiopia produced something unique for a fossil-rich African nation: a generation of indigenous scholars.

In 1988, Berhane finished his doctorate at Berkeley. Some American colleagues doubted he would return to Ethiopia once he had his Ph.D. in hand. Why would anybody choose to work in the world's poorest nation under a brutal regime—especially somebody already tortured by that government? The suggestion offended Berhane in a way that most westerners could not fathom. He was *Ethiopian*—an identity born of pride in the country's heritage and tempered by suffering. Indeed, if *anything* united the diverse peoples of polyglot Ethiopia, it was their shared history of suffering. He would not abandon his country nor betray the faith invested in him. In America, he could only teach. In Ethiopia, he could *serve,* and there was no better place for a scholar of early humanity. "Our parents didn't want us to come back," recalled his wife, Frehiwot. "They felt, 'Oh my god, now they're safe. They don't have to die.'" But Berhane never wavered. "In his field of study," explained Frehiwot, "he knew there was no professional there who could do the job."

In 1989, Berhane was appointed head of the fossil lab at the Ethiopian National Museum and later acting director of the museum itself. He trained young workers and helped the most promising ones secure scholarships to doctoral programs in the United States and Europe. "He put in the first core for paleoanthropology in the national museum," said Yonas Beyene, who won a scholarship to study archeology in France. "He wanted to build institutions, build infrastructure, and build the human capacity." Berhane became the father of human origins research in his country—and an example of the Ethiopian proverb: *The fruit that matures first becomes the pasture of birds, and the one who expresses himself first exposes himself to the hatred of others.* Berhane held views similar to his mentor Tim White: science was not a consensus-building exercise and there was a right way to do things. In one case, the storekeeper hid an important fossil and for years refused to tell Berhane where it was. Some became resentful: why did he favor certain employees? Suspicious bureaucrats questioned his intentions: why did his museum need a computer? Why should the Derg government allow students to study abroad? And, given that Ethiopia had banned antiquities expeditions, just what was he doing out in the countryside?

Shortly after his return, Berhane convinced his government to authorize a countrywide survey for new antiquities sites. The survey provided field experience for the emerging generation of Ethiopian anthropologists, paleontologists, archeologists, and

During the ban on foreign fossil-hunting expeditions, Berhane Asfaw launched a countrywide reconnaissance for new fossil and archeology sites. The mostly African team developed new techniques to use NASA imagery to locate fossil-rich sediments. The survey launched the careers of several prominent African scientists and found sites that produced major discoveries. Here the crew poses in the Ethiopian lakes region with a banner from the National Geographic Society, one of its funders. (*From left to right:*) Alemu Ademassu, Yohannes Haile-Selassie, Yonas Beyene, Berhane Asfaw, Gen Suwa, Pelaji Kyauka (a Tanzanian graduate student at Berkeley), Sileshi Semaw, Giday WoldeGabriel, and Tim White.

geologists—several of whom went on to become internationally known scientists. Only two foreigners were invited: Tim White and his Japanese student, Gen Suwa. In 1988, White secured a $29,000 grant from the U.S. National Science Foundation; the agency hoped to foster more goodwill toward American researchers when Ethiopia lifted its moratorium. (Before funding arrived from the NSF and the National Geographic Society, White donated personal funds to kick-start the project.)

Even in the Great Rift Valley, fossils are only found on a tiny percentage of the land surface, so success begins with figuring out *where* to look. The challenge is like trying to catch glimpses of the play of evolution by finding a few pinholes in the theater wall; the windows are rare and usually provide only episodic glimpses of the full story. Armed with declassified images taken from NASA satellites and space shuttles, the survey developed new techniques for locating fossil-bearing geological layers. By measuring the intensity of reflected sunlight, the scientists could distinguish different types of rock and identify geological exposures likely to contain petrified bones and artifacts. Space images pinpointed targets for boots-on-ground inspection, and field observations refined the interpretation of images. This heavens-to-earth approach enabled the crew to locate the rare haystacks where they might find the rarer needles. Later, these methods would lead to major discoveries—and pioneer techniques employed around the world.

Such an expedition dominated by Africans was a novelty on the continent of human origins. As the chief geologist, Berhane recruited his old university friend, Giday WoldeGabriel. It was not the first time Berhane had asked him for a favor. When being hunted by security forces in 1977, Berhane asked Giday to buy his bus ticket so he could flee the capital.

Giday's biography provides a glimpse of the trials faced by scientists of his generation. Born a peasant in the remote highland province of Tigray, he lost his father when he was a child. His family tilled a subsistence farm of chickpeas, barley, and teff, and plowed the ground with sharpened tree branches. One day, he saw huge clouds of locusts darken the sky and descend on his village. The locusts reduced crops to stalks and trees to bare branches. To avoid starvation, the family walked to a distant region and worked for another farmer until the next growing season. He was schooled by American missionaries under a sycamore tree in his village and later won a scholarship to a Lutheran boarding school. He earned another scholarship to attend Haile Selassie University. In the twilight of the emperor's regime, Giday entered a vibrant university with an international faculty and watched it all gutted by the revolution and the Red Terror. He was briefly expelled for joining a protest against the old regime. After the revolution, the university was closed for two years and the students were dispatched into the countryside for an ostensible development campaign—but Giday and others eventually concluded the whole thing really had been a strategy to disperse opposition to the Derg regime. The emperor's name was stripped away and the school was renamed Addis Ababa University. One of his favorite geology teachers was murdered by the militia. He started with a class of eighteen geology students, and only two remained by the time he graduated in 1978; the rest had been killed, been arrested, gone home, or fled into exile. Giday himself steered clear of politics and just tried to finish his degree, support his family, and stay alive. A cadre of communist bloc academics arrived, and Giday completed a master's degree under a Soviet geologist. "He didn't speak English, I didn't speak Russian," recalled Giday. Upon graduation, he became a geology instructor at the university and met a few foreign scientists who persisted working in Ethiopia. Jim Aronson, a kindly, bespectacled geologist on the Lucy team, recruited Giday to Case Western Reserve University in Cleveland.

Then came one of the great ironies of his life: after surviving famine, revolution, and terror in his homeland, Giday arrived in an ostensibly safe country and was nearly beaten to death. When he took a field trip to examine classic geological formations of the American west, robbers snuck into his campsite at night, clubbed him with firewood, stole his money, slashed his car tires, and stuffed him into the trunk. "Then they left me there to die," Giday quietly recounted. He wrote his Ph.D. thesis on the Ethiopian rift system (the research moratorium did not apply to geologists, so he was free to conduct field surveys). Nevertheless, he decided not to pursue a career in his homeland and instead took a position as a geologist at the U.S. Los Alamos National Laboratory (where he worked on projects such as underground nuclear waste disposal and geothermal energy) and became an American citizen with a security clearance.

When invited to join the inventory project, Giday returned to

Ethiopia with a sense of patriotic duty—and anxiety. Civil war raged in the countryside, and Tigrayan guerillas had turned his home province into a combat zone. At checkpoints, government soldiers eyed the geologist suspiciously because his name instantly identified him as Tigrayan. In the ministries, people whispered nervously about the growing insurrections in the northern provinces. As the inventory finished its work, the country fell apart. By the late 1980s, the Soviet Union—then in its own final years of disintegration—cut military and economic support to Ethiopia, and the Derg regime crumbled. Several ethnic-based rebel armies with captured weaponry pushed toward the capital. Deserters from the Derg conscript army tried to flee across the Afar Depression, and many were massacred by the local tribes. During the

Photograph © 2002 David L. Brill

Giday WoldeGabriel.

inventory, crews looking for fossils stumbled upon more recent remains of dead soldiers.

In the project, Giday began working with the sole Caucasian scientist—Tim White. "Tim never delegated anything," said Giday, "and then complained about not getting help." Once White disapproved of the manner others had loaded gear atop a car, and he climbed up to fix the job himself. "He went up and started throwing things in all directions," recalled Giday. "People were running away from being hit. He was wild and uncontrollable." Yet over time, Giday recognized an Ethiopian virtue in his American colleague: *selflessness*. White asked much for the science but little for himself. He worked harder than anybody and pushed himself to the point of exhaustion. He *suffered*. White habitually disregarded academic colleagues who urged him to restrain his criticism, but he learned to listen when his African friends warned, *This is not Ethiopian*. "He is more African than Africans," observed Giday, "and more Ethiopian than most Ethiopians."

Between 1988 and 1991, the survey team spent 127 days in the field and found new sites that later produced major discoveries. But the Ethiopian government barred the explorers from entering existing project areas, so all the fieldwork brought White and Berhane no closer to the promising fossil fields they had glimpsed in 1981—and whatever came before Lucy. "There were in fact papers in Gewane instructing the local authorities to arrest us if we tried to go," recounted White, "signed by one of the highest guys in the country during the Derg. Shows you the kind of influence that our enemies had at that point."

In 1990, the government finally lifted the ban on field research. At last, the Middle Awash team could resume the search for fossils in the Afar Depression. Nine years earlier, they had piled stone cairns atop animal fossils because the museum had no space to store them. The cairns were still standing—and infested with snakes. Downstream at Hadar, Don Johanson joined a separate expedition back to the place where he had found Lucy. Both teams found more *afarensis* fossils, but still nothing older. A headline in the journal *Science* announced: "First Hominid Finds from Ethiopia in a Decade." But the momentum was short-lived. A few months later, the civil war reached Addis Ababa and threatened to destroy the little museum and Ethiopia's precious record of early humanity.

IN MAY 1991, THE REBEL ARMY MASSED ON THE OUTSKIRTS OF THE CAPITAL. The Derg regime collapsed, and international observers feared Ethiopia would descend into anarchy like its neighbor Somalia. President Mengistu boarded a plane and flew into exile. Generals lost control of their troops, and bands of heavily armed soldiers roamed the city. Jail doors were thrown open and prisoners streamed into the streets. In London, U.S. diplomats brokered negotiations between the government and several ethnic-based rebel movements. Following reports of chaos in the capital, the chief American negotiator gave his blessing for one rebel army, the Tigrayan People's Liberation Front (TPLF), to enter the city and restore order. Western diplomats had been impressed by the discipline of the peasant force and the political sophistication of the TPLF leaders, particularly an intellectual named Meles Zenawi, a former university acquaintance of Giday and Berhane.

In Addis Ababa, 2 million inhabitants huddled in their homes and awaited the climactic battle of the civil war. Berhane and his staff locked down the museum. In addition to fossils like Lucy, its treasures included the throne of Haile Selassie, jewels, and crowns of past monarchs. A breakdown in civil order raised the specter of looters. The main building of the museum was an Italian mansion built for the viceroy during the fascist occupation of the 1930s. It stood inside a walled compound with verdant gardens and a fountain. Behind the old villa stood the little fossil lab and storeroom. As director, Berhane had ordered the installation of several fireproof vaults in case of artillery barrage. Throughout the city, police and soldiers abandoned their posts to avoid being shot by the advancing rebel army. Berhane went to his museum guards, a dozen men with mismatched weapons and backgrounds (one was an out-of-work surgeon), and begged them to remain on duty. At his request, they stayed.

That evening, Berhane drove home through near-deserted streets. Addis Ababa was a sprawling city where corrugated metal roofs stretched for kilometers across a mountain valley dotted with eucalyptus trees. Tanks veered through the streets as if the drivers didn't know where to turn. Government troops wandered the capital with rocket launchers and automatic weapons. Soldiers armed for battle begged for food and water from unarmed civilians—an image of Ethiopian humility that remained with Berhane for the rest of his life. He arrived at home to find his wife waiting anxiously with their two young children.

The fighting began that night. Bursts of gunfire echoed over the rooftops. Berhane phoned the museum every hour and the guards reported looters pillaging the government offices next door. "Don't allow anybody to get into the compound," Berhane ordered. "You have the right to shoot and I am responsible."

But the looters never touched the museum. By morning, the rebel army had seized the capital. State-run media instructed the populace to stay home and prohibited cars from the streets. Berhane announced he would walk the eight kilometers to work. "What?" gasped his wife, Frehiwot, fearing for his safety. "How can you go?" He set off carrying a bag of food for his museum guards, who had remained at their posts all night and would be hungry. He made his way through checkpoints, passed rebel tanks with kids clambering aboard, and found his guards waiting outside the museum gates in civilian clothes. They had stripped off their uniforms and surrendered their weapons. Berhane peered through the gates and saw the compound occupied by rebel fighters with big afros, shorts, and AK-47 rifles. The TPLF guerrillas were sometimes derided as *woyanes*, an Ethiopian term for "country hicks." But the

Tigrayans wore the slur with pride: they represented the *peasants*, not the elites who had dominated the country for the last century. They refused to admit the director into his own museum.

The next day, Berhane and an assistant walked to the local military command post and asked to speak to the officer in charge, a wiry man who looked like he had marched all the way across the country on meager rations. Berhane explained the situation and asked the officer to rearm his guards. The rebel looked at him quizzically. *What's a museum? What's a fossil?* If these things were so valuable, why had not Mengistu taken them when he fled? Berhane led the commander back to the museum for a tour. He explained: *These treasures belong to all humanity. We are Ethiopians, and it is our duty to curate these bones for the world. If we destroy them, history will hold us responsible.* The commander ordered his soldiers to return the weapons, donated an extra AK-47, and supplied an armful of ammunition. Ethiopia's senior scholar of human origins returned to his museum with a bundle of bullets slung in a shirt.

Lucy and the antiquities remained safe. Whether the anthropologists would remain safe was another question.

THE NEW GOVERNMENT BANNED FIELDWORK AGAIN BUT RELENTED LATE IN the year. Only those most tolerant of risk ventured into the postwar chaos—like Tim White. In December 1991, White departed with a small crew. The highways remained unsafe due to bandits, so the fossil team joined a truck convoy escorted by military vehicles and pickups mounted with machine guns. They traveled over pockmarked roads and past burned-out hulks of tanks, abandoned rocket launchers, and armored personnel carriers, and finally pulled off the highway in Gewane, a small town beneath the two-kilometer-tall Ayelu volcano. It was the place of the last phone, the last pavement, last gas, last supplies, and last outpost of state authority before entering tribal territory. The expedition rendezvoused with its next escort of five soldiers from the Ethiopian

People's Revolutionary Democratic Front (EPRDF), the former rebel army that had morphed into the fledgling government. They also hired three Afar workers, who were nervous and fearful of what waited ahead.

The Middle Awash fossil fields straddled the borderlands between two hostile tribes. For generations, the Afars and Issa had waged a blood feud whose origins nobody could remember. Warriors raided back and forth, stole livestock, and killed enemies. Until recently, the Derg army had enforced a tribal border along the main highway through the Afar desert. The road linked Addis Ababa to the Red Sea port of Assab and served as the country's economic lifeline for fuel, arms, food, and supplies. Soldiers had kept Issa to the east and Afars to the west. But once the Derg regime vanished, so did border control. The Issa pushed westward across the highway to the Awash River for water and grazing land. The fossil fields became a battleground.

The caravan motored into a badlands dotted with acacias and thorn bushes. Here and there, stone crypts of Afar warriors stood atop ridges. Over time, the tombs collapsed and bones spilled down the slopes. A fossil hunter had to be careful not to mistake modern remains for ancient fossils.

The cars crossed a basalt plateau and came to a place called Sibahkaietu. The Afars had warned that the first Issa would appear there. Sure enough, they did, identifiable by the white skirts wrapped around legs so gaunt they looked like sticks. The Issa men slung assault rifles over bare shoulders. Soviet weapons had flooded the country, and the AK-47 had become as ubiquitous as the traditional daggers that the tribesmen strapped to their waists. White and the others watched as the Issa drove huge trains of cattle and goats across the river floodplain.

The fossil team and soldiers stopped to greet some Issa elders and explained that they had come to look for bones that had turned to stone. As a token of goodwill, an elder presented the visitors with two goats. Alemu Ademassu, the museum fossil-casting technician, shoved the goats into the backseat and they drove on. White recorded in his field notes: "Many women, children and men were seen, all waving happily."

That would soon change.

A young museum employee named Yohannes Haile-Selassie drove the lead car. Two soldiers sat beside him in the front seat, and the Afar workers in the back grew nervous as they ventured deeper into Issa territory. The Afars wore civilian clothes because traditional attire might have gotten them shot on sight. Near the Awash, the grassy floodplain turned into a thick forest. The cars wove through the floodplain shrubland—and into an ambush.

Gunfire erupted and tracers whizzed through the air. A tire burst. "All we could hear was flying bullets," recalled Yohannes. "They were shooting from both sides. We were trapped in the middle."

As the cars spun around to flee, another Issa charged directly toward them. It was the old man who had given them the goats a few minutes before. He shouted at the gunmen to stop shooting.

The soldiers leapt out of the cars and scrambled to defensive positions for a tense standoff. Soon about twenty Issa gathered in a noisy bunch. The shooters seemed unfazed by the presence of the soldiers, and utterly unapologetic about nearly killing the fossil team. The Issa elder explained that some kids had gotten spooked by the unfamiliar sight of vehicles. "They thought it was some kind of an Afar army coming to get rid of them," recalled Yohannes. Bullets had hit White's trailer and narrowly missed its cargo of fuel tanks. Miraculously, nobody had been hurt.

The old Issa tried to smooth things over. Could the visitors please take their goats, drive ahead, and forget this little misunderstanding? "We were terrified—very, very terrified," recalled Zelalem Assefa, an antiquities officer. He looked nervously at White and feared the hard-charging American would insist on pressing ahead. But even White was spooked: *too fucking dangerous.*

Around the next bend might be another band just as trigger-happy as this one—and perhaps they would be better shots. The team dared advance no further. As they drove away, more gunfire echoed from behind. Whatever fossils lay ahead would remain hidden yet another year.

THE TEAM WITHDREW TO THE TOWN OF GEWANE, CAMPED BESIDE AN ABANdoned military installation where they felt safe, and spent a few miserable days searching nearby outcrops until calling it quits due to rain, illness, and lack of sleep from incessant truck traffic, barking dogs, and hyenas. They watched Afars and Issa trading potshots in the distance. A shitty field season—literally, since everyone got sick with diarrhea. Zero hominids.

Then Berhane lost his job. In the transition to the new government, opposition mounted against his leadership. "The museum was just a one-man house," said Mulugeta Feseha, a former employee. "Any office key, scholarship, the rights of people in there—all were dictated by Berhane." Another critic later complained: "The problem is that Berhane has convinced himself that he can do everything and that nothing, in this country, goes without his blessing." In 1992, the new government informed Berhane and a number of officials from the old regime that they would be dismissed. The next day, Berhane departed on a scheduled trip to attend a scientific conference in Las Vegas, where he planned to deliver a report on the recent discoveries from Ethiopia. Rumors spread that fossils had disappeared and Berhane had fled into exile—or that he had not left the country at all and really was hiding at home. Some employees protested being locked out of the workplace. "We stayed outside of the office for three days," said Mulugeta Feseha. "Every office was locked because the keys were with Berhane. . . . The minister was upset and he said, 'OK, you break the door.'" In Las Vegas, Berhane received a fax summoning him back to Ethiopia. He delivered his scientific presentations, flew home, and found government inspectors conducting an audit and investigation at the museum. He quarreled with them, locked his lab, and went to see the minister of culture. "The minister told me that he heard that Lucy is no more in the safe. I told him it is not true," recalled Berhane. "I had coffee with him and he told me to go to the vice minister." The next day Berhane went to see the vice minister—the man who had faxed him in Las Vegas. "He had arranged for me to be arrested," said Berhane. "He told security that I entered into the laboratory without permission and interrupted government work." Berhane spent two days in jail and was fired. Upon his release, he became just another civilian scientist pleading for permission from a bureaucracy now controlled by adversaries. The government inspector who had quarreled with Berhane at the museum became a journalist and published critical articles about his tenure as director. There was one consolation—his team held the permit for one of the richest fossil territories on earth, a place where a scientist could spend a lifetime discovering human history back to its deepest roots. By the end of that year, they would pick up the trail they had sought for so long—the creatures who came before Lucy.

BUT MORE MACHINATIONS WERE BREWING. THE ORIGINAL PERMIT HOLDER of the Middle Awash territory had not given up his claim.

There was a long backstory. In the 1970s, geologist Jon Kalb had been one of the original members of the first fossil-hunting team to explore the Afar Depression, but he fell out with his partners and was ejected from the group. As it happened, he had narrowly missed the biggest discovery of his generation. "Jon Kalb—his footsteps were within five feet of the Lucy skeleton!" recalled Johanson. "He walked right by it the year before!" After being cast out, Kalb convinced the Ethiopian government to award him a portion of territory that had not been explored by his former teammates—roughly eight thousand square kilometers upstream from Hadar that later became the Middle Awash. After securing a permit for the territory in 1975, he spent three years conducting surveys, marveled at its abundance of fossils and stone tools, and called attention to its promise to yield fossils back to the dawn of the human family.

Kalb cut an unusual figure among fossil hunters. A fun-loving adventurer who spoke in a Texas drawl, he dropped out of the geology program at Johns Hopkins University without finishing his Ph.D., lived permanently in Ethiopia, and staged expeditions throughout the year (most western researchers had Ph.D.s, worked at universities or museums, and traveled to Africa for a single annual expedition). Kalb was no academic. Once he delivered a lecture at the National Academy of Sciences with a raging hangover after a night of heavy drinking and suddenly realized he was standing at the podium with no memory of how he got there. In another episode in Ethiopia, he announced finding an ancient hominid that turned out to be a fossil baboon. After the revolution, many westerners fled or were expelled, but Kalb continued to run expeditions to the Afar and encouraged training of Ethiopians at a time when almost no indigenous scientists worked in the profession. He warned that Ethiopians were growing resentful toward foreign scientists. "If the Government decides to throw us all out of the country," he said in 1977, "I can understand their reasons." As it turned out, Kalb himself would be expelled—but for very different causes.

Rumors circulated that Kalb might be a CIA agent. In that era, a climate of suspicion was fueled by post-Vietnam revelations about CIA assassination plots and destabilization campaigns against foreign governments. Kalb received funding from an obscure foundation, and he employed an aerial photo specialist who also happened to be the son of the CIA station chief at the U.S. embassy in Addis Ababa. According to Kalb, in 1974 Ethiopian antiquities official Bekele Negussie visited his office to discuss a delicate matter. "He said Don Johanson has accused you of being in the CIA," recalled

Kalb. "He warned me to be very careful." (Johanson denied spreading the rumors.) Kalb disavowed any connection to the intelligence agency and attributed the charge to blackballing by rivals. During the terror, such an accusation could be very dangerous—especially for Ethiopians who associated with that foreigner. "The Derg people did not make distinctions of individuals," recounted Aklilu Habte, the former minister of culture. "If you were American, you must be with the CIA. If you had an American friend, people would suspect *you* were in the CIA."

Lacking his own credentials, Kalb tried to set himself up as a scientific impresario in the style of Richard Leakey by recruiting academics (in 1976, he invited Tim White, who was busy elsewhere and declined). In 1977, Kalb submitted three proposals to the NSF seeking $500,000 for fieldwork, and all were rejected. The NSF typically circulates proposals to scientific peers for evaluation, and three were from Berkeley. At least one of them, archeologist Glynn Isaac, warned the agency about the CIA allegations—and he was not alone. "It is widely rumored that Kalb works for the CIA," NSF official John Yellen wrote in an internal memo in 1977. "The grad students who work in Ethiopia told me they thought it was true." One of Kalb's senior academic collaborators, archeologist Fred Wendorf from Southern Methodist University, complained of a smear campaign. "What would he have reported?" Wendorf scoffed about the idea that Kalb worked as an intelligence operative. "That the water tasted like hippo urine?" To salvage the Middle Awash mission, Wendorf invited a trusted old friend—Desmond Clark, a "damned good man" who commanded respect from both academics and Ethiopian antiquities officials. But Kalb refused, explaining "I no longer trusted anything connected with Berkeley."

And the NSF did not trust Kalb. "It would be well worth our while to have another Leakey in Addis but Kalb is not the man and he should not be given any encouragement," Yellen wrote in the NSF memo. Soon it became a moot issue. In August 1978, Ethiopian security agents expelled Kalb. About ten days later,

Desmond Clark arrived in Ethiopia and asked to take over the orphaned territory, and his Berkeley team acquired the huge concession. (Wendorf later withdrew to pursue projects elsewhere and gave his blessing for Clark to assume command.) With Clark in charge, the NSF became more enthusiastic about funding research in the Middle Awash—and that launched the 1981 expedition of Clark, White, and Berhane and all the events that followed. Kalb suspected a conspiracy to steal the territory, but the charge did not pass muster with his former partner. "It was because he lacked formal academic credentials," Wendorf insisted, "not because they heard he was a CIA agent, that the NSF openly sided with the Berkeley group."

White joined the Berkeley expedition after the Kalb drama but accepted the judgment of the Ethiopian government—and the valuable territory that came with it. "It's very unlikely that Ethiopian security would expel somebody on the basis of baseless rumors," White told a reporter in 1983. "If in their judgment he's a CIA man, that's good enough for me."

It remains unclear whether Kalb's expulsion from Ethiopia really resulted from the CIA rumors, which Ethiopian officials had been hearing for years. Kalb blamed the whisper campaign for his banishment. But Desmond Clark later recounted hearing a very different version of events from antiquities official Berhanu Abebe: Kalb was ordered to leave the country because he failed to uphold his agreement to bring professional scientists into the field, store his collections in the national museum, and raise funds to improve the facility—and then tried to enlist political allies to circumvent those requirements.

Enraged, Kalb waged a long campaign to prove he had been sabotaged by scientific rivals—particularly the Berkeley team and Johanson. (The accusations levied by his Ethiopian allies were aswirl when the government suspended foreign expeditions in 1982.) For a decade, the NSF denied the CIA allegations played a role in the rejection of Kalb's funding applications, but Kalb obtained documents though the Freedom of Information Act, tracked down members of the review panel, and discovered the rumor played a larger role than acknowledged. In October 1982, an internal NSF inquiry confirmed that the CIA allegation did indeed arise during deliberations and noted that three reviewers of Kalb's proposals were from Berkeley. "Clark's eventual move into what had been Kalb's territory make it all seem a bit collusive, even if all is innocent," wrote NSF official Charles Redman. In the urgency to secure the valuable research area for an American team, the NSF bypassed its normal channels of peer review and allowed the Berkeley group "freer rein than normal projects and negative comments have been explained away too readily." (The situation became even more tangled when Yellen later joined the Middle Awash team and conducted an NSF-funded archeology project there with his wife, Alison Brooks.) Unfortunately, Redman concluded, there was no good remedy and the only recourse was to "try to soothe Kalb's resentment." In the end, the NSF did exactly the opposite. Redman told Kalb he had been the "victim of very unfortunate circumstances" but denied any impact of the CIA rumors. Kalb responded with a handwritten note: "You know and I know who, how & when the allegations were used at NSF." He warned that the NSF was making a grave error—and vowed to prove it.

In 1986, Kalb sued the NSF seeking $1 million in damages. One year later, the two sides reached a settlement, and the NSF acknowledged discussing the CIA allegation, apologized, and agreed to pay Kalb $20,000. (The NSF continued to maintain that Kalb's proposals were declined for valid reasons.) The dispute came to symbolize the cutthroat competition of paleoanthropology, and in 1992 the muckraking book *Impure Science* described it as a case study of "the breakdown of peer review at the National Science Foundation." Kalb later wrote his own memoir, which again portrayed the Berkeley takeover in a Machiavellian light—and likened the courtly Desmond Clark to King Leopold in the scramble for Africa. In fact, the correspondence shows that Clark took pains to avoid any perception that he sought to "muscle in." He invited Kalb's Ethiopian students to join his new team and even tried to find a role for Kalb—until the Ethiopian government explicitly barred any further participation by the expelled geologist. Kalb and his supporters in Ethiopia remained bitter, and for decades he would haunt his successors in the Middle Awash like a ghost. With the fall of the Derg, the door opened for him to return to Ethiopia.

Chapter 6
BADLANDS

Oftentimes the hardest part of finding fossils was just getting into the field. So it was again when the Middle Awash team staged its next expedition. Another round of bureaucratic battles stalled the expedition in Addis Ababa in late 1992. Government officials invited Kalb to return and retake portions of his former territory from the Berkeley group. Berhane and his team fought to keep their project area intact and spent weeks marching from one office to another pleading for letters of permission. "They tried to intimidate Tim—'if you take Berhane, you will lose your permit,'" recounted Berhane. "Finally, the minister intervened and gave us a permit." But then the team couldn't get antiquities officers to chaperone them as required by law. "Three people declined in a row," recalled archeologist Yonas Beyene. One senior government official would issue an order, another senior official would contradict it, and low-level employees tried to steer clear of trouble with either faction. "The only way to protect yourself was to come up with some excuse not to go to the field," recounted Zelalem Assefa, an antiquities officer at the time. "The priority was to keep your job." After one month of delays, the team finally secured the requisite permissions, personnel, and supplies (they even needed a permit to purchase toilet paper) and set off in late November. Until then, the team had concentrated on the better-known east side of the Awash River, but it had become too dangerous since the Issa takeover, so the expedition headed to the west side, an expanse they had barely explored. They did, however, know that it contained older sediments—and better chances of finding fossils older than *Australopithecus afarensis*.

On the first morning in the wilderness, the fossil team awoke in camp as the sunrise illuminated the underbellies of clouds in pastel. The massive cone of the Ayelu volcano came into view in the south. Nearby lay an Afar village of loaf-shaped huts constructed from gnarled branches and grass mats. As day broke, Afar children took livestock out to forage, spotted the strangers and cars, and ran back to their village.

The Afar positioned their settlements in open ground or atop hills so enemies could not approach undetected. Any intruder caused alarm. During the civil war, the Afar had killed Derg soldiers and plundered their weapons. Recently, there had been clashes with soldiers of the new government and more gunfights between Afars and Issa. The villagers woke an Afar warrior named Elema, a

Photograph by and © Tim D. White 1992

Fierce pastoralists: the Afar were armed with military weaponry and regularly clashed with Issa over territory, water, and livestock. Here Elema tends cattle and goats at Bouri Village.

wiry man with a thin mustache and a wide gap between his incisor teeth. He rose, shouldered a long assault rifle, and strapped on a military belt with extra ammunition clips, a knife, and a Soviet-made pistol plucked from a Derg soldier. Later, when he gained enough English to describe himself, he would proclaim, "Elema was fucking bad ass!"

He set off across the badlands. He found the strangers camped by a line of vehicles and trailers, their cargoes concealed by olive-colored tarps. He saw a handful of Ethiopian highlanders and a couple of foreigners, one with skin the color of a cow's udder and hair the color of desert grass. Elema had seen a few white men before in his travels to Djibouti, but never here in his homeland. Two highland Ethiopians—scientists, he later discovered—stood off to the side, watching the scene like hyenas.

Elema was enraged to discover the intruders had come with Afar guides from other clans, who had no business escorting strangers into his territory. Elema's father was an Afar *balabat*, a chief of the local Bouri-Modaitu clan—and nobody entered their domain without permission. He unslung his rifle and demanded the interlopers leave—immediately.

The Afar guides explained that the scientific expedition had come looking for bones. They presented letters from the central government in Addis Ababa. The visitors even had permission from the sultan of the Afar people.

Elema ripped up their letters. Here *he* was boss—not the government in Addis, not Afar from neighboring clans, and not even the sultan. The Afars argued loudly in their language, indecipherable to the Amharic-speaking highlanders. Giday WoldeGabriel watched the Afar guides. They looked uncomfortable—not a good sign. As the argument became heated, one Afar tried to grab Elema's rifle. The expedition leaders shouted to back off, fearing the confrontation might escalate into another shooting. Eventually the Afars walked away from camp, sat in the desert, and argued all morning.

Off on the side, White watched yet another frustrating drama unfold. All year long, he'd studied maps and satellite images, picked targets, prepared gear, and attended to everything he could imagine that might be required. Years of experience had taught him one near certainty of fieldwork: unforeseen complications always reared up and threatened to derail the entire endeavor—shootings, bureaucratic roadblocks, denunciation by "colleagues," political attacks against Berhane, flat tires, blown engine radiators—and now this motherfucker with a rifle. "Africa," he once groused, "always wins."

The entire mission depended on a few weeks of fossil collecting. By now, he had hoped to be many kilometers farther down the trail. Every hour spent dicking around was sixty minutes squandered in a preciously short field season. Was this expedition going to turn into another abortive cock-up like the last one?

A few hours later, the Afar negotiators stood up. A deal had been reached. Elema had a job on the team. The man who began the day threatening the scientists would become their new guide.

NOT UNTIL AFTERNOON DID THE EXPEDITION REACH ITS DESTINATION along the bank of the Hatayae River, where the crew pitched its base camp for the season. White's father, the highway superintendent, had taught him the importance of logistics: every excursion—whether a road camp or family outing—depended on planning, supply, and organization. He never rested until the camp was in military order. He decreed how tarps were laid, tents staked, cars parked, tools organized, and ropes coiled. No detail was too trivial to escape his attention or trigger a tongue lashing if he didn't like what he saw.

As the camp was being readied, an Ethiopian knocked deadwood from a tree and a branch crashed onto White. He fell to the ground with blood streaming from his scalp, reached up, and felt a nail-size thorn driven into the temporalis muscle of his head. He pulled out his Leatherman all-in-one tool and asked Giday WoldeGabriel to extract the thorn with the pliers. And that was that—back to the mission. Over the years, White racked up injuries and illnesses like passport stamps: malaria, dysentery, giardia, hepatitis, pneumonia, and more. Once, sitting under the dinner tent, he bit a piece of goat meat, hit a bone splinter, and shattered a tooth. He screamed in agony, cursed, lay his head on the table, and instructed colleagues to patch the broken tooth with super glue. But the tooth kept breaking apart and White finally ordered them to just cover the exposed root with a glob of adhesive. He would get it fixed after the field season.

The Afar men hired as "security" didn't make anybody feel particularly secure. Firearm safety was a concept as alien as written language. The expedition leaders ordered the guards to click on the safeties of their rifles, but when the scientists turned their backs the gunmen flipped those safeties off again. Many of the Afars had never ridden in vehicles. They squeezed into the crowded seats and swung their rifles with no concern about where the muzzles might be aimed. The scientists insisted all passengers point guns straight up—"towards Allah," White gestured—and forbade grenades in the cars. (There had been close calls when a *bombe* thumped onto the floor and rolled on their feet while everybody counted *one, two, three* and braced for an explosion.) Afar men tried to spit out of the car without realizing that windows were covered by a transparent substance called *glass*. Some became carsick. "You're all packed in the back of a Land Cruiser and then one of them just vomits out camel milk, which is what he had for breakfast," said anthropologist Bruce Latimer. "I can't even describe how bad it is."

The scientists explored a wilderness where few outsiders had ventured. One day they came upon a camel herder who leveled his AK-47 at the windshield. The crew ducked behind the engine block while Elema exited to parley. The gunman explained that he had mistaken them for some kind of strange, terrifying animal. He had never seen a car before.

Every new location required diplomacy and *dagu*, the Afar ritual greeting and sharing of news. The scientists called upon clan leaders, presented gifts of sugar and coffee, and requested permission to search for fossils. It consumed precious field time, but there was no way around it. Some clans erected barriers and demanded tolls. Few Afar people had seen foreigners or government representatives, and the expedition always drew crowds—Afar women in colorful skirts, naked children, gaunt elders with dignified beards and *hata* walking sticks, and stern men with guns. Did these strangers really seek worthless bones turned to stone? Some suspected an ulterior motive. Did the explorers really desire gold, minerals, or oil? Had the government come to arrest and kidnap their people? Elema proved an invaluable ambassador and helped negotiate passage, scout trails, and find stream crossings. "Making roads, talking to people," Elema later recalled. "Fucking politics, every day."

WHITE SOMETIMES COMPARED THE TEAM TO THE CORPS OF DISCOVERY, THE Lewis and Clark expedition that explored the Louisiana Purchase of North America. In the Middle Awash, the paleonauts not only mapped unknown lands but also sought windows into the deep past—geological exposures where the remains of ancient life eroded to the surface of the badlands. (In eastern Africa, most fossils come to the surface naturally by erosion and are found during foot surveys. In southern Africa, by contrast, they are typically found embedded in limestone caves.) The geological time scale is divided into a series of epochs and most of human evolution is confined to three—the Miocene (23 million to 5.3 million years ago), Pliocene (5.3 to 2.6 million years ago), and Pleistocene (2.6 million years ago to 11,700 years ago). The Middle Awash preserved sediments from all three.

The first task was reconnaissance. The huge territory remained mostly unknown, so the explorers spent weeks making surveys and skimming the land before deciding where to focus attention. Day

The Geological Time Scale

	EPOCH	BEGINNING DATES	EVENTS
YOUNGER	Holocene	11,700 years to present	• Begins after last ice age • Natural warming trend • First human civilizations • Agriculture
	Pleistocene	2.6 mya (million years ago)	• *Homo sapiens* populate globe • Stone tools ~2.5 mya • 3 mya—2.5 mya *Homo* genus appears
	Pliocene	5.3 mya	• 3.2 mya Lucy • ~4 mya *Austalopithecus* genus appears • Oldest substantial fossil record in human family
	Miocene	23 mya	• Almost no human or ape fossil record 4 mya—8mya • Separation of human lineage from African apes in late Miocene (date estimated from biochemistry but no fossil evidence) • Diverse Miocene apes in Africa and Eurasia in early and mid-Miocene • Relatively warm global temperatures
	Oligocene	34 mya	• 30 mya split between ancestors of modern apes and modern old world monkeys • 30 mya East African Rift System begins to form
	Eocene	56 mya	
OLDER	Paleocene	66 mya	• Squirrel-sized proto-primates

after day, they combed the badlands from the Awash River to the western slopes of the rift margin. They passed extinct volcanoes, crumbling Afar crypts, and cliffs pocketed with hyena dens. They slogged up channels of ephemeral rivers that sliced into ancient layers of geologic time and over hills of chocolate-colored clay that once had been the bottoms of primeval lakes. They wandered over landscapes paved with innumerable Stone Age tools and blades. "We kept on moving," said Giday WoldeGabriel, "because there was nothing to hold us in any one place."

Eroding outcrops opened windows on deep time—5 million years here, 2.5 million there, 1 million years somewhere else. Broken fossils hinted at lost forms of ancestors that seemed parodies of their descendants. Giant otters. Giant hyraxes. *Sivatherium*, extinct relatives of modern giraffes, with short necks and stout bodies. *Hipparion*, horses with three toes instead of hooves. *Deinotheres*, ancient elephants with tusks that dangled downward like twin ivory goatees. *Gompotheres*, elephants with four tusks. *Agriotherium*, a bear nine feet tall. The fearsome pig *Notochoerus*, weighing as much as a grand piano, and other pigs as tiny as lapdogs. *Chalicotheres*, horse-size herbivores with stubby hindlimbs, and long forelimbs upon which they knuckle-walked like apes. They had teeth like rhinos and clawed feet like anteaters, and for decades naturalists assigned their heads and bodies to two different species—a reminder that extinct creatures might resemble nothing alive today. To grasp the past, one must abandon the expectation that it will resemble the present.

Somewhere amid this strange menagerie lived another bizarre group of animals—bipedal primates who gave rise to modern humans. Unfortunately, they were among the rarest of all. At Bouri, the territory of Elema's clan, they found the first remains of ancient humanity—two broken thighbones attributed to *Homo erectus*. But those pieces were only about one million years old and the searchers prioritized older fossils. The expedition crossed the rift floor to the western margin, searching the oldest rocks whose age approached the estimated split time of humans and apes. Those long treks failed to yield a single hominid fossil.

ON THE THIRD WEEK, THE TEAM EXPLORED A SERIES OF HILLS WHOSE heights commanded sweeping views across the rift floor. The basalt massif of an extinct volcano known as Dulu Ali loomed over ranges of hills and gullies. Space images revealed its true geological identity: an enormous dome. In the deep past, underground eruptions formed a giant cyst of lava that pushed up a 250-meter-thick stack of sediments from the Pliocene and Miocene epochs. (Geologists would later precisely date the strata between 3.9 and

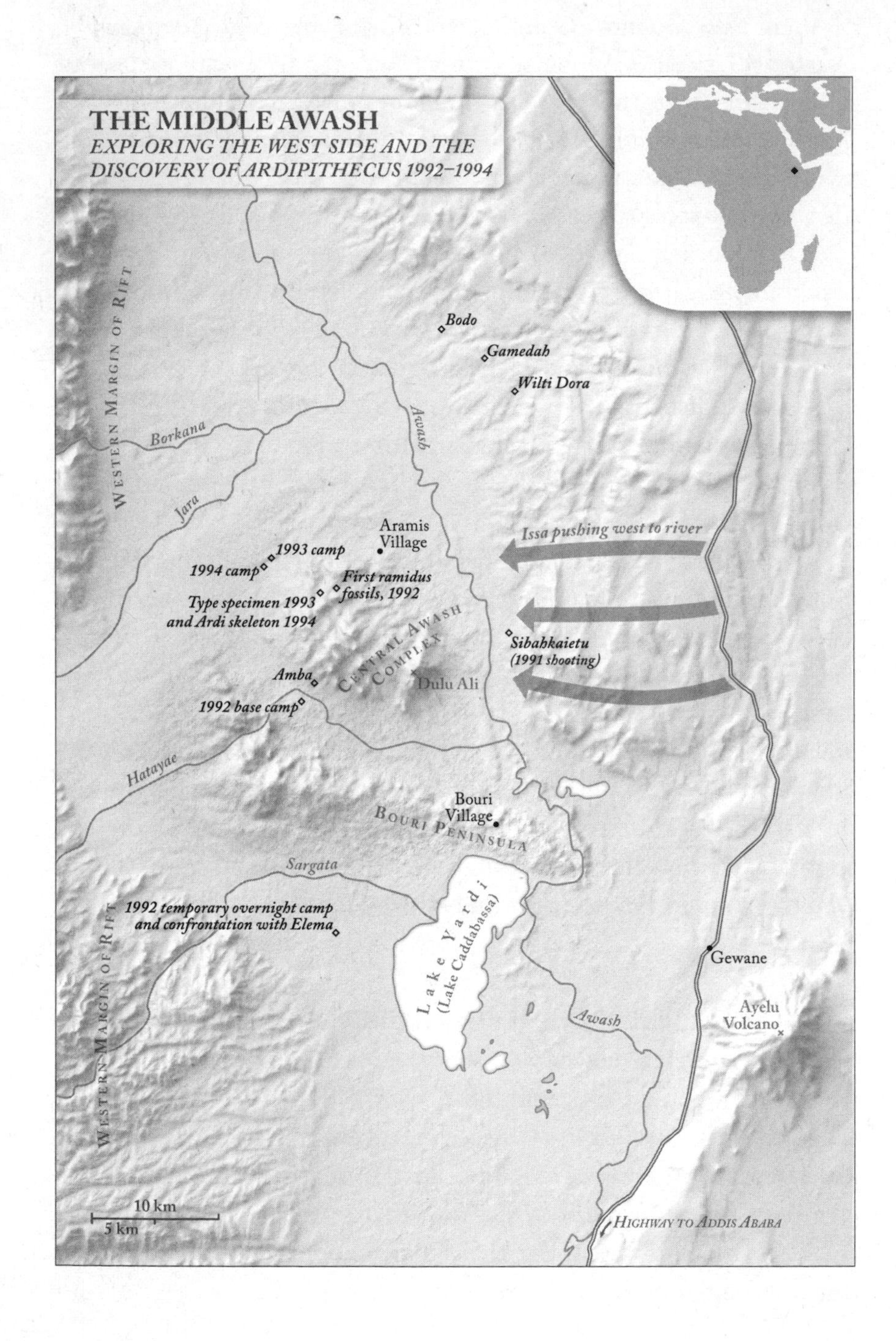
THE MIDDLE AWASH
EXPLORING THE WEST SIDE AND THE DISCOVERY OF ARDIPITHECUS 1992–1994
Western Margin of Rift
Borkana
Jara
Awash
Bodo
Gamedah
Wilti Dora
Aramis Village
Issa pushing west to river
1993 camp
1994 camp
First ramidus fossils, 1992
Type specimen 1993 and Ardi skeleton 1994
Central Awash Complex
Sibahkaietu (1991 shooting)
Amba
Dulu Ali
1992 base camp
Hatayae
Bouri Village
Bouri Peninsula
Sargata
1992 temporary overnight camp and confrontation with Elema
Lake Yardi (Lake Caddabassa)
Gewane
Ayelu Volcano
Awash
Western Margin of Rift
10 km
5 km
Highway to Addis Ababa

5.6 million years old.) Called the Central Awash Complex, the formation sprawled over five hundred square kilometers and was sliced by drainages whose eroding slopes exposed the remains of creatures that lived millions of years ago, making it a prime target to illuminate the dark ages before Lucy.

Without geological context, fossils are meaningless. One of the oldest disciplines in geology is stratigraphy, or the study of the layering of the earth. In 1667, Danish anatomist Nicholas Steno proposed the law of superimposition: sediments form in succession from older below to younger above. In the late 1790s, British canal engineer William Smith added what became known as the principle of faunal succession: certain types of fossils predictably appeared in certain layers of rock. Fossil-bearing layers stack from older below to younger above.

Sediments settle horizontally, but over time they may tilt with geological faulting—and this part of the Afar Depression was a train wreck of tectonic movement. The Central Awash Complex dipped a few degrees to the northeast, allowing scientists to predict that the oldest layers would be tipped up on the backside on the southwest, which was where they began searching. At a place called Amba, they found one of the richest fossil sites White had ever seen, with exquisitely preserved bones of horses, hippos, and other fauna. It would take months before the geology lab at Berkeley could provide precise ages for the rock samples, so until then the explorers would have to make a best guess based on "biochronology," or the fossil fauna. Among the cache was *Nyanzochoerus kanamensis*, a pig species well-known from other sites in Africa. White knew his pigs and this one spoke to him: 5 million years old. That meant the fossils in this layer were roughly 2 million years older than Lucy and tantalizingly close to the then-estimated split time between the ancestors of humans and chimpanzees. Somewhere in this field, he figured, there *should* be fossils near the deepest roots of the human lineage. It was the right time, right geological window, right guild of animals, right everything—except right then no damn hominids, no matter how hard they looked.

Finally, the team gave up and moved higher up the complex to younger deposits near an Afar village called Aramis. White had low expectations because he had made a fruitless pass through the area in 1981. Back then, what he saw seemed so barren that he wondered if Kalb's expedition had already stripped all the fossils; unbeknownst to him, his predecessors left just as frustrated. This time they would check more thoroughly. White sent the surveyors on a long walk down a grassy plateau. Instead of intact bones they found fragments. Instead of jaws they found teeth—or pieces of teeth. Some fossils bore scars from gnawing or signs of being etched by digestive acid. The remains at their feet had been mauled, munched, and shit out by ancient carnivores. That was apropos: the fossils seemed like crap.

Then somebody cried out, "I've got one."

The eyes of fossil finders are as idiosyncratic as personalities. Some people excelled at finding skull pieces, others rice-size rodent bones, and Gen Suwa was a wizard of teeth. A meticulous Japanese scientist, Suwa had studied under a famous primatologist back home. Japan has a long tradition of primatology, but Suwa aspired to study ancient humans. No such programs existed in Japan, so he headed overseas to graduate school at Berkeley. There he quickly established a habit of working late in the lab and poring over drawers of teeth, which he found both peaceful and mesmerizing. One night, his reverie was broken by the arrival of a stranger: Tim White, freshly returned from a trip to the Cleveland Museum of Natural History, encountered a like-minded soul who haunted the lab in the late hours. The next day, White invited Suwa to his office and pulled out some bone fragments to quiz the new student. To simulate the jumble of fossil sites, the professor typically mixed pieces of hominid bone with other stuff such as a raccoon molar, baboon incisor, coconut shell, or shard of pottery. Suwa identified every bone. In fact, Suwa found his first human ancestor fossil not in the field but in the Berkeley lab—a tooth that other experts had misidentified as a monkey.

Gen Suwa studies a newly discovered hominid tooth near Aramis in 1992.

Teeth are time capsules. As the cutlery and pulverizers of our digestive system, they are engineered for durability. Dental enamel is the hardest substance in the body, so teeth survive long after bones have turned to dust and are the most common fossils. The older the species, the more its fossil record is dominated by teeth. Some are known *only* by teeth. All apes, including humans, share the same basic formula of top and bottom tooth rows: four incisors, two canines, four premolars, and six molars. Variations in tooth size, geometry, and function provide clues about how lineages have evolved through time.

For his dissertation, Suwa had proposed a study on hominid teeth collected from the Berkeley expedition in the Omo Valley of Ethiopia. His advisors enthused: *Splendid idea!* Ever precise, Suwa decided the topic was too broad: he would focus on just premolars. *Wonderful,* said his professors. Eventually Suwa came back a third time. The topic still demanded tighter focus, he explained—the *lower* premolars. Suwa's exactitude rivaled that of his mentor, Tim White. But while White was volcanic, Gen was stable as bedrock.

On that day in the field, Suwa had been following an Ethiopian surveyor who had just picked up a monkey fossil. A monkey always was a promising signal, since it revealed that primates were present—and human ancestors were big primates often found alongside monkeys. Suwa walked slowly through the little gully and scanned the ground. Amid a thin cover of pale grass swaying in the breeze, he spotted a familiar shape. He picked it up. *Upper right third molar. The wisdom tooth.* The upper chewing surface formed the distinctive bowl shape of a molar and two whitish roots formed a V-shape. *Hominid.*

Once again, a pig provided a clue about age. That morning, somebody had picked up a broken mandible of *Nyanzachoerus jaegeri*, a pig that had gone extinct by 3.8 million years ago. Nearby

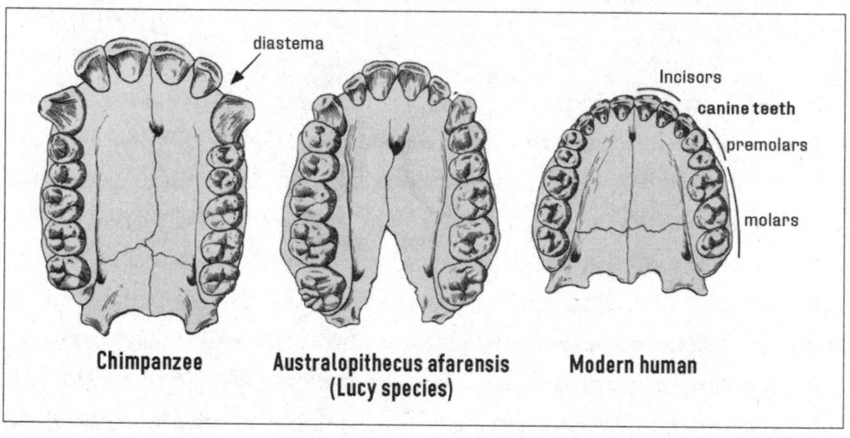

Humans and other apes share the same basic formula of teeth. Each half of the tooth row contains *(from front to back)*: two incisors, one canine, two premolars, and three molars. The characteristics of these teeth differ between species and provide important signals of evolutionary change.

Chimpanzee: U-shaped dental arcade, wide incisor teeth, large honing canines, diastema (gap between canines and incisors), single-cusp premolars, and thin enamel coating.

Australopithecus afarensis (the Lucy species): U-shaped dental arcade, reduced canines, small diastema, and big chewing teeth with thick enamel. The *Au. afarensis* teeth were often described as intermediate between humans and chimps.

Human: Parabola-shaped dental arcade, canines small and diamond-shaped, no diastema, two-cusp premolars (also known as bicuspids).

lay the fossil of *Anancus*, a primitive, mastodon-like species that walked the earth 4 million years ago. Those two fossils had come from a layer above, indicating Suwa's tooth must be older.

Suwa had found exactly what the expedition wanted, yet couldn't hide his disappointment. "Too bad it's only a third molar," he grumbled. In this scattering of crushed and mauled fossils, the piece might be the only human ancestor they found, and it happened to be the *least* informative tooth in the mouth because third molars do not reliably distinguish between species. Wisdom teeth offered little wisdom. But the tooth did reveal human ancestors were there. The question was, could anybody find more?

The crew dropped to hands and knees and spotted fragments they had missed when walking upright. A few feet away, they found an incisor tooth and upper arm bone. For the next few days, they returned to crawl, sieve, and pick up more isolated pieces, including teeth and a shard of skull. At 4 million years, such scraps might be all anybody could expect to find of these creatures, whoever they were. Such was the nature of the business: paleoanthropology is the aggregation of broken remains to compose a mosaic of the past.

A few days later, an Ethiopian named Alemayehu Asfaw added another important piece. Alemayehu was a taciturn, slender man with thick glasses and a quiet lisp. He had begun his career as an antiquities officer and often picked up better fossils than the foreigners he was supposed to be supervising. He had found the first *afarensis* jaws for Johanson at Hadar and a skull for Kalb in the Middle Awash. He also had an independent streak and a habit of roaming away from the group—what White diagnosed as a terminal case of "hominid fever." Although no longer employed as an antiquities representative, he couldn't leave the bone trade. In group surveys, he edged ahead and White yelled at Alemayehu *Asshole* to slow down and remain with the team.

A brief primer on the taxonomy of assholes: In the Tim White classification, the word carried many shades of meaning. A snarl of *Asshole!* might condemn an enemy as morally or scientifically bankrupt. Down a notch, he sometimes used it to vent irritation that evaporated as quick as sweat in the desert. But *You asshole* could express sardonic affection, even grudging admiration. In moments like this, White adored that asshole.

Alemayehu had found a fragment of lower jawbone of a hominid child who had died before losing its milk teeth. The teeth looked so primitive that Alemayehu had initially believed he had found another monkey. But Suwa knew better. The fragment, about the size of a face of a watch, still bore its baby first molar—which *was* the kind of diagnostic tooth that Suwa sought. The first molar clearly distinguishes humans from other apes. (In fact, the first molar had enabled Raymond Dart to correctly identify the first human ancestor ever found in Africa in 1924.) In humans, molars are large, four- or five-cusped teeth specialized for grinding (the name comes from the Latin *mola,* or "millstone"). Two things immediately distinguished these new discoveries from any known species. Lucy's genus, *Australopithecus,* bore big molars adapted to eating coarse foods. In contrast, this creature had *small* molars shaped more like apes. Second, *Australopithecus* was known for thickly enameled chewing teeth. But this creature sported thin enamel like chimpanzees, our closest ape relatives. The little jaw also preserved an incisor tooth that had not fully erupted when the child died. A chimpanzee has broad front teeth and a toothy grin. "This one is narrow," observed White. "That's a hominid incisor."

The scientists felt a thrill of discovery. Molecular studies estimated that humans and chimps shared a common ancestor only 5 million or 6 million years ago, and these new fossils brought the corps of discovery a large step closer to that point. A more precise calculation of their age would have to await analysis of volcanic rock samples back at the geochronology lab. In the meantime, the pigs and other fauna provided a rough indication that these fossils were around a million years older than Lucy. After eleven years of searching, the fossil team exulted at the first traces of what they sought—the roots of humanity.

As the fossil trail got hotter, so did tensions with the local Afars. One day, a young American anthropologist named Scott Simpson was scanning the ground and looked up to see an odd-looking Afar man bearing down on him with a big rifle and scowl. The hunched figure stalked up with a strange gait and snarled in a high-pitched, grating voice. His front incisor teeth had been filed into sharp points like fangs. He wore a bracelet on his wrist—a traditional Afar kill trophy—and a strange necklace made from zippers. Simpson had no idea what the man was trying to say and shrugged. The gunman turned away and went looking for somebody who could appreciate his tirade.

Scott Simpson *(center)* celebrates a fossil discovery while Elema *(left)* and Gadi *(right)* brandish weapons at Aramis, 1992.

Aramis—the place with the promising concentration of fossils—lay inside the territory of a clan known as the Aliseras, who had had a fearsome reputation even among the Afar. Earlier, White and Bruce Latimer had been confronted by the same little man while surveying on the edge of Alisera territory. The two skeleton experts quickly diagnosed his odd posture: *pathological scoliosis*. He was a hunchback. They also took note of the zippers around his neck. In the old days, Afar warriors took testicles from slain enemies. More recently, they collected zippers from their victims—mostly Derg soldiers—and this little guy had adorned himself with them. Latimer dubbed him *Zipperman*. Later they would learn his name was Gada Hamed, or Gadi.

Zipperman was back—and angry.

The expedition had brokered an alliance with Elema and his Bouri-Modaitu clan, but that partnership held little value in Alisera territory. Not long before, warriors from the two Afar clans conducted a joint raid against the Issa on the east side of the Awash. On their way home, one Bouri fighter accidentally shot an Alisera. Or was it really an accident? Some Aliseras suspected the killing might have been intentional. The agitated hunchback was the brother of the slain Alisera, and he was enraged to find the man who had shot his brother among the employees of the fossil team. Zipperman seemed ready to collect another zipper.

The Aliseras also resented that the expedition had not hired more of their clan. The situation grew more tense as the fossil team continued working at Aramis. When the sun climbed overhead at midday, the badlands became too hot and bright to search for fossils, and the crew rested over lunch. The expedition leaders belatedly discovered an Afar tradition—any passerby was welcome to share the meal. Suddenly the desolate badlands spontaneously generated many passersby. Some brought children and umbrellas. "They forced themselves into our lunch spot," recalled Berhane. "They wanted us out—and at the same time they wanted to eat our food!" But breaking bread together failed to generate goodwill. "One day instead of bringing their *hata* walking sticks they brought their guns," said Simpson. "They were really belligerent, just reaching in and grabbing food. What they were doing was establishing their dominance."

Photograph by and © Tim D. White 1992

Berhane Asfaw takes a volcanic ash sample under the eyes of Gadi and another Afar.

On the fourth day at Aramis, White and Berhane went to collect a sample of a volcanic ash that they hoped would pinpoint an age for the new fossils. They walked to a ravine that exposed a cross section of the ash layer as Gadi followed a few steps behind and kept up a gravelly rant. White glanced at his gun: the selector was flipped to FIRE. They could no longer ignore all the warning signs. The scientists consulted their Afar guides. Elema and another guide offered the same advice: *Get out!*

The team did not wait to finish sieving. They packed unsifted soil into sacks and drove away. At last, they had picked up the trail of whatever came before Lucy. But that promise, like much in the Afar, proved short-lived.

Chapter 7
THE ZIPPERMAN'S ASH

Somewhere in ancient Africa—no one knows exactly where—a volcano erupted and blasted huge volumes of ash into the atmosphere. At Aramis, cold ash blanketed the ground like a blizzard one and a half meters thick. The eruption possibly triggered torrential storms and hellfire lightning and turned a fertile valley into an apocalyptic scene of desolation.

Over the years, life returned. Spiral-horned antelope browsed on the leaves of floodplain forests and woodlands. Groundwater springs irrigated thickets of palms, bayberry, and fig trees amid meadows of grasses and sedges. Colobine monkeys scampered through the branches. A large primate foraged on the ground and took shelter in the trees. Big cats stalked prey and packs of hyenas chewed carcasses to splinters. Crocodiles, hippos, and fish plied through rift lakes and rivers. Year after year, rivers overflowed banks, and fine silt entombed the broken bones of this ancient menagerie. Derived from the weathered basaltic rocks of the volcanic highlands, the silts bore a slightly reddish tint from oxidized iron. Centuries later, another catastrophic eruption left its mark on deep time when a volcanic vent exploded somewhere near Aramis and blasted out a coarse-grained basaltic ash that settled in a darker layer above the earlier one.

Those two events of the Pliocene epoch remained visible in the geology in two bands of ash above and below the fossil-rich, reddish silts. The scientific team labeled ash layers after Afar words for animals. The lower ash became *Gàala*, or "camel," and the upper one *Daam Aatu*, or "baboon." For ancient life, volcanoes brought disaster. For scientists, eruptions were manna from heaven. Volcanic rocks contain potassium, and its radioactive decay allowed geologists to calculate how much time had passed since the eruptions. Until that point, the field team could only venture educated guesses about the age of the fossils based on fauna such as pigs and elephants. The ash Berhane had collected under the watch of the Alisera gunmen would begin to answer an essential question: just how old were these mysterious ancestors?

THE HISTORY OF LIFE IS WRITTEN MOSTLY IN SEDIMENTARY ROCK, BUT ITS timeline is written mostly in volcanic rocks. Until the mid-twentieth century, the age of the earth remained a guessing game. Geologists estimated the ages of rocks based on their relative position in the strata or the types of fossils embedded within them. The discovery of radioactivity put dates on the timeline. In the early twentieth century, British physicist Ernest Rutherford suggested that radiation might be used to calculate ages of rocks. Every element has a fixed number of protons but may have variable numbers of neutrons. These variants are known as isotopes. Some isotopes are radioactive, unstable, and decay into stable "daughter isotopes." The decay occurs at a constant rate, unaffected by environmental conditions, and provides an absolute measure of the passage of time. After World War II, Willard Libby pioneered a new method of dating based on the decay of carbon 14. In 1949, Libby and J. R. Arnold of the University of Chicago showed that the carbon 14 method accurately predicted the ages of artifacts, such as relics from tombs of Egyptian kings, a linen wrapping from the Dead Sea Scrolls, and a scrap of bread from Pompeii. But the carbon 14 method only worked on objects less than fifty thousand years old (newer advances have extended the range to about seventy thousand years). For older rocks, scientists turned to other forms of radiometric dating. In 1956, Clair Cameron Patterson of the California Institute of Technology used a technique involving uranium and lead to calculate the age of the earth at 4.5 billion years. In the fossil field of East Africa, geologists turned to potassium, a common element in volcanic eruptions. Potassium-40 decays into argon-40, and by comparing the ratio between the two, scientists could measure the passage of time since the rock formed. Before a volcanic eruption, all argon-40 leaks out of the hot magma and effectively resets the stopwatch to zero. The accumulation of argon-40 within the cooled ash or lava provides a measure of time, like sand within an hourglass. The more argon, the older the rock.

Volcanoes commonly occur near the edges of tectonic plates and rise along the axis of the Great Rift Valley. Over the eons, they have layered East Africa with ashes and lava flows. Scientists had no means to calculate the ages of the fossils themselves, but could date the volcanic rocks above and below—and thus obtain age brackets of minimum and maximum ages. In 1959, Berkeley scientists Garniss Curtis and Jack Evernden used the potassium-argon method to calculate the age of the the *Zinjanthropus*, or Olduvai Man, skull found by Mary and Louis Leakey in Tanzania. It turned out to be 1.75 million years old—three times older than what experts had guessed. Radiometric dating revolutionized the study of prehistory. Without it, said Desmond Clark (who had spent the first decades of his career in the pre-radiometric dark ages), human origins research "would still be foundering in a sea of imprecisions sometime bred of inspired guesswork but more often of imaginative speculation."

By the 1990s, the old potassium-argon technique had given way to a newer method known as argon-argon (so named because it analyzed two isotopes of argon, one stable and the other unstable). This technique could be applied to single mineral crystals, allowing researchers to avoid contaminants, a helpful advance because violent eruptions often obliterate older rocks whose fragments become mixed with the new ones and thus skew age results. Argon-argon became the preferred method, and in 1993 it finally resolved the age of Lucy at 3.2 million years (her date had remained uncertain for two decades after her discovery). In November 1993, geologist Paul Renne of the Berkeley Geochronology Center gave White the newly calculated results for another ash from the Afar—the sample collected under the Zipperman's gun. The *Gàala* tuff was 4.4 million years old—1.2 million years older than Lucy. (Later, Renne also calculated an age for the higher *Daam Aatu* ash—and it, too, proved to be 4.4 million years old, revealing that the two layers sandwiched a very narrow window of time.) The new fossils were even older than expected and took the fossil hunters closer to the estimated split time between humans and our closest ape relatives.

WHERE DO HUMANS SIT IN THE TREE OF LIFE? THIS QUESTION POSED ONE OF the great mysteries of twentieth-century evolutionary biology. Primates span an array of hundreds of species, including apes, old-world monkeys, new-world monkeys, lemurs, lorises, and tarsiers. They share common traits, such as grasping hands and feet with opposable thumbs and toes (with a few exceptions such as humans, who have lost opposable toes), flat nails instead of claws, forward-facing eyes, stereoscopic vision, large brains, and mostly single births.

Humans belong to one particular subgroup of primates—apes. Laypeople often confuse apes and monkeys, but the two branches of the higher primate tree separated around 30 million years ago and are biologically quite different. Apes lack tails, have more upright posture, broad chests, mobile shoulders, larger brains, greater intelligence, and more complex social lives. Compared to monkeys, apes live longer, breed more slowly, and have longer intervals between births. Apes and humans are more closely related to each other than either is related to a monkey. You may think an ape looks like a monkey, but in reality the ape is genetically closer to you.

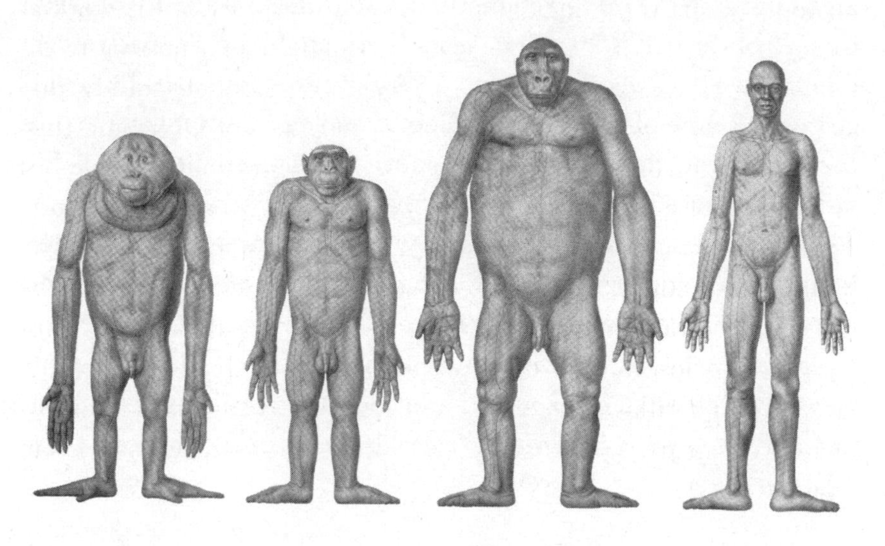

The great apes drawn by anatomist Adolph Schultz (1891–1976). (*From left to right:*) orangutan, chimpanzee, gorilla, and human.

The modern great apes include four genera: *Homo* (humans), *Pan* (chimpanzees and bonobos), *Gorilla*, and *Pongo* (orangutans). For a century after Darwin, the evolutionary relationship between these branches remained uncertain. Until the 1960s, taxonomists assigned places in the family tree based on anatomical comparisons. Chimps, gorillas, and orangutans were classified into the single taxon *Pongidae*, known colloquially as *pongids*. Humans (and fossil human ancestors) were classified under the family *Hominidae*, known as *hominids*. This old view held (mistakenly, as it turned out) that the chimps, gorillas, and orangutans were closely related to each other and more distantly related to their human cousins. In other words, the pongids perched together on one branch of the family tree and we hominids sat by ourselves on another. Many scholars assumed that humans split from the other apes at least 20 million years ago—near the very beginning of ape history.

This view was shattered by a revolution in molecular biology. In the late 1950s, scientist Morris Goodman began researching blood proteins of the the human immune system as part of his work at a cancer institute in Detroit. At the time, there was no way to directly compare the genetic code of DNA, but scientists could study the structure of proteins—which were products of DNA and thus served as a proxies for genetic relationships. Goodman, a professor at Wayne State University, found little to no differences in blood serum protein between humans and the African apes (chimps and gorillas) and slightly more differences between humans and the Asian apes (orangutans and gibbons). His work indicated that humans were closely related to the African apes. It also suggested chimps and gorillas were *not* such close kin to orangutans. Surprisingly, African apes were more closely related to humans than Asian apes. (Goodman was not the first to make such an observation. Around the turn of the century, biologist George Henry Falkiner Nuttall made similar findings with blood proteins.) Goodman asserted the traditional family tree was wrong and that humans were just another branch of the African apes.

Such unsettling results were not easily accepted—especially by experts. In 1962, Goodman attended a scientific conference at a castle in Austria and proposed that the African apes be removed from the family *Pongidae* and placed in the human family *Hominidae*. As a molecular biologist, Goodman had no allegiances to traditional classification systems based on physical appearance or vague concepts of evolutionary "grades." His proposal was rejected by eminent authorities such as ornithologist Ernst Mayr and paleontologist George Gaylord Simpson. To traditional evolutionists, it seemed perfectly obvious that the other apes should form a separate taxonomic class united by features such as opposable toes, long forelimbs, big canine teeth, and arboreal habits. Likewise, it seemed self-evident that humans stood in their own taxonomic class—one believed to have separated from the other apes tens of millions of years ago.

But another attendee at the castle summit did take Goodman seriously. Sherwood Washburn of Berkeley became excited by the prospect that molecular science might resolve long-nagging questions about the family tree. Others already had envisioned how: In the mid 1960s, scientists Emile Zuckerkandl and Linus Pauling of the California Institute of Technology posited that molecular-based methods could reconstruct family trees—and even serve as an "evolutionary clock" that told when the branching of the family tree occurred. Back at Berkeley, Washburn encouraged his graduate student Vincent Sarich to research blood proteins with a newly arrived biochemist from New Zealand named Allan Wilson. Until then, molecular scientists could measure genetic difference but not rate of change and thus remained unsure when the ape lineages split. Wilson and Sarich posited that the rate of molecular evolution accrued at a steady rate known as the "molecular clock." If the rate remained constant between species, investigators could measure the *relative* difference, compare it to a split time in the fossil record (the divergence of apes and old-world monkeys about 30 million years ago), and figure out a timeline for the branching of the lineages. By analyzing the blood serum protein albumin, Sarich and Wilson estimated that humans shared a common ancestor with chimps and gorillas only 5 million years ago—far more recently than previously assumed. (By contrast, there was a sixfold greater genetic difference between the African apes and old-world monkeys.)

In 1967, Sarich and Wilson published their results in *Science*, and it was a Big Bang for a new paradigm of recent human origins from the African apes. The new findings upset the traditional family tree with its startlingly late split of humanity from the African apes. Many anthropologists and paleontologists reacted with disbelief. "Mostly they thought we were idiots," recalled Sarich. Over the next two decades, more molecular studies affirmed the findings of Sarich and Wilson and firmly established that humans were closely related to the African apes—and all three were only distant cousins to the Asian apes. The molecular findings disowned *Ramapithecus*—an Asian fossil species once purported to be the oldest hominid—from the human family because it was too old and in the wrong place. The quest for human origins in eastern Africa gained renewed fervor. Skeptics responded: if humans were so closely related to chimps and gorillas, why didn't fossils of human ancestors look more like apes?

Then Lucy arrived.

The Lucy team interpreted the *Australopithecus afarensis* fossils within the new paradigm and emphasized the apelike aspects of Lucy's species. After the publication of the fossils, White recalled Sarich bouncing into his office "happy as a lark." For years, molecular scientists had been fending off criticism from anthropologists and, at last, molecules and morphology no longer seemed at odds—the oldest members of the human family seemed to be converging on an ancestor similar to chimps and gorillas. "The atmosphere changed because of Lucy," recalled Sarich. "They saw how apelike Lucy was. It would appear one didn't have to go back very much further to get to something that was not distinguishable from an ape. From there on, it was clear sailing."

Again and again, the Lucy team emphasized that the human fossil record became increasingly similar to the African apes as it went back deeper in time. Johanson called the chimpanzee "the general-purpose ape, the one to which we look with increasing confidence for anatomical characteristics it may have inherited from a hypothetical common ancestor to all apes."

Even so, for the next two decades, molecular scientists wrestled with the so-called trichotomy problem: what was the genetic relationship between humans, chimps, and gorillas? The methods of that era could not sort out the order of branching between the three African lineages. Some studies hinted that humans were particularly close to chimpanzees. For example, in 1975, Mary-Claire King and Allan Wilson analyzed several lines of molecular evidence and concluded that humans were more than 99 percent identical to chimps. The three-way split finally was resolved in 1989 by Adalgisa Caccone and Jeffrey Powell of Yale University. Despite the longstanding *perception* that African apes were so similar to each other, chimps actually were even genetically closer to humans than gorillas—yet another finding that blew the minds of the experts. By the 1990s, the argument about the family tree had been settled. The genus *Pan* (both common chimps and bonobos) was the nearest human relative, and our genetic code was estimated to be 98.4 percent identical. Many scholars came to share an expectation: the older the human ancestor, the more it would resemble a chimpanzee.

The molecular revolution coincided with a boom of research on wild apes. In 1960, Jane Goodall went to the forests along the east side of Lake Tanganyika in Tanzania to observe chimpanzees near a stream called Gombe. She witnessed a primate reality show of behaviors eerily similar to humans with courtships, alliances, rivalries, and murders. One day Goodall saw a chimp she called David Graybeard dip a stick into a termite mound to catch insects for food—the discovery of apes using tools, something previously assumed to be the exclusive province of "Man the Toolmaker." When Goodall reported the news, Louis Leakey cabled a famous reply: "Now we must redefine tool, redefine Man, or accept chimpanzees as humans."

Goodall was the most celebrated pioneer of a broader explosion of research on primates. Western and Japanese researchers set up observation posts at ape habitats across Africa. Old notions of human distinctiveness gave way to new visions of humans as made-over apes. The line blurred between human and simian. In 1964, one psychotherapist adopted an infant chimp and tried to raise her like a human daughter (coincidentally, she also was named Lucy). Washburn, the son of an Episcopal minister, took to the bully pulpit with a new vision of the earliest human ancestors—long-armed, big-fanged knuckle-walkers like modern African apes. In lectures, he flashed a photo of Hollywood starlet Dorothy Lamour sitting

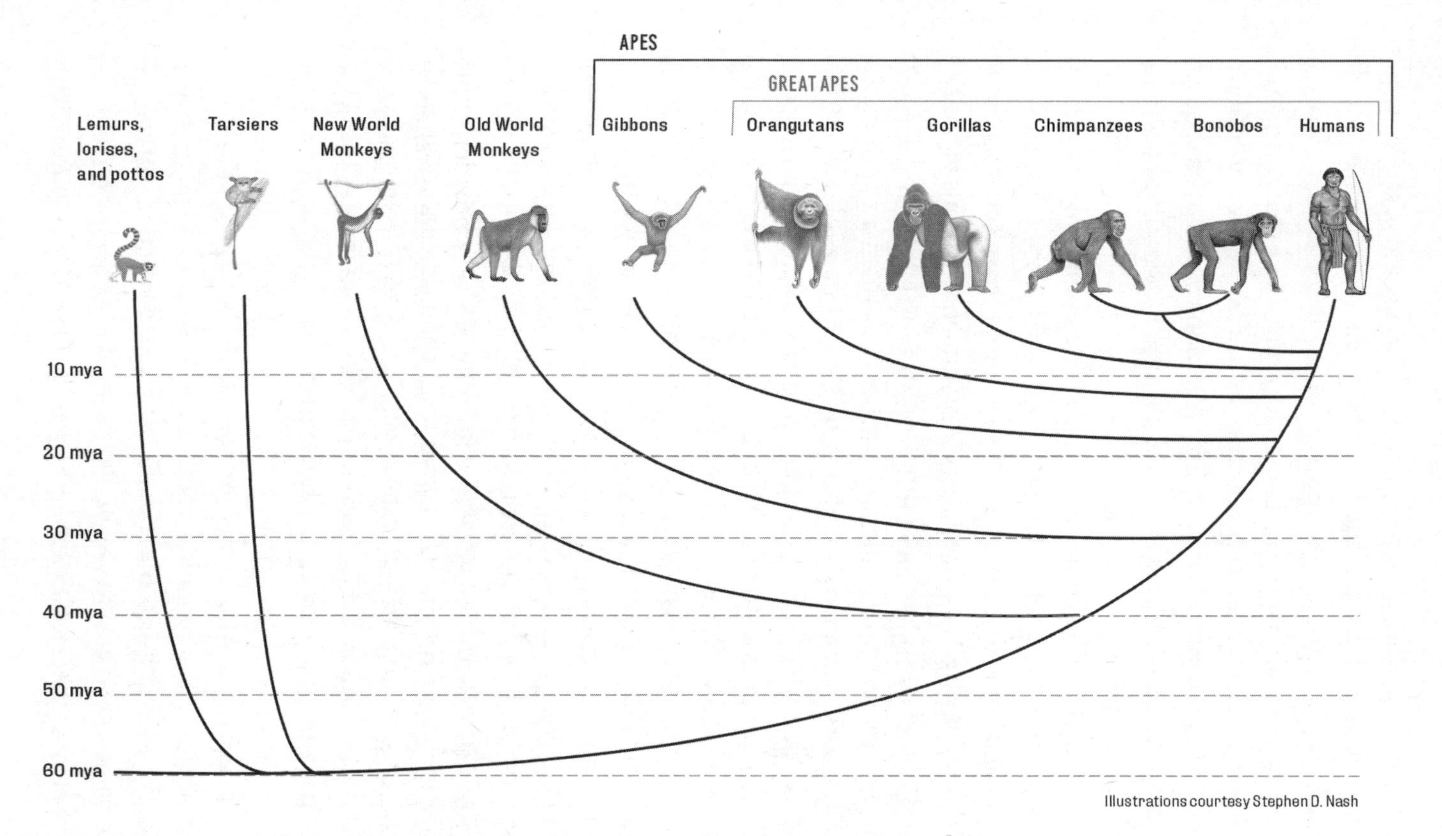

Illustrations courtesy Stephen D. Nash

The Primate Family Tree. Split time estimates are approximations based on molecular studies circa the 1990s.

beside a chimp who apes her posture. The message: there wasn't much difference between the two. But even Washburn eventually found himself dismayed by the readiness of so many of his colleagues to assume that modern apes had remained nearly frozen in time. "The chimpanzee is becoming the ancestral ape," Washburn warned at one conference in 1983. "We don't know what the ancestral form was like. We're better off if we use chimpanzee and gorilla plus a big X."

Nevertheless, the chimpanzee became a commonly used proxy for the still-undiscovered last common ancestor of humans and apes. This chimp-focused view of the human past became known as "chimpology" or "*Pan*thropology"—a pun on the genus name *Pan*. Lacking fossils near the ape-human split, scholars made inferences from our ape relatives. Paleoanthropologists filled in the missing pieces of fossil skeletons like Lucy to fit an image of a chimplike ancestor. Scientists tried to understand the origin of language by teaching chimps how to talk. They tried to retrace the innovation of tools by teaching chimps to knock together stones. They modeled the beginning of upright walking by inducing chimps to waddle on treadmills. They tried to demystify human sexuality by watching promiscuous chimps screw each other. They looked for the origins of human violence and warfare in chimp aggression. The debate narrowed to which of our closest relatives offered the better model of the protohuman—the aggressive common chimp (*Pan troglodytes*) or the mellower bonobo (*Pan paniscus*), sometimes known as the pygmy chimp. In 1992—the same year the fossils were discovered at Aramis—author Jared Diamond labeled humans as "the Third Chimpanzee" in a book by the same name. He wrote:

> Just imagine taking some normal people, stripping off their clothes, taking away all their other possessions, depriving them of the power of speech, and reducing them to grunting, without changing their anatomy at all. Put them in a cage in the zoo next to the chimp cages, and let the rest of us clothed and talking

people visit the zoo. Those speechless caged people would be seen for what we all really are: chimps that have little hair and walk upright.

That ethos widely permeated the study of human origins. A prime example is the evolution of the brain—the most notable human distinction as shown by the fact that you can read this book and an ape cannot. Since the time of Darwin, evolutionists had viewed the human brain as a reproportioned version of the ape brain. Chimp brains are roughly the size of the oldest human ancestors, and many scientists interpreted human cognition as a more sophisticated version of patterns observed in our closest relatives. By one line of thinking, humans and chimps possessed the same basic neuroanatomy, but we simply had more of it—particularly in the neocortex—and could stack processes on top of each other to attain more complex cognition. But we differ greatly from our ape cousins in one respect—imagination. Chimps have a limited capacity for abstract thinking while humans excel at it. This powerful attribute enabled our ancestors to envision how rocks might be chipped into tools, communicate with language, construct narratives, and formulate systems of belief. Humans fight not only over mates and territories but *ideas*. The downside: the human mind systematically filters its own perceptions. We notice what we *expect* to see and become blind to inconvenient facts. Fossils, being nothing more than mute bone, are blank slates for human interpretation. The chimplike ancestor became a prediction to be tested—with the confirmation bias that often attends such expectations. For many, a protohuman like a chimp or bonobo became the null hypothesis, or the default answer until proven otherwise.

MOLECULAR STUDIES ONLY COULD REVEAL THE SEQUENCE AND TIMING OF the branching of the family tree, not reconstruct ancestors. The only way to know *for sure* was to find fossils—and the hottest trail led into a dangerous valley of bones where a seeker could lose a zipper or worse. The Afar remained as perilous as ever: in 1993, an Ethiopian antiquities officer with the team of Maurice Taieb was shot and killed at a militia roadblock elsewhere in the Awash Valley. When the Berkeley expedition returned that year, the leaders took the risk of moving smack into the domain of the hostile clan that had chased them away the previous season. The Aliseras had issued an ultimatum: *To work in our territory, you must camp in our territory.* The clan envied all the benefits enjoyed by the rival Bouri clan and wanted more for themselves.

The fossil team pitched camp in Alisera land beneath the shade trees beside a seasonal river known as Ganduli, a short drive from the outcrops where they had found the hominid specimens the previous year. Under the guns of their former tormentors, the fossil hunters began to rake up bones like there was no tomorrow—because, given the precariousness of working in Ethiopia, there might not be.

Chapter 8
UNDER THE VOLCANO

On the third day of the 1993 field season, White stood in the hardscrabble terrain near Aramis wearing his bush hat decorated with a band of rattlesnake skin. "In this operation," he announced to his crew, "every single piece of bone will be picked up off the slope and will go into those metal pans."

The team had returned the day after Christmas and immediately began collecting more hominid fossils: a molar on the first morning, a canine in the afternoon, another molar on the second day. On the third day, White himself found part of a skull—a chunk of the right temporal, the section around the ear, that had been gouged by carnivores chewing the temporalis muscle. Virtually every fossil bore scars from gnawing, revealing that ancient carnivores had ravaged carcasses down to bone. White hoped to find more pieces of the creature with the broken temporal bone and organized a crawl—an up close search on hands and knees.

Ten people people knelt shoulder to shoulder, including a young American professor, an Ethiopian museum worker, a rich American benefactor, and a bare-chested Afar man. The crew began inching across the desert, and the air echoed with the *plunk!* of hard objects dropping into metal pans.

"You guys are moving a little bit too fast for my taste," White called after them. "Slow down. There's only one of these in the whole world. We don't want to lose a piece just because we were in a hurry."

These scraps of the human family were among the oldest ever found, and their rarity obliged the team to treat the area like a crime scene. White ordered a 100 percent collection: take *everything,* no matter how broken, tiny, or unidentifiable—bone shards, mouse teeth, fossil wood, seeds. Unidentifiable? *Collect!* Might be a rock? *Take it anyway!* The key to success was efficient workflows; disciplined teams outperformed any glory-seeking individual. He lined up searchers shoulder to shoulder to ensure no gaps in their fields of vision and sometimes laid down nylon cords to keep people in lanes. He only let his crews move *up* slopes; moving downhill put their eyes too far from fossils and risked trampling evidence. At most fossil fields, a 100 percent collection strategy would produce truckloads of fossils, many worthless to science. At Aramis, however, the fragmentary remains and narrow time window kept the volume manageable—and the scarcely known period of Pliocene Africa made it all valuable.

Day after day, the crew picked the desert clean. By the time the first light illuminated the Ayelu volcano, White was up and moving, planning, checking maps, checking equipment, and checking off whatever needed doing—because there was *always* more to be done. Every morning before driving out of camp, he pulled down the sun visor on his bush car and scanned his checklist: *tire pump, car jack, tow tope, ice axe, air photos, lunch cooler, water, bush clippers, shovels, tarps, crowbar. . . .* The list never changed yet he checked it every day, just to be sure. At night, team members spotted his lanky figure stalking through the darkness and attending to details that kept him from sleep. "He just keeps going," observed Yohannes Haile-Selassie, then an Ethiopian graduate student. "How come this guy doesn't get tired?" At night, White reviewed each day's collections piece by piece and dictated his field diary into an audio recorder because he didn't have time to write. When he transcribed the recordings back at Berkeley, the exhaustion in his voice made him bone-tired all over again.

Some paleoanthropologists, notably the Leakeys and Johanson, celebrated their luck at finding fossils. White didn't trust in mystical notions like luck. What mattered was efficient supply, organization, preparation, and teamwork. Once he snuck away from the crew and scattered plaster fossil casts over the ground so he could measure the efficiency of his surveyors when they returned years later. Every detail counted, so he micromanaged without apology. People endured his whipcracking for a simple reason: no one was more productive. "Tim is without question the best field worker there is," said anthropologist Bruce Latimer. "His science, logistics, and efficiency are phenomenal. If you want somebody to find a fossil, there's nobody better on the planet earth than him."

The distinctive sandwich of sediments between the two volcanic ashes appeared repeatedly over several kilometers in three drainages that the Afar called the Aramis, Adgontole, and Sagantole. Giday WoldeGabriel trekked across the badlands mapping the stratigraphy and collecting volcanic samples for radiometric dating, and whenever he encountered the familiar fossil-rich layers, he called on the radio and the fossil crew checked it—and often found more bones. "This is a remarkable layer," White recorded in his field notes, "widespread, and always with the same signature."

The Aliseras, the former tormentors, were trained to become fossil hunters. The Afar had been stepping over fossils for centuries and even had a word for bone-shaped stones: *gohola*. The paleontologists taught the locals to recognize different types. They compared bovids to cattle—animals readily recognizable to the herders. Crocodiles were familiar to the Afar because the *dobado* still lurked in the Awash and occasionally snatched people (one worker had a stubby thumb that had been bitten off). Monkeys they compared to the *Daam Aatu* baboons that ranged over the land. And hominids? There was no word for "ancient human" in the Afar language. The only ancestors they knew were the ones buried in the stone crypts that dotted the landscape—and the scientists wanted to avoid any suggestion of grave robbing. Nor could they mention apes, which were unknown in the Afar desert. So the scientists explained that they sought the bones of the *Kada Daam Aatu*—the "big monkey." The name was not strictly accurate taxonomically, but it got the point across, and in the field practicality mattered more than anything.

On the fourth day, White took the crew to a new place, a parched badlands of salmon-colored earth where a few wispy strands of knee-high brown grass swayed in the breeze. Ledges of gray rock ringed the edges of the shallow valley like crumbling pavement. It was the familiar *Daam Aatu* basaltic tuff, marking the area as a priority target.

He told his crew to walk—*slooowly*—toward the far side of the gully. Following his orders, they would pick their way over a dusty kilometer of dry grass, thorn trees, and animal dung while looking for fossils. White drove to the other side and parked his car as a target for the crew. He brought along Gadi, the hunchbacked Zipperman. Other members of the expedition were terrified of the wild gunman (one American described him as "psychotic"), but White was growing fond of him. At lunch, they sat together under the shade trees, exchanged fragments of Afarina and English, and broke into wild laughter. "Somehow Tim got attached to that guy," said Giday. "He was very aggressive and closer to Tim than any of the other guys." As they warmed to their visitors, the Afars calmly recounted battling Issa foes and slaughtering Derg soldiers during the civil war. White admired their ability to survive in such a hostile place, and sometimes he showed more respect for illiterate Afars bedecked in kill trophies than a certain class of pompous academics.

Gadi's cordiality marked a turnaround in relations with the Alisera clan. He became an all-purpose worker, guard, guide, tracker, geographer, camel herder, and enforcer. "He was a guy who got respect because of his talent—his gun talent," said Berhane. In his own peculiar way, Gadi proved more reliable than some of the self-serving local leaders, who had a habit of forgetting previous agreements and conniving for more of this or more of that. Gadi remained a steadfast friend—and so utterly fearless that even his own *balabat* seemed scared of him. Some pathology had left him sterile, so he remained a bachelor warrior of his clan. He wore a rifle strapped over his shoulder and extra banana clips dangling from a harness. Sometimes his rifle sprouted plumage: like many Afars, he stuffed ostrich feathers in the muzzle to keep the weapon free from dust.

So it was that Gadi became a regular sidekick in White's car. While waiting for the crew, White climbed out of the vehicle and checked some nearby outcrops for fossils. Gadi followed.

Photograph © 1997 David L. Brill

Tim White and Gadi enjoy a laugh at the site where Gadi found the type specimen of the new species. The crew marked the site with a boulder and cement slab bearing Gadi's handprint. Gadi later suffered a mortal wound in a gunfight against his Issa foes.

"Doktor Tee, amee."

Doctor Tim, come here. Gadi pointed to something on the ground by his sandals.

White bent down and saw a small white tooth amid a scatter of rocks. *A hominid molar.* His unschooled bodyguard had just outdone all the Ph.D.s.

It turned out to be another red-hot clue. When the crew arrived and began searching the area, they found a canine tooth from the other side of the mouth—a signal that the tooth row had separated recently and more pieces might lie nearby. "There is a chance to get an entire dentition here," White said as he held the pieces in his hand. "This is left canine and this is a right first molar. That probably means the whole palate was here."

The crew crawled, scraped up loose soil, sieved, found more fossils, and soon beheld the finest specimen found so far—a set of ten teeth from one individual. It was enough to prove these creatures were not only older than Lucy, but something biologically distinct. Zipperman earned another trophy—the teeth that would become the type specimen of the oldest species of human ancestor.

THE SUBSEQUENT RECOVERY OPERATION TURNED INTO A GRIND OF CRAWLing, sieving, and picking through pans. One tall man on the periphery kept gazing into the distance. Alemayehu Asfaw, infected as ever with his terminal case of hominid fever, longed to be somewhere else—wandering free, doing what he did best, and finding fossils. Sieves were torture. Sometimes he complained that crawling hurt his knees. Alemayehu lacked patience to be stuck in one place. His boss lacked patience for layabouts, and finally White told Alemayehu to get the hell out and go survey on his own. Good riddance.

Alemayehu wandered off and found a fine fossil of a kudu, a spiral-horned antelope. He duly reported the fossil to White, who was busy recovering Gadi's tooth set and in no hurry to chase after another damn antelope. When Alemayehu finally led him to the spot, the boss spotted something his surveyor had missed—the upper arm bone, or humerus, of a human ancestor buried in the dirt beside the kudu. Even so, he credited the hominid discovery to the Ethiopian. A small excavation uncovered two more bones of the forearm, the radius and ulna—and raised hopes that more might be buried in the same spot. By then, however, time in the field had run out. For the following year, the scientists could only wonder: What more might be hidden below?

BY THE TIME THE TEAM BROKE CAMP, THEY HAD ENOUGH MATERIAL TO DEclare a new species. In October 1994, eight months after returning from the field, the Middle Awash team published a paper in *Nature* announcing the new species *ramidus,* after the Afar word for "root," *ramid.* As the name implied, it was portrayed as an early member of the human lineage—"a long-sought link in the evolutionary chain of species between humans and their African ape ancestors." The announcement described "the most apelike hominid ancestor ever known." The searchers had envisioned finding the early stages of the human lineage and the taxonomic name suggested they had done just that. In one television interview, White declared, "We think this species is at the root of the human side of the family tree."

The new species seemed to affirm predictions—an ancestor not far removed from our ape cousins. The small, thinly enameled molars were likened to chimpanzees. Other clues hinted at nascent humanity. Both chimps and gorillas have daggerlike projecting canines that sharpen themselves by rubbing against the lower premolars. But *ramidus* had smaller, blunter, diamond-shaped canines with no self-sharpening wear—the earliest example of canine reduction, a diagnostic feature of the human family.

The skull fragments displayed what the discovery team described as a "strikingly chimp-like morphology" while an upper arm bone displayed a "mosaic of characters" from both apes and human ancestors. (The new species initially was put into the existing genus *Australopithecus*, but within a year additional discoveries would cause the team to create a new genus *Ardipithecus*.) "Were *Ardipithecus ramidus* any less humanlike," *Nature* editor Henry Gee later wrote, "it would be hard to decide whether it is a closer relative of a chimpanzee than humans." In short, it was depicted as exactly what the profession expected. A mysterious early member of the human family was obliged to fulfill the hopes of its last surviving relatives. As Gee added: "If *Ardipithecus ramidus* hadn't been discovered, we would have had to invent it."

Yet the actual haul remained meager—only seventeen fossils, so little that all could almost be held in a pair of cupped hands. Much remained unknown. Did this creature walk upright? Only indirect evidence suggested it did—the geometry of the skull fragment suggested it rested atop a vertical spine. The team had found no bones from below the waist, so upright locomotion—long assumed to be another defining trait of the human family—remained only an inference.

Twenty years had passed since the discovery of Lucy and the long scientific drought in Ethiopia finally had ended. The *Nature* headline blared "Earliest Hominids." In a commentary that accompanied the announcement, Professor Bernard Wood, an anthropologist at the University of Liverpool, extolled, "The metaphor of a missing link has often been misused, but it is a suitable epithet for the hominid from Aramis." The *New York Times* front-page headline exclaimed, NEW FOSSILS TAKE SCIENCE CLOSE TO DAWN OF HUMANS. The London *Times* announced, SCIENTISTS FIND THE 'MISSING LINK.'

In one television appearance in 1994, White emphasized the similarity of *ramidus* to our closest relatives and even invoked Jared Diamond's contention that humans were the "third chimpanzee." As he told the interviewer, "It's a woodland or forest dweller, something closer in habitat to a chimpanzee, something closer in anatomy to a chimpanzee."

In time, he would regret those words.

Chapter 9
THE WHOLE THING IS THERE

The announcement meant that the secret was out—the Middle Awash expedition had picked up the trail of the oldest human ancestor, the biggest news in paleoanthropology since the naming of Lucy's species sixteen years earlier. That triggered all the excitement, hype, and skullduggery that had come to afflict the science of old bones. Journalists begged to join the expedition. All requests were refused: too dangerous and too distracting. One reporter from the *Sunday Times* of London hounded the team about the tension brewing between White and his former partner, Don Johanson. "I will not participate in a media-invented and orchestrated feud," White snapped at the reporter. "There are more important things to be involved in." The journalist took particular interest in one member of his expedition—the billionaire Ann Getty, who had ferried the team to Ethiopia on her private Boeing 727 jet. Getty herself joined the field expedition (and had been present when it came under fire in 1991). During Berhane's tenure as museum director, the Gettys had chartered another jet, this one a DC-10, to import a massive donation of lab equipment, fossil cabinets, and even carpenters to install them. In Ethiopia, Berhane's opponents turned her philanthropy into a weapon and had charged that Getty money bought improper influence (and the vice minister of culture also had complained that guests drank alcohol at a party to celebrate her donations to the museum). Such were the paleo politics in Ethiopia.

In November 1994, the expedition departed the capital. "So paranoid are they about media attention that they deliberately put out false information about their whereabouts," reported *Times* journalist Mary Anne Fitzgerald. The reporter hired a car, pursued the Middle Awash team into the desert, followed the tire tracks, and intercepted some team members at a stream crossing. "They snubbed me," Fitzgerald recounted, "refusing even to say hello." The Middle Awash team abandoned the pursuers at the crossing, and the reporter never got within fifty kilometers of their camp, wherever it was.

The caravan had vanished deep into the wilderness. In Alisera territory, some veterans of the expedition felt a pang of fear when a hunched figure appeared with a gun. *Zipperman*. Who knew what his mood might be this time? Gadi rushed forward to greet *Doktor Tee*. "The emotional connection between Gadi and Tim was very real—*very* real," recalled Doug Pennington, a Berkeley student on his first trip to the Afar. "The emotion on Gadi's face was clear—absolute delight. It was really touching, actually."

Once again, the visitors camped by the edge of Alisera territory, but at a different site because the well they had dug in the dry streambed the previous year had proved to be a mistake with disturbing consequences. "Finally, we learned the Afar guys take their camels to drink from that water so they can clean their intestines," said Berhane. "It's really a laxative."

AFTER FIVE DAYS OF TRAVEL AND PITCHING CAMP, WHITE SATISFIED HIMself that the mission was well enough established to let the fossil hunters burn off some energy. As usual, the team leaders had insisted on logistics first—make a safe and sanitary camp, build roads, establish security, dig wells to secure good water so the field season wouldn't turn into another shit-fest, get the geologists to the right places, help Berhane manage all the drama of hiring the locals, and basically herd a bunch of goddamn cats. The crew got hominid fever and nagged White: *When can we search for fossils?* The top priority for the season was a major excavation at the arm site found by Alemayehu the previous year. Begin there? *Fuck no.* Better to warm up with a lower-priority target, burn off some energy, and get eyes calibrated for the real work that lay ahead. White took the crew to locality six—pronounced "loc six" for short—a section near Aramis where Gadi had found the type specimen one year earlier. The little valley already had been picked over multiple times. They could search until sundown and had two hours. He left them to their work.

White shouldered twin impulses—multitask *and* micromanage. Sure as the sun was going down, something that needed to be done wasn't getting done without him. Somewhere somebody might be fucking up. Before climbing into his car, he paused atop a hill to snap a photo of the crawlers. Already the line was disintegrating. *Damn Alemayehu creeping ahead again! Why can't people follow instructions?* Years later, his frustration welled up anew as he reviewed his archival photos. "I'm really reluctant to call it a disciplined crawl," he growled, gesturing at the imperfect line, "because there's no FUCKING discipline with these assholes."

Those assholes—actually a line of students, professors, Ethiopian museum workers, and Afars—inched across the scrubland basin. Among them crept Yohannes Haile-Selassie. He was lucky to be there after having almost been shot by the Issa three years earlier. He had proven himself as a star pupil of Berhane at the museum, and White recruited him to the Ph.D. program at Berkeley, where students called him "Johnny." He had risen high enough in White's esteem to be trusted crawling in a fossil hot zone.

He straightened up. He held a broken bone about the size of a pencil stub. He served as a teaching assistant for White's human osteology class and knew his bones. *Second metacarpal*. A bone located in the palm below the pointer finger. *Hominid*.

Following protocol, the team halted, and carefully retraced steps as if backing out of a mine field. Nobody wanted to invoke the wrath of you-know-who by crushing a fossil with a klutzy step. As the sun went down, they returned to camp with the first fossil, found on the first day of the season. At the time, it seemed just another isolated fragment in a ravaged assemblage. Unbeknownst to them, it would prove to be the first hint of something much bigger.

Searching continued elsewhere. Night and day, Afars showed up, pleaded for jobs, and asked to move into camp. Berhane hired as many as possible, but there were never enough jobs or tents to meet the demand. One night, gunfire broke out over camp. People awoke and saw tracer bullets streaking over their tents.

"What's that?" one American student called out from his sleeping bag.

"Bullets! Turn off your light!"

It was the Afar version of a workplace shooting. Disgruntled job-seekers vented their rage by firing over camp and stealing a rope. Suddenly it became the Aliseras' turn to be scared: if the expedition fled, the boom economy would vanish. Gadi and another Afar tracked down the shooters, tied them up with the purloined rope, and delivered them to the tribe elders for punishment.

THE HAND BONE FOUND BY YOHANNES WAS THE FIRST *KADA DAAM AATU* OF the season. Even so, nobody was getting excited just yet. In the carnivore-mauled assemblage, scraps were the rule, and not a single intact bone had been found.

A few days after the initial discovery, the crew returned to locality six, crawled again, and found a finger bone and another chunk of metacarpal. Inside White's mind, a switch tripped: *There's more here than one isolated hand bone*. With dustpans and brushes, they scraped away several inches of loose soil to the underlying hard layer of ancient sediment. With a sieve, they shook away clouds of dust to a gravelly scree and dumped it onto a tarp. In the shade, workers picked through every piece—and found toe bones. Another switch: *multiple elements of a skeleton*.

Late in the afternoon, they uncovered what appeared to be a shinbone embedded in the ground. *An intact long bone!* Instead of trying to free the fossil, they extracted the entire block of soil and wrapped the sides in plaster bandages. Cleaning could wait until time allowed—usually back at the museum lab. In the process of excavation, they uncovered another small white object—a thumb bone. Another switch: *fossils in situ*.

The discoveries stopped. Locality six had been picked clean and the oldest ancestor vanished again. The crew took the shinbone and other fossils back to camp and stored them in the fossil tent. There they sat for the following month while the crew systematically crawled one locality after another.

Then it rained. Out-of-season storms turned the desert into muck, and dry drainages became chocolate-colored torrents. The river beside the camp overflowed its banks and swamped the tents. Vehicles could not move in the mire. Stuck in camp, White's mind flew back to the inaccessible desert. Could there be more out there at loc six? The sieve had been disappointing, but maybe that was good news. "It may mean that most of the material is still in situ in the bank up above," he recorded in his field notes, "so there is still hope for this site."

One tent served as the fossil storage room and field lab, known as *gohola bota*, the Afar term for "fossil place." White examined the tibia collected weeks earlier. Half hidden in the chunk of hard earth and wrapped in a plaster jacket, it almost resembled a tree root. Did it really belong to a human ancestor? The same individual as the previously discovered hand and toe bones? Or was it some other primate? White literally wrote the book on bones: in 1991 he published the 662-page *Human Osteology*, a classic textbook for medical and anthropology students around the world. He spent days cleaning the tibia, carefully dripping chemical hardener onto the brittle bone with a syringe, and flicking away bits of reddish earth with dental picks. He repeated the process multiple times and checked his work with a portable microscope. When the rains cleared, everybody returned to the field—except White, who took the unusual step of remaining in camp all day to finish cleaning the fossil. By the night of December 20, he beheld a clean shinbone—in his judgment, now unquestionably a human ancestor.

The next morning, the crew hurried back to loc six.

IT WAS THE SHORTEST DAY OF THE YEAR. THE SILHOUETTE OF THE AYELU volcano loomed in the morning haze, and already White felt the press of time. Over the next few hours, the morning haze would turn to broiling heat and the most productive hours of the day would be gone. This day—winter solstice in the northern hemisphere—granted the least daylight all year. In truth, the length of days didn't change much ten degrees north of the equator, yet the date underscored the time crunch: the field season was running short. In the desert, time, daylight, and manpower were precious commodities to be rationed as carefully as water.

Up on the ridge, Gadi stood with his assault rifle casually thrown over his humped shoulder, keeping watch over the place named for him, Gadi Bota. Down in the gully basin, a boulder marked where he had found the type specimen a year earlier.

White stalked through the badlands, his rangy frame clad in denim, and issued commands in a terse baritone, with scientific terminology for precision and profanity for emphasis. He was about to bet against very long odds that the corps of discovery might find something that no human had ever seen before.

The rains had washed the surface clean and caused more erosion. In storms, torrents of runoff carried fossils downhill. Likewise, sometimes mini tornados known as dust devils, or *ho hos*, could fling a small bone far from its original source. One could never be sure where a fossil came from—unless you had evidence in situ, like that tibia. Most fossils are found loose on the surface as isolated pieces whose point of origin remains unknown. A fossil still embedded in the sediments raised hope of finding a concentration of bones.

The crew had planted small yellow flags to pinpoint the locations of fossils recovered so far. The scatter pattern hinted how bones might have spread from the point of origin and broken apart on the surface. White likened this kind of detective work to the "placer" mining of the gold rush in his native California: figure out the pattern of fragments, follow it upslope, and look for the mother lode. In most instances, of course, there would be no bonanza. Contrary to popular belief, most fossils in the eastern African rift system were found lying on the surface—a tooth here, a finger bone there—and digging rarely produced anything. The idea of hitting a jackpot skeleton was mostly fiction. "Ninety-nine out of a hundred times you never find anything in situ," he once cautioned. "Including Lucy—not a piece in situ."

A painful reminder lay about two hundred meters away at locality seven, the place where the arm bones had been found in situ the year before. At the beginning of this field season, Berhane Asfaw and Yonas Beyene began an excavation with a large team and high hopes. Despite weeks of digging, they found only a few more wrist and hand pieces—weird shapes unlike anything the anthropologists had ever encountered before. They found nothing more, and eventually concluded the arm had been an isolated piece, perhaps dragged away from the rest of the carcass by an ancient carnivore. Except for those few bones, the excavators had dug what wildcat oil explorers called a dry hole—but the rains had turned this one into a pond.

An excavation was a time-consuming process with a high risk of failure. The arm site already had diverted workers who could otherwise survey kilometers of ground each day and collect fossils by the bucketful. One dig was a costly drain on manpower. But *two*?

The yellow flags marked precisely where each fossil had been found in the gully wash and suggested a pattern: the scatter narrowed toward a small mound where the tibia and a finger bone were found embedded in the ground. That meant there was a chance—a very small chance—that more fossils might remain hidden in the same little hillock.

Deep cracks had opened in the mound. *Bad news.* Clay soils expanded and contracted with each wet-dry cycle and could destroy delicate bones. Typically, only a few small, hard bones got squeezed to the surface and survived—like the hand and foot fossils found in the gully weeks earlier.

White raked his geological hammer across the side of the mound and exposed a cross section of layers. "As we excavate down through this horizon here, down through this hill, we'll find more specimens," he predicted as an assistant documented the moment on video. The sediment was laced with white lines: fossilized roots of plants that grew 4.4 million years ago. If those roots survived this long, maybe bones could, too.

"All we can do now," he said, "is excavate and hope for the best."

IN PALEOANTHROPOLOGY, AN EXCAVATION BEARS LITTLE RESEMBLANCE TO the popular conception of that word. Excavation is more like surgery, except the incision is measured in square meters, the operation drags on for days, and the patient is already dead. To rescue an antiquity, the discoverer must destroy its context. White never forgot the words of one of his graduate school mentors: *We kill our informants in the process of studying them.*

The only remedy: document meticulously. There was just one chance to collect evidence before it became lost forever. The crew pounded a datum stake into the ground, a fixed point of reference for all subsequent measurements. They recorded distances, elevations, slopes, and precise locations of every piece found thus far. They placed a video camera on a tripod and kept it rolling all day.

Millimeter by millimeter, the excavators reduced the mound with scrapers, dental tools, porcupine quills, and dust brushes. Each person spent an entire day picking away at a section not much bigger than a laptop computer. By sundown, they had gone down only one or two centimeters. All brushed soil was carried to a sieve station to catch every scrap.

Brushing, brushing, brushing . . . shapes slowly emerged from the dust. *Bones!* When a crew member uncovered a fossil, White often took over. He sprawled on his belly with his nose inches above the bone and carefully extracted it with dental picks. From the reddish earth came the skeletal grin of the long-sought *Kada Daam Aatu*—the jawbone of an ancient human ancestor.

"Look at that canine! Whoa!" White sang out. "We got this individual at *exactly* the right time of its life. It died before the teeth wore out. The third molar is fully erupted. All the teeth are here!"

The excavation moved in slow motion, but by the standards of paleoanthropology it seemed a torrent—bones of the jaw, hand, foot, pelvis, and skull appeared day after day. To White's surprise, many long bones had survived—barely. Fossils that looked solid proved to be powdery and brittle. Once exposed, they dried quickly in the heat and crumbled. White hurriedly applied chemical hardener before the pieces disintegrated. A student with a backpack sprayer misted the excavation area with water, and the team laid down plastic to slow the evaporation.

They dared not expose the most fragile pieces in the field. Instead, the crew dug up blocks of soil and wrapped them in orthopedic bandages, the kind that doctors use to make casts around broken limbs. White would finish cleaning the fossils in the museum lab. Until then, whatever lay inside remained a mystery.

The skull came out of the ground in a pumpkin-size block of

Gadi stands watch while the excavation team slowly uncovers the bones of the skeleton. Each flag marks the location of an individual bone.

sediment. Ann Getty knelt on her kneepads and cut free the block with a knife. Suddenly the pedestal cracked open.

"Need a bandage," she called out.

White whipped around.

"Aaaggh!" he yelled. "Go away!"

Startled, Getty blinked as White loomed over her. Yes, the Getty family had donated millions of dollars to human origins research and loaned its private jet, but donor relations and VIP treatment meant nothing to White when *the world's oldest fucking hominid skull* had been jeopardized. He called for an orthopedic bandage. He would finish the job himself.

"Don't touch it," he snapped, "unless you want to be bandaged in with it."

The team set up an awning and worked through the midday heat. The car stereo blared Grateful Dead songs from the album *Skeletons from the Closet*. Atop the ridge by the cars, Gadi kept watch with his rifle yoked over his shoulder. The strange *farenji* obsession with the *Kada Daam Aatu* apparently left him bored.

He unslung his gun and aimed at the crew—his idea of a joke. Henry Gilbert—Getty's nephew, whom White had taken on as a graduate student—aimed his video camera at the Zipperman, who aimed back with his gun. White strode up the hill toward his car and found himself facing the barrel of Gadi's rifle.

"Get that gun out of your hand!"

Gadi lowered his weapon.

THE SUN SETS QUICKLY NEAR THE EQUATOR. WHEN THE BLAZING BALL SANK toward the western rift margin, White repeated two Afar words: *Ayro korte.* "Short sun." Hurry.

It was dangerous to stay out after sunset. Hostile Afars or Issa could sneak up unseen. Packs of hyenas and big cats began to prowl. The tooth-scarred fossils of Aramis provided a stark reminder of what big carnivores did to helpless primates—then and now. A car could roll into a ravine while returning to camp. But the task was too urgent to stop, and the crew finished excavations by lantern and car headlights. New Year's Eve passed with White sprawled on his belly under the glow of headlights. The end of the year underscored the deadline: *Get it out of the ground and safely back to the lab.*

White lifted a finger bone and a piece came off in his hand. He cursed, "Son of a bitch." He dripped Glyptal chemical hardener onto each piece under his nose. The fumes gave him headaches. He finished excavating the last finger bone by flashlight.

"You can't take shortcuts at this site, man," White muttered to himself as darkness fell. "You just can't take them. These bones are just way too fragile to mess around with."

Every night, the crew cleared all equipment and piled heavy stones atop the excavation area. The labor had to be undone every morning and consumed precious time when they had none to spare. But a fossil couldn't be left vulnerable overnight. Jackals loved the chemical smell and came sniffing around in the dark like glue junkies. These crumbly fossils had barely survived careful archeological extraction and wouldn't survive an animal's pawing. Indeed, in the morning, amid all the paver stones and yellow flags, they found fresh piles of jackal shit. Even after 4 million years, carnivores still threatened to ravage the precious bones.

NOT A SINGLE BONE REMAINED ARTICULATED WITH AN ADJOINING ONE. Hand bones, skull shards, and leg bones all lay in a jumble. But there was no duplication of parts, and a revelation slowly dawned: *it's all one individual.* On the final day of the year, ten days after beginning the excavation, White acknowledged it aloud: *a skeleton!*

His assistants were less restrained. One shouted, "The whole fucking thing is in there!"

White spread-eagled on the ground, excavating the mandible. He pressed his face close to the fossil with his baseball cap turned backward and glasses on the ground. He slowly dripped Glyptal onto the specimen and flicked soil with his dental tools. "Nobody has ever seen one of these before," he murmured. "Quite a privilege, huh?"

One day, a familiar bone came into view. "It's a middle finger!" White called out. "Lucifer's!"

It was the digit that the Romans called *impudicus,* the impudent one, the modern signal of "fuck you." This phalange was long and curved, and more like an arboreal ape than any human ancestor ever seen before.

White instructed one of his assistants to document the moment with a photo. "There are going to be people in the field who say, 'Hey, what finger is that?'" White explained. "I'm going to show them."

He laid his hand next to the bone as the camera snapped away. Then he gave the finger.

ALL TOLD, THE TEAM RECOVERED MORE THAN 125 PIECES OF THE SKELETON, including virtually all of the most important elements—skull, teeth, hands, arms, pelvis, legs, and feet. It turned out to be a female, the oldest woman in the human family, more than a million years more ancient than Lucy and more complete. She had come to rest in a shallow swale of an ancient floodplain, and there she had lain for 4.4 million years until the seekers of the *Kada Daam Aatu* came along at just the right time.

A single fossil is like a fragment of ancient writing; a skeleton is like the Rosetta stone. It contains clues to decipher the entire message: the body plan, proportions of limbs, size ratio of brain and body, style of locomotion, even behavior and environmental adaptations. Over the following decade, the team collected parts of thirty-six other individuals of the same species and the skeleton gained context: a *population*. From all of that, it became clear that this animal marked not only a new species but a previously unknown stage of evolution from the Dark Ages of the human past.

The discovery placed a heavy onus of responsibility on the scientific team. Such a skeleton might never be found again, so they would take as long as necessary to decipher it—no shortcuts. In that moment, the discoverers had only a vague notion of how the skeleton would change our understanding of human ancestry. They had no conception of the length of the journey upon which they had embarked, nor the revelations that lay ahead. They would spend three years on the excavation alone and spend many more years gathering more fossils nearby. They would take fifteen years to agonizingly reconstruct the skeleton and make sense of it. Nearly fifty scientists from around the world would be enlisted to study thousands more fossils of extinct animals, reconstruct the ancient environment, and build the timeline of geology. Together, they would reveal new truths, send old notions to their graves, ignite hatreds, and cleave the scientific community. But this much immediately appeared certain: it was the most important discovery for early humanity since Dinknesh. Eventually, this skeleton not only would prove as revolutionary as Lucy, it would rewrite the story of Lucy herself.

Chapter 10
A POISON TREE

About seventy-five kilometers to the north, a rival expedition wondered what was happening upstream in the Middle Awash. Down the meandering river, Don Johanson had returned to Hadar, the place that had produced Lucy and so many *Australopithecus afarensis* fossils. Earlier that year, Johanson and his team announced their own major discovery—the first complete skull of Lucy's species. But older fossils still eluded him.

The journalist who had been spurned by the Middle Awash team found herself welcomed in Johanson's camp and described him in the field: "Johanson steps out of his four-wheel-drive vehicle into the 100°F heat of the Afar badlands. Dressed in a Ralph Lauren shirt, shorts and sandals, he looks like a businessman decked out for a day-trip in southern Spain." Johanson had mastered the art of paleo chic. Another writer, from the London *Observer*, described him as "Indiana Jones in Armani." Much had changed in the two decades since he wandered over the Hadar moonscapes as a shirtless, bushy-haired scholar of teeth. He had become a best-selling author and sophisticate who lunched with journalists over foie gras salad and scallops, debated the wine selection, and hobnobbed with the donor class. Recently he had been disturbed because one of his biggest funders, the Gettys, had abandoned him, and Mrs. Getty had joined his rivals in the Middle Awash. What were they up to?

Two camps jostled to position themselves around the base of the family tree. That year, Johanson had narrated a major *Nova* television documentary in which he declared, "We believe Lucy's species was the root of the human family tree. She's our earliest ancestor, the missing link between ape and human." Not long after those words were broadcast, his former friends in the Middle Awash found a much older species in the human family—and even named it after the word for *root*. "Clearly, here we have the ancestor of Lucy," White said in a television interview. One estranged scientist at Johanson's institute at Berkeley sensed that the boss was upset. "Johanson is insanely jealous that his Lucy is no longer the oldest human ancestor," said geologist Paul Renne. Hard feelings ran both ways. In the Middle Awash, Berhane Asfaw and his colleagues seethed with animosity against Johanson—and soon would put him out of business in Ethiopia. A paleo cold war had begun.

TIM WHITE CONSIDERED DON JOHANSON ONE OF THE FINEST MORPHOLOgists he had ever met. One night back in the Lucy era, the two anthropologists walked on the beach, and White reached into his pants pocket and discovered the cast of a fossil tooth the way laypeople find spare change. He handed the tooth to Johanson and challenged him to identify it in the dark. Relying only on touch, Johanson named the fossil and described its notable features. "There are very few people in the world with those capabilities and insights," recalled White. "That was evident to everybody who worked with Don." Over time, White sensed that something had changed, and his friend had become more enamored with fame than science. In his second book, Johanson boasted: "I was introduced to royalty, given awards, and came to be regarded as a public spokesman for American anthropology." Their joint expedition to Olduvai Gorge in 1986 marked a turning point. "We published it and the usual Johansonesque escapades were involved—sending his photograph out," recalled White. "That was what was important to

Don—being a celebrity." White had abetted the hype, and finally decided he had enough. Johanson's ghostwriter, Jaime Shreeve, summed up the rift between the two men: "the style of Don and the substance of Tim finally came to a clash." White quietly divorced himself and redirected his energies elsewhere, including his work in Ethiopia.

Berhane Asfaw also had joined the Olduvai expedition and had worked at Johanson's institute during his years as a graduate student. "I never liked Johanson from the first day I saw him because he was not supportive of anybody in Ethiopia to be trained in this field," said Berhane. "He was scared to see somebody working in the Ministry of Culture and in antiquities who knew what was going on." He branded Johanson a "neocolonialist"—a foreigner who sought to exploit African antiquities while investing little in return. When Berhane became director of the Ethiopian National Museum, he made it clear that foreigners—Johanson in particular—would have to comply with stricter regulations. "Berhane was always quick to remind me," Johanson remembered, "that any failure to follow ministry guidelines might result in cancellation of the Hadar Research Project."

Relations deteriorated after Ethiopia ended the research moratorium in 1990. When Johanson flew into Addis Ababa, Berhane suggested he get out of town to avoid hostile questions from the Ethiopian press about the *Lucy* book. Johanson hid in a cockroach-ridden hotel on the highway to the Afar until his colleagues came to transport him to the Hadar field camp. "That was all a smokescreen," Johanson alleged years later. "They just wanted to make life difficult for me." Soon afterward, Johanson's team at Hadar was visited by Berhane Asfaw (then still head of the National Museum) and Tadesse Terfa, the head of the Ethiopian Center for Research and Conservation of Cultural Heritage (the agency that regulated antiquities research). They discussed removing a portion of the Hadar territory called Gona and transferring it to Ethiopian archeologist Sileshi Semaw, a protégé of Berhane. At that point,

Gona was only known for producing ancient stone tools, which held little interest for the Hadar fossil team. As it turned out, however, Gona later was found to have deposits from the Pliocene and Miocene epochs as old as 6.4 million years (and fossils of *Ardipithecus*), making it prime hunting grounds for the earliest members of the human family—the ones Johanson had dubbed "Lucy's grandparents" and had dreamed of finding.

White also came along on the 1990 visit to Hadar and recalled tense exchanges. After the field season, Tadesse Terfa rebuked Johanson for exporting geological samples and videos without proper clearances from the Ethiopian government. The Hadar team complained of unfair treatment. During the tumultuous transition to the new government after the civil war, Berhane's supporters tried to rally the research community behind him, but Johanson remained reluctant. "I said, 'You know, it's difficult for me to do that after what's happened to me,'" Johanson confessed. "I had talked to people in the ministry who told me Berhane was very openly saying, 'Johanson is never going to work here again.'"

White slammed the door on his friendship with Johanson. He would remain loyal to his Ethiopian colleagues and his own purist vision of science. Taken aback by the abrupt snub, Johanson approached mutual friends—couldn't they work this out? No, they could not. White refused to comment publicly, but his icy wrath was clear enough. "Tim is a very black-and-white person," said Johanson. "There are not a lot of shades of gray in his life. We had a long-term, very productive relationship, and then it was just terminated."

By 1994, the rift between the two camps had become an open feud and part of the on-the-ground reality of human origins research in Ethiopia. "They were always bitterly complaining about each other, trying to slice each other's throats," recalled archeologist Steve Brandt, who remained friendly with both sides. "Cut them out of this, cut them out of that—I heard that all the time, particularly from Berhane and Tim." In the national museum, bystanders and administrators were swept into the drama. "That place was like a battlefield," remembered Zelalem Assefa, a former Ethiopian antiquities officer. "For some of us, that gave us a very bitter experience. . . . Instead of collaboration for the sake of science, you'd see big names fighting each other—for glory and fame."

Johanson began to suffer other consequences of becoming a scientific celebrity. Back in California, he had conceived of his institute as a nexus of interdisciplinary scholarship, and it also housed the Berkeley Geochronology Center, a world-leading center on radiometric dating. The staff was divided into two teams: paleontologists under Johanson and geochronologists under Garniss Curtis, a rumpled, elderly scientist who had pioneered the science of dating volcanic rocks. According to one inside account by Institute for Human Origins geologists, "both men sport considerable egos and . . . explosive tempers." Johanson busied himself as a masterful popularizer of his science. As he later recounted, he followed "the Louis Leakey School of Anthropology: To keep yourself in the field 10 percent of the time, keep yourself on the stump the other 90 percent." He churned out books in the Lucy franchise and devoted a large portion of his time to his *Nova* series, produced by his then wife, Lenora Johanson. The geochronologists wrote 90 percent of the institute's scientific publications, generated 70 percent of its grant money, and complained that Johanson failed to respect their work. Long-simmering tensions finally blew up in 1994 when Johanson spotted the geochronologists lunching with one of his prospective donors at the Chez Panisse restaurant in Berkeley. Johanson accused Curtis and his staff of trying to poach his philanthropists, and the two men yelled at each other back at the office. News reached board member Gordon Getty, the IHO's largest benefactor (and Ann Getty's husband). Alarmed by Johanson's pattern of behavior, Mr. Getty decided the IHO founder had to go and threatened to pull his $1 million annual support. The IHO board held an emergency meeting, and Getty motioned to terminate Johanson. "I think he could leave with a minimum of embarrassment, if all were of a mind to minimize embarrassment," Getty reportedly told the board. The motion failed by a 9–4 vote (Johanson was opposed by Getty, Curtis, Paul Renne, and F. Clark Howell). Then the tables turned: one of Johanson's supporters motioned to fire the entire geochronology group, and that motion passed by the same margin. Those who sought to depose Johanson wound up being deposed themselves. Gordon Getty resigned from the IHO board and pledged $1 million per year to the newly independent geochronology group (which was deeply immersed in the Middle Awash work). After a messy lawsuit, the two sides reached a settlement, and Johanson moved his institute to Arizona State University three years later.

In Ethiopia, Johanson faced additional troubles. In 1994, Berhane Asfaw, Giday WoldeGabriel, Yonas Beyene, Yohannes Haile-Selassie, and Sileshi Semaw levied a series of allegations against Johanson and the IHO, including grave robbing (which had come to light only because Johanson dramatically described it in his book), commercializing Ethiopian antiquities, and distributing fossil casts in violation of Ethiopia's new antiquities laws. Four of the accusers worked on the Middle Awash team. Berhane later explained, "We found ourselves no longer able or willing to tolerate the blatant exploitation and abuse of privileges granted to these researchers for over twenty years by this country." Gona became a battleground, too: Sileshi Semaw accused the IHO of trespassing on his new territory and "claim jumping." The IHO denied the accusation and dismissed it as part of a sabotage campaign by rivals. "They are trying to destroy our organization," charged Bill Kimbel, the scientific director of the IHO. The Ethiopian attorney general began an investigation into the complaints, and the government suspended the IHO research permit. It would take five years for Johanson, Kimbel, and the IHO team to regain permission to work at Hadar.

Johanson blamed the Middle Awash team for his troubles. He would not forget it.

• • •

FOR DECADES AFTERWARD, THE TWO CAMPS ADVANCED COMPETING NARRAtives. The Middle Awash team believed Johanson maneuvered to undermine Berhane in order to secure a more foreign-friendly antiquities administration in the Ethiopian government. "When he heard I was removed from the position, he had a party at the Ethiopia Hotel," said Berhane. "Can you believe that? He had a party!" When an Ethiopian journalist aired a critical report about him, Berhane noticed that it included a photo of him taken at an IHO event during his graduate school days. The Middle Awash team saw the conflict in stark terms: they were doing science on behalf of the developing country and Johanson was a "safari anthropologist" working for himself. In turn, Johanson accused his rivals of making false allegations in order to establish a monopoly as the scientific superpower in Ethiopia. He cried alarm about the expanses of fossil-rich lands controlled by Berhane, White, and their associates—not only the rich Middle Awash but additional territories acquired by veterans of Berhane's inventory project. "Obviously, Tim is the one person who seems to be the thread through all of this," Johanson said years later, "although he is very clever and Machiavellian in the way he's approached this."

"As Tim told me many times, the power is in the fossils," added Johanson. "If you control the field sites and you control the discoveries, you control the field. That's how Tim looks at this. He was to be supremely in control of this entire field. And one of the ways of doing that, of course, was to try to control one of the greatest fossil fields ever—the Afar triangle."

White dismissed such talk as "Johanson's spin" and more self-aggrandizement. To White, Johanson became another antihero—a model to be shunned. When literary agents later approached White about writing a popular book, he declined and directed them to his Ethiopian colleague Berhane Asfaw. White also distanced himself from his profession: in 1994, he left the Berkeley anthropology department—which he felt had fallen under the sway of postmodernist social anthropologists who did not believe in science—and moved to integrative biology. Just as the team hunted fossils beyond the edges of civilization, they would work on the outskirts of their science.

Johanson sometimes became irritated when people mistakenly credited White—his partner in naming the Lucy *species*—for being codiscoverer of the *skeleton*. Likewise, he grew annoyed that many people in France mistakenly believe Yves Coppens found Lucy. In truth, *who* discovered a fossil matters little to science—but it matters greatly to some scientists. "It's very odd that one of the leaders of a paleoanthropology expedition has ever made the discovery," Johanson later explained, "so I held that very dear in my career and my life."

Indeed he did—to the point where some peers privately questioned his account. Even among the Lucy team, some suspected that the famous skeleton actually had been found by Johanson's graduate assistant, Tom Gray. Even Maurice Taieb thought so. The rumor was false—on that Gray and Johanson both agree—but it does reveal the dynamics of jealousy, possessiveness, and suspicion. The nearest witness was Gray, and he confirmed that Johanson spotted the first piece of the Lucy skeleton on that day in 1974. But seconds after the initial discovery, Gray recounted that he himself pointed out the second piece—a bit of skull beside Johanson (the *Lucy* book claimed Johanson spotted that piece, too). Gray felt the discovery should have been credited to both men. Within a day of the find, Gray recalled having a sinking realization: "I was going to be cut out of this." He also disputed Johanson's account of the grave-robbing episode, which described Gray reaching into the Afar stone tomb to plunder the bone. Not so, insisted Gray—Johanson stole it. "I thought, if I stick my hand in there I'm going to get bit by a goddamn snake," recounted Gray. "There was no way on god's green earth I was going to do that. I can guarantee you that was not me." So why did the *Lucy* book pin the theft on him? "My guess is that Don knew it was wrong and wasn't willing to take the responsibility for it," said Gray. When asked about these episodes, Johanson responded that Gray had a chance to review a draft of the book for accuracy. Gray recalled raising mild objections but did not press the issue. "Donald can be incredibly vindictive," said Gray, who later dropped out of the profession. "If I had raised too much of a stink, I could have seen him trying to torpedo my whole dissertation defense." To be fair, anthropology has a long tradition of exhuming graves, and some comparative collections are composed entirely of bones from ancient cemeteries. Even so, Johanson's enemies seized the grave-robbing episode as a symbol of exploitation and pummeled him with his stolen bone.

In retrospect, Maurice Taieb concluded that Johanson simply had been too young to manage the sudden temptations of fame. Taieb found himself dropped from the IHO team—despite being the person who'd invited Johanson to the rich Afar fossil fields in the first place. The geologist said in his French-accented English, "Donald Johanson had ego—he wants to have this honor only for himself."

TWENTY YEARS AFTER LUCY, THE AWASH VALLEY PRODUCED A SECOND blockbuster discovery—a skeleton 1.2 million years older than the one that made Johanson famous. The Middle Awash team already sensed this would be the biggest discovery since Dinknesh, and much would be different. There would be almost no publicity until the scientific work was complete. A closed circle would operate in secrecy. African scholars would be equal partners.

Once again, newly found fossils were laid out on the camp table. On the final nights in January 1995, the crew readied the fossils for transport back to the museum. With broad grins, the anthropologists packed fossils in crumpled newspapers and toilet paper. To document the moment, they turned on the video camera, and Berhane and Yohannes narrated an account of the discovery in Amharic for Ethiopian television. In the background, White stood silent and kept working. One key to his success was knowing when to step forward and take charge of operations and when to fade into the background and defer to Ethiopians.

Yohannes Haile-Selassie and Berhane Asfaw break into broad grins as they pack bones of the skeleton at the end of the field season.

Once again, the discoverers quibbled about who deserved credit for discovering the new skeleton. The videographer asked Yohannes to describe how he found the first piece, but the student demurred and looked uncomfortable. Self-promotion seemed too immodest, too un-Ethiopian.

"I don't want to say *I* found it," protested Yohannes.

"Well," piped up White, "say *we* found it then."

Chapter 11
THE PLIOCENE RESTORATION

The next big surprise came in the lab.

Trailing plumes of dust, the expedition caravan slowly wound its way out of the badlands at the end of the field season. For all the excitement, the discoverers still didn't know exactly what they had found. The skeleton arrived at the Ethiopian National Museum with many bones still encased in pedestals of ancient sediment wrapped in orthopedic bandages and aluminum foil. A passerby might have mistaken the skeleton for a bunch of crudely potted bulbs that had yet to sprout. The team checked in their fossils at the museum and made a cursory announcement of their discoveries in a press conference in Ethiopia. The field season had been repeatedly extended to rescue as many fossils as possible before calling it quits, so White had time to clean only a single bone before catching a plane out of Addis Ababa. Acting on a hunch, he picked specimen number 88, a small foot bone known as the medial cuneiform.

The medial cuneiform reveals whether or not a primate has an opposable toe. In apes, the far end of the bone bears a rounded, cylinder-shaped joint for the first metatarsal bone of the foot, which allows the grasping toe to open and close. In humans the joint is flatter and holds the big toe in line. The loss of the grasping toe represented a watershed when our ancestors sacrificed climbing for walking upright on the ground. Had *ramidus* crossed that Rubicon?

In the closing hours before his flight, White sought to answer that mystery.

FEET ARE RARE IN THE FOSSIL RECORD. EXTREMITIES ARE FILLED WITH tasty ligaments and muscles, like a handout for scavengers: *eat me.* "Feet are carnivore hors d'oeuvres," explained White. "They're out on the end of the limb and easily chewed off. They're one of the first things to go." Case in point was Lucy: her lower extremity had only two toe bones and an ankle bone. Similarly, the discovery of at least thirteen individuals of Lucy's species from the "First Family" site failed to yield a single intact foot; the best was a partial foot missing its front half. When White and Gen Suwa sought to reconstruct the feet that made the Laetoli footprints, they cobbled together a composite from *four* different fossils that spanned nearly 2 million years. This paucity of fossils gave rise to ferocious debate.

The *ramidus* fossils were an extraordinary exception—the team picked up lots of 4.4-million-year-old foot bones. Some of the metatarsals—the long bones of the foot—resembled *Australopithecus afarensis* and hinted bipedality. So did the bit of ravaged skull base, whose shape suggested it sat atop an upright spine. On the other hand, the fingers and toes seemed long and curved, like tree-climbing apes. Specimen number 88 might reveal whether this creature bore the telltale sign of arboreality, a grasping toe.

White sequestered himself in the hominid room at the national museum. He was not a man prone to sentimentality, but something about fossils left him in awe. One of his greatest pleasures—a privilege, honestly—was witnessing the reappearance of extinct species. "It really is a kind of time machine," he once remarked.

"As you take those sand grains away from a fossil, you're able to see an organism nobody has ever seen." In his eyes, damaging a fossil wasn't just a mistake; it was a crime against humanity. One clumsy move could destroy the last remains of an extinct creature never to be found again.

He bemoaned the carelessness of peers writ all over the fossil record—fossils damaged with dental drills, high-pressure air guns, acid baths, and more. White believed he could recognize the microscopic signatures of certain offenders. The very first hominid fossil found in Africa, the Taung Child, was pockmarked by indentations where discoverer Raymond Dart had chiseled away rock matrix with his wife's knitting needles. One Neanderthal skull had been smoothed with sandpaper; other scholars mistook the scratches for a mortuary ritual. "The preparer damage across those fossils is *obscene,*" White hissed. Sometimes he almost wished those specimens remained in the ground rather than suffer such abuse at the hands of amateurs.

A shape appeared beneath his microscope—a rounded surface on the far end of the bone. On the medial cuneiform, that could

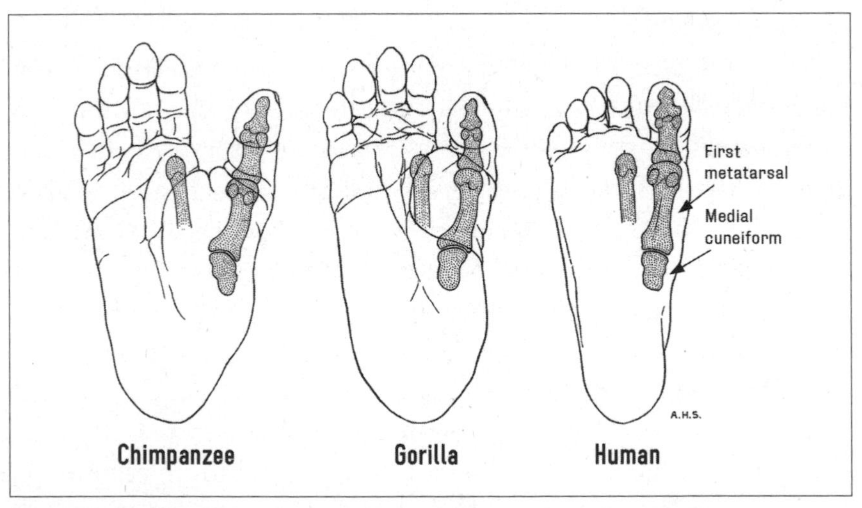

The medial cuneiform bone articulates with the first metatarsal. In apes, the joint bears a cylindrical surface that allows mobility in the grasping toe.

mean only one thing—the mobile joint of a grasping toe, the mark of a creature that climbed trees.

White's boyhood, thumb-worn copy of the book *Early Man* showed an illustration of an upright ape-man with prognathous snout and divergent toe. As he grew older, he came to realize that the beautiful illustrations were the fanciful imaginings of an artist. No such member of the human family had ever been found.

Until now.

Here it was: a member of the human family with an opposable toe—something entirely new to science. White loathed the myth of the eureka moment, a trope long abused by writers and publicity-pandering peers. In retrospect, however, he had to admit this bone *was* a genuine moment of revelation—perhaps the greatest surprise of the entire skeleton. He didn't have time to remove all the rock matrix but had exposed just enough to reveal something very important. As psychologists have demonstrated, the source of expertise is no secret: it is simply the ability to recognize patterns based on years of experience.

White had written the book on bones, but on questions of locomotion he listened to one man. In the final hours before catching his plane out of Ethiopia, White walked down to the casting lab and ordered a duplicate of the medial cuneiform. He would send the cast to the one expert he trusted on such matters. At that moment, his old friend lay in self-imposed exile, reeling from personal tragedy, and forgotten by much of the profession.

Chapter 12
STANDING UPRIGHT

Owen Lovejoy was a bearded, middle-aged anthropologist who had begun his career as a gravedigger, except he dug *up* skeletons. Back before such things became impossibly controversial or illegal, Lovejoy exhumed ancient cemetery graves by the hundreds, each skeleton revealing its own peculiar quirks and pathologies that filled him with wonder. Where did *that* come from? A Bible college dropout, he embraced the heretical notion of evolution, became a scientist, and earned a reputation as an expert of singular insights.

Lovejoy was known as the man who put human ancestors on their feet. Two decades earlier, he had served as the chief locomotion expert on the Lucy team and showed that *Australopithecus afarensis* walked upright more than 3 million years ago, far earlier than many scholars had expected.

Lovejoy spoke with a deep, smoky accent of his native Kentucky that sounded like a voice-over for a bourbon commercial. It sometimes became a growl of disdain when he encountered ideas he didn't like—and there were many. "There are probably more myths in human evolution than any other science," he said. "And they're just perpetuated!" He didn't just politely say he disagreed; he dismissed such things as nutty, preposterous, hopelessly lost in space, or the stupidest thing anybody ever heard. Such rhetoric did not endear him to what he called "the profession."

"It's too bad anthropologists don't have patients," he said, "because all my enemies would have dead ones."

In the eyes of the profession, he, too, was as good as dead.

It was January 1995, and snow covered the ground around Lovejoy's home in Ohio. The fifty-one-year-old scientist had arrived at the bleak midwinter of his life. Once he had been regarded as one of the most creative and incisive minds in his field, but years of silence left friends and foes alike wondering if he had decided to quit. Two months earlier, his wife had died from brain cancer. The bereaved anatomist remained intellectually groggy and barely worked. He had promised some papers to journals but just couldn't force himself to write them. He avoided visiting the bone collection at the Cleveland Museum of Natural History because the journey would take him past the hospital where he watched his wife slip away. He was sick of academia and all its damn politics, and he toyed with the idea of saying the hell with it and taking early retirement.

Then a package arrived from Tim White.

White was one of the few colleagues with whom Lovejoy still bothered to keep in touch. They had known each other since the 1970s when White, then a graduate student at Michigan, drove his pickup truck down to Ohio to examine the bone collections at the Cleveland museum and slept on Lovejoy's living room couch. "We were both equally suspicious of everything anthropologists did—and have remained so for the rest of our lives," Lovejoy recalled.

They liked each other because they were so unlike each other. White was a field man who spent much of his life in the desert amid scorpions, snakes, and nomads with rifles. In truth, Lovejoy had only a vague notion of what his friend did out in Ethiopia. All he knew was that White would get irritable in the fall as he prepared for his next expedition, disappear into the wilderness for a couple of months, and come back with crazy tales and lots of interesting fossils. Lovejoy avoided working outside and no longer got his hands dirty, except when law enforcement asked him to assist in forensic cases. White was an empiricist who demanded hard data; Lovejoy was a theorist whose breakthroughs were intellectual. White was literal and combative; Lovejoy was speculative and reclusive.

Inside the package Lovejoy found a cork-size chunk of white dental plaster, the kind of cast that lab technicians make when they want to dash off a rough copy in a hurry. It was the medial cuneiform bone from the oldest foot in the human family. Because the original fossil remained encrusted with earth, the cast looked more like a blob of plaster than a bone, and the exposed joint surface was even smaller than a pinky fingernail. White allowed Lovejoy to form his own impression then followed up by e-mail to reveal the punch line.

It's got a joint for an opposable big toe, said White.

Bah. A big toe that could grasp like a thumb? Lovejoy didn't buy it. If there was one thing that distinguished human ancestors, it was big toes that faced forward, not off to the side. (In anatomical parlance, other primates have *abductable* big toes that open and close, and humans have *adducted* ones.) Anatomists had long recognized that our big toe was unique in the animal kingdom. It was called the *hallux*, derived from the Greek word for "I spring forward." Every other known human ancestor had forward-facing toes. Oh, sure, some of the "profession" argued otherwise, but Lovejoy dismissed them as hacks who wouldn't know an abducted hallux if it kicked them in the gluteus maximus.

Lovejoy had spent the last decade fighting people who wanted to turn human ancestors like Lucy into quasi-apes. He was tired of hearing about all the arboreal features his opponents saw in the Lucy species—grasping toes, walking like circus apes with bent-hips-bent-knees, blah, blah, blah. Lovejoy got so fed up he quit going to conferences. To him, opponents seemed like hunting dogs who instinctively ran everything up a tree. And here his oldest and most trusted colleague spouted another such argument. *Et tu, Tim?*

• • •

LOVEJOY WAS RAISED AS A CREATIONIST. HE GREW UP IN A DEVOUT METHodist family in Lexington, Kentucky, where his father served as song leader in the local church and the children attended weekly Bible school. Scriptural verses still leaven his speech, and he sometimes echoed a passage from First Corinthians. "When I was a child, I believed everything my mentors taught me," Lovejoy later recounted. "I have spent the last forty years unbelieving just about everything that I learned." The family spent summer vacations in the woods of northern Michigan, and as a boy he explored the woods, mucked about in the ponds, and took an interest in biology. When Lovejoy was thirteen, he caddied a golf game for his father, and on the way home they stopped at the auto mechanic where Lovejoy saw his dad drop dead from a heart attack. Lovejoy figured he had gone to heaven, just as a devout Christian should.

Lovejoy duly attended evangelical Wheaton College in Illinois, the alma mater of Billy Graham. The great books curriculum inspired critical thinking and his conversion to atheism. (It also inspired pranks like dropping Ping-Pong balls in the chapel organ pipes so the balls would rattle in the tubes during services.) Lovejoy dropped out of Wheaton and transferred to Western Reserve University in Cleveland (the school later merged with the nearby Case Institute of Technology). For a time, he fancied himself as a poet. He studied psychology but found the workings of the mind too vague for his tastes. On a lark, he took a summer job on an archeological excavation of an ancient Native American burial ground led by archeologist Olaf Prufer. Prufer was a squat cannonball of iconoclasm with a bushy gnome beard, German accent, bayonet intellect, and disregard for political correctness or academic niceties. Prufer had an unusual pedigree: his father had been a German spy during World War I who played a geopolitical chess match against Lawrence of Arabia and the British in the Middle East. As a child during the Nazi era, Prufer joined the Hitler Youth, and his father served as the Nazi ambassador to Brazil.

After the war, Prufer became estranged from his father, went to work for the government of newly independent India excavating prehistoric sites, earned a Ph.D. at Harvard, and became an archeologist who specialized in the Native American cultures of Ohio. That improbable biography stamped him with a fierce intellectual independence and little patience for fools. He liked Lovejoy.

By Lovejoy's account, Prufer hired him for reasons that had little to do with archeological qualifications. Lovejoy played guitar and could entertain the drunken crew around the campfire at night. He proved a diligent student, and in 1967 Prufer appointed him "chief shovelhand" on a bigger excavation at the Libben site in northwest Ohio. "We dug up 1,500 skeletons, and that changed everything," Lovejoy recalled. "Skeletons were going to be my life."

Libben held the remains of a fishing village beside the Portage River more than one thousand years ago, when northwestern Ohio had been covered by the Great Black Swamp. It was the largest cemetery excavated in the Americas. Whenever the crew hit a skeleton, orthopedic surgeon King Heiple dropped into the hole and delivered an impromptu tutorial about whatever oddity they had found—and they *all* had oddities. One skeleton of a young man wore a headdress made from a wolf's tail, and his body had arrowheads embedded in his spine and ribs; his skull bore gashes suggesting he might have been scalped. Nearly every skeleton exhibited some pathology such as arthritis, cancers, evidence of healed breaks, kidney stones, bladder stones, or bones that had improperly fused together. There was no such thing as a typical skeleton; instead, *normal* spanned a large range of biological variation and imperfection.

Lovejoy became fascinated by the nuances of bone. When Prufer moved to the University of Massachusetts, Lovejoy followed. He earned a Ph.D. there in biological anthropology and wrote his dissertation on the Libben remains. At the time, most anthropologists only had cursory training in anatomy and none in biomechanics. Lovejoy apprenticed himself to orthopedic surgeons and biomedical engineers who designed artificial hips and knees, and he spent his weekends solving problem sets from engineering textbooks. He became the first scholar to apply biomechanical engineering principles to human evolution.

When Prufer moved to Kent State University in Ohio, Lovejoy again followed and took a job teaching anthropology in 1968 even before completing his Ph.D. For a time, he became a seer of doom who predicted the imminent extinction of humanity due to pollution. Meanwhile, the campus was swept up in the protest movement against the Vietnam War. In 1970, Lovejoy heard a commotion outside and stepped onto the roof of his department building to watch demonstrators facing off against the Ohio National Guard. "I heard the shots," Lovejoy recounted. "I thought they were firecrackers, until all hell broke loose." He had witnessed the climactic moment of the antiwar movement—the Kent State shooting. Troops had opened fire on student demonstrators, killing four and wounding nine. In that turbulent era, the distrust of authority played out in the classroom, too. The young professor taught from textbooks at a time when students questioned everything. If a teacher couldn't defend orthodoxy, something was wrong.

One of those things was the prevailing wisdom about the evolution of human locomotion. At the time, the early fossil record was dominated by *Australopithecus* from South Africa, which many scientists dismissed as inefficient bipeds capable of only semi-erect shuffling. Some prescient scholars such as British anatomist Wilfrid E. Le Gros Clark had recognized as early as the 1950s that *Australopithecus* stood erect, but this remained far from the consensus view. British anthropologist John Napier described them as waddling with an inefficient "jog trot." Sherwood Washburn portrayed them upright with a "shambling half-run." Some prominent authorities deemed *Australopithecus* a dead-end branch doomed to extinction—"just bloody apes," according to British scholar Lord Solly Zuckerman.

But Lovejoy was not trained by anthropologists, and much of

the literature on locomotion struck him as fanciful storytelling. He conducted a biomechanical comparison of human bipeds and African apes, then applied his insights to the then-oldest hominids, mostly South African *Australopithecus* fossils believed to be about 2 million years old. The best evidence was a femur and pelvis, and Lovejoy determined they bore the major hallmarks of bipeds. By 1973, Lovejoy had concluded that bipedality was far more ancient than big brains or tools, and *Australopithecus* walked with a striding gait like modern humans.

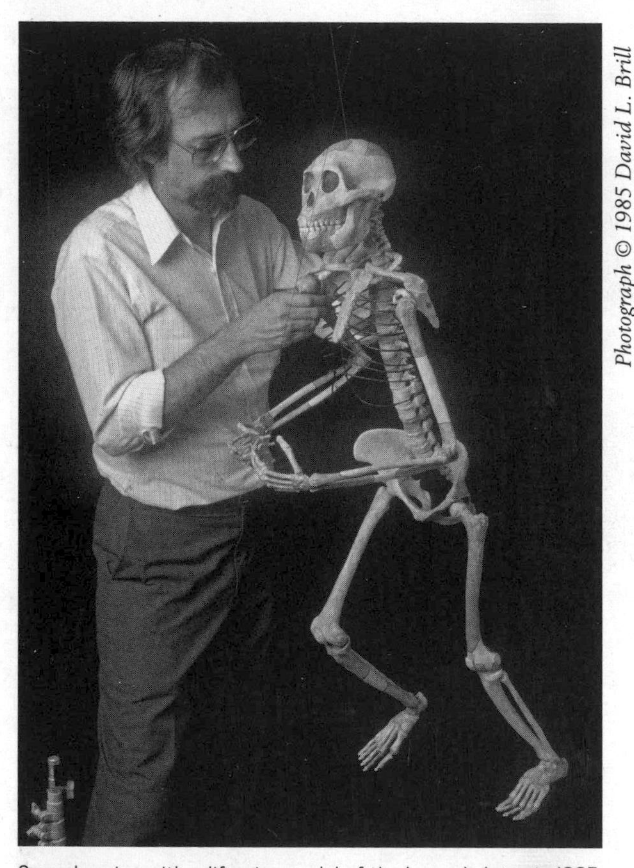

Photograph © 1985 David L. Brill

Owen Lovejoy with a life-size model of the Lucy skeleton in 1985. His belief that *Australopithecus afarensis* was an advanced biped aroused fierce opposition from critics who insisted that Lucy's species was arboreal.

Then confirmation arrived at his door.

In 1973, Don Johanson showed up at Lovejoy's house in Kent with a box labeled "primate." Johanson had just returned from the first fossil expedition to the Afar Depression of Ethiopia and was keen to show Lovejoy what he had found. Inside the box, nestled in foam, lay a fossil knee—the one that prompted the plundering of the Afar grave. Johanson specialized in teeth and jaws and needed the counsel of an expert on locomotion.

The two men sat on the floor of the living room, and Lovejoy examined the fossil knee. "It was gorgeous," he recalled, still moved by the memory decades afterward. "Beautifully preserved." Despite the tiny size—the creature would have stood only about three feet tall—it bore the telltale traits of a bipedal knee. The biggest surprise was the age, then estimated to be at least 3 million years old. Upright walking had just grown a million years older.

The following year Johanson returned with Lucy, whose knee was virtually identical to the first one he had shown Lovejoy. Even better, her skeleton had a partial pelvis, which Lovejoy quickly diagnosed as further evidence of bipedality. Four years later, the Laetoli footprints provided yet more confirmation of the ancient roots of upright locomotion—and left Tim White greatly impressed by the insights of his friend.

Lovejoy's reputation soared. Johanson's popular books and television documentaries lionized him as an oracle of human evolution and locomotion. Within a few years, however, critics turned against Lovejoy. In the 1980s, an opposition emerged led by scientists at the State University of New York at Stony Brook. They believed Lucy's species remained arboreal and dismissed Lovejoy's interpretation as a "fairy tale." Critics sniped that Lovejoy portrayed the apelike hominids as "little people" and exaggerated their similarities to modern humans. One Stony Brook anthropologist rebuilt Lucy's pelvis to resemble an arboreal ape. Lovejoy told him he didn't know what he was talking about, and then stopped talking to him entirely. Similarly, a group of French scholars claimed Lucy and her species remained in the trees. A pair of Swiss scientists made another apelike reconstruction with a new twist: Lucy was actually a male and should be renamed *Lucifer.* Colleagues who formerly sided with Lovejoy drifted toward the new view—that Lucy and her species were climbers.

Lovejoy became dismayed by what he read in the journals. In his eyes, his opponents were obsessed with quantification and anatomical reductionism. They would identify an anatomical feature, take measurements showing it fell below the human average, and—*eureka!*—conclude arboreality. Lovejoy allowed that Lucy might have occasionally climbed trees but insisted her species had given up too many arboreal adaptations to have spent much time up there. "She was not just capable of walking upright," he insisted, "it had become her only choice." Lovejoy lacked the stomach for the debate. He purged offensive literature from sight by tossing journals into a hole in his office ceiling. "He used to stand on top of his desk, pop open the acoustic panel, throw the journal in, and put the panel back," recalled former student Bill Kimbel, bellowing with laughter. "He was worried it would one day collapse and kill him!"

Colleagues begged Lovejoy to answer his opponents. He grumbled that it wasn't worth his time to argue with people whom he dismissed as clueless. He stopped attending conferences, and his publications on locomotion slowed to a trickle. He wrote an influential article in *Scientific American* on the evolution of human walking but refused to even acknowledge the alternative point of view. It would take him twelve years to directly confront his critics, but he did not deign to do so in an anthropology journal and instead published his magnum opus on human gait in an *orthopedic* journal. By withdrawing, Lovejoy effectively ceded the debate to opponents, and the pendulum of professional opinion swung toward the arborealist view for the next twenty years. In the 1970s, *National Geographic* illustrated Lucy walking upright. A decade later, the magazine showed her in the trees.

That wasn't the only point of controversy. In 1981, Lovejoy proffered a controversial theory about the origins of bipedality—it was a mating strategy that allowed human ancestors to become monogamous by freeing the hands so males could provision children and mates. He suffered the academic equivalent of a biblical stoning.

For a time, Lovejoy sank into a funk and absented himself from the university. A connoisseur of fine automobiles, he spent days working on his sports cars and hanging out with the mechanics at the local auto shop. The garage owners gave him a mechanic uniform with a breast patch that said OWEN and dispatched the professor to make runs to the auto parts distributor. When students had questions, they called the garage.

Lovejoy got divorced. A few years later he remarried, and his second wife, Lynn, started waking up with headaches. Doctors initially diagnosed migraines. By the time they realized it was cancer, the disease had metastasized throughout her body, and she lived only four more months. Lynn died on November 2, 1994, at University Hospital in Cleveland, a few blocks from the Cleveland Museum of Natural History, where Lovejoy had done his groundbreaking research in the Lucy era. Only days after her death, the fossil crew in Ethiopia found the first piece of the new *ramidus* skeleton. In the dead of winter, the grieving anatomist held its most intriguing bone.

LOVEJOY STARED AT THE MEDIAL CUNEIFORM FOR DAYS. HE COULDN'T UNderstand why White kept calling the thing an opposable toe.

Lovejoy believed that upright walking was *the* original evolutionary innovation that separated the human lineage from the other apes. To be a member of the human family was to be a biped—with a forward-facing toe, period. A biped couldn't also be arboreal. Oh sure, they might occasionally have gone up in trees to escape predators or obtain food but not often enough to be called real climbing specialists. Upright walking, he figured, required such a drastic reorganization of the body that once our ancestors left the trees they could never really go back. The idea of an arboreal-bipedal hybrid seemed as ludicrous as a horse-drawn automobile. Inconceivable to anyone who understood biomechanics. When people argued otherwise, Lovejoy could only shake his head and repeat, *Thank goodness you don't have patients.* When our ancestors stood up, those opposable toes would have disappeared in a heartbeat of geological time.

Lovejoy could almost hear White's impatience growling between the lines of his e-mails as they squabbled for a week. Back at Berkeley, White compared the fossil to the same bone on a chimpanzee foot and could barely tell them apart.

"This guy is a footgrabber, Owen," he insisted.

Lovejoy wouldn't budge, and White grew exasperated. *Typical Lovejoy—a stubborn son of a bitch.* Having cleaned more fossils than most colleagues see in their lifetimes, White had a knack for taking bearings from small landmarks of bone. He told Lovejoy to find the lateral facet, the smooth surface where the medial cuneiform joined with the neighboring bone of the midfoot. Got it? OK, now lay the bone on the tabletop and look for the cylindrical surface of the toe joint. Grumbling, Lovejoy obliged.

Then, to the best of his recollection, he swore.

LOVEJOY NEVER COULD PREDICT HOW LONG HE WOULD HAVE TO STARE AT A bone before it revealed some deep truth. With his glasses halfway down his nose, he would study it from different angles. Turn it over. Put it in articulation with the adjoining bones and consider how the animal moved. Do something else for a while and let the questions percolate in the back of the mind. One day, revelation would arrive as unexpectedly as a letter in the mail, and he'd finally recognize what lay in plain sight all along—and become disgusted with himself for not seeing it earlier.

Suddenly, he saw the convex joint surface that could mean only one thing. An opposable toe.

After one week of resistance, Lovejoy surrendered. This oldest known species of the human family had a grasping foot. "Now ready to admit defeat," he wrote White. "I didn't really want to but, hell, you're right."

Chapter 13
THE WHOLE WORLD WANTS TO KNOW

What the hell was this thing? At first, nobody was quite sure what to call it. In anatomical discussions, White and Lovejoy dubbed it ML—an inside joke poking fun at commentators who had exhumed the old cliché about a "missing link" when pontificating about the discovery. Lovejoy took to calling it *the beast*.

At night, he lay awake and tried not to think about Lynn, so he thought about the new skeleton. Maybe he wouldn't retire after all. Maybe he'd stick around and share the wonder of revealing the beauty of the beast. Never did he expect to encounter such a marvel—a creature so primitive that it still bore an opposable toe, with hints of anatomy unique to the human family. Already he sensed that the skeleton would prove his critics wrong. Lucy was no missing link of locomotion between bipeds and apes, as his opponents had claimed. *This* must be the transitional biped—or at least a large step closer. Upright locomotion came much earlier than Lucy, and the new species promised to show them how it happened.

The thrill of discovery stirred in Lovejoy again. What else awaited in the Ethiopian museum? What more might White and the field crew find in the Afar badlands? "From your description of this beast," Lovejoy told White, "there's a lot more surprises where this one came from!"

• • •

IN MAY 1995, A BRIEF ANNOUNCEMENT APPEARED IN *NATURE*. WITH LITTLE explanation, the Middle Awash team moved *ramidus* into a newly named genus, *Ardipithecus*. As a result, the earliest species of human ancestor ceased to be *Australopithecus ramidus* and became *Ardipithecus ramidus*. The new genus name derived from the Afar word *Ardi*, for "ground," and the Greek word *pithēkos*, for "ape" or "monkey" (apropos for fossils, it also means "trickster"). The name suggested both *ground ape* and *root species* of the human lineage. Major discoveries often take on nicknames, and the oldest woman in the human family became known as Ardi.

The revised classification appeared in an unlikely place—the journal's corrigendum section for rectifying errors. Never before had such a blockbuster announcement been published as a *correction*. Colleagues were left to to read between the lines. Compared to its younger successor *Australopithecus*, the new genus *Ardipithecus* was distinguished by its lack of big chewing teeth, thinner enamel, and larger canines. The notice briefly mentioned the skeleton (which also had been revealed in a few news stories) but provided no details. The opposable toe and climbing adaptations would remain secret for another fourteen years.

THAT SPRING, WHITE RETURNED TO ETHIOPIA AS SOON AS HE FINISHED classes at Berkeley. He felt the weight of humanity on his shoulders. "It is the most important specimen in the world right now," he told the *San Francisco Chronicle*. As he later explained to *National Geographic*, "*Ramidus* is the first species this side of our common ancestor with chimpanzees. It's the link that's no longer missing." He planned a blitz cleaning of all the skeleton bones—many still encased in blocks of sediment—but first he took time to document their original condition. One day he stood outside the museum photographing and videotaping the fossils in daylight when an Ethiopian official approached and questioned him.

"When will it be ready for display at the museum?"

"Probably three or four years would be my guess," answered White.

"Why so long?"

White explained the ordeal of restoring the delicate fossils. He showed one example—the sacrum, a triangular section of fused vertebrae at the base of the spine, still buried inside a chunk of earth.

"Is this closer to chimpanzee or human?"

"We don't know yet because it's covered still," answered White. "That's one of the things we'll be working on."

"You're going to clean it?"

"Oh, I'll clean everything eventually. But eventually is a long time."

"That's what they are wondering," pressed the official. "Are you going to finish in two or three months?"

"No," answered White. "In some ways it will never be finished."

In that moment, he revealed an attitude that would pervade the saga of *Ardipithecus*. In his purist vision, science was an endless task, and he refused to rush this most important specimen to satisfy somebody else's curiosity or impatience. He was a stickler who valued accuracy far more than quick delivery or popularity. "Many people would work on a fossil and reach a level they would call *publishable*," said geologist Paul Renne, a longtime member of the Middle Awash team. "Tim doesn't stop at that stage. He goes well beyond it to the absolutely ironclad stage."

THE INVESTIGATION WOULD ADVANCE IN STRICT SECRECY. IN ITS 1982 APplication to the National Science Foundation, which funded the expeditions that ultimately found Ardi (plus many other fossils), Desmond Clark and Tim White's team promised a spirit of collaboration for their fossil fauna: "Fossils collected by the Middle Awash research team will not become the exclusive domain of any person or group but will be available to the entire scientific community." The timeline for such access remained unspecified. By the time Ardi arrived at the museum a dozen years later, their position had hardened—no access for anybody outside the team. New antiquities laws had come into effect, the science had become more fractious, and the team bunkered down. Ethiopian regulations required scientists to "keep every discovery in secret" and follow strict protocols for the release of information. An unauthorized leak could provide a pretext to cancel a research permit.

It was common for discoverers to protect exclusive rights for a certain period of time and generally agreed that people who performed the hard work of finding fossils deserved the first crack at describing them. But the profession lacked standards, so practices varied between research teams and became flashpoints of conflict. Fossils, territory, and money were the raw materials of the discipline, and scientific tribes fought over them accordingly. Some teams offered sneak previews or special access to friendly colleagues, or simply valued outside input to enhance their own understanding. The Middle Awash team gave everybody the same answer: sorry, nothing would be revealed until the team finished all its work. White defended the policy as a matter of equity. "There was none of this favoritism—we're going to give you a sneak peek but the other guy is not going to get a sneak peek," he later explained. "I saw that as a graduate student and that's why I was dead set against it. I saw the special access Richard Leakey would give to 'the chapel.' If you were the Prince of Holland or Stephen Jay Gould, you were given special treatment." During the Lucy era, Johanson allowed some outside researchers to examine the *Australopithecus afarensis* fossils if they agreed not to publish before his team. The Ardi team would offer no such embargoed reviews (with very few exceptions) and would hold off making anatomical descriptions until they were not only sure of the facts but ready to declare what it all *meant*. Lovejoy said: "We decided, the first time this thing gets published it's going to be the *truth*." Eventually colleagues and journalists labeled the tight-lipped Ardi team as "the Manhattan Project of paleoanthropology."

For fifteen years, nothing leaked. White hammered his team with regular reminders to keep their mouths shut. Once the garrulous Bruce Latimer blabbed about an unpublished fossil to a reporter and suffered a memorable thrashing from White—a cautionary example that made an impression on his teammates. "I've been taken to the woodshed," said Latimer. "Then I shut up."

WHITE SHUT HIMSELF IN THE CLOSET-SIZE HOMINID ROOM OF THE MUSEUM. For weeks, he worked eighteen-hour days with dental tools, syringes, and a microscope, and cleaned every piece of the skeleton himself. He refused to delegate the task.

He continued the excavation in miniature, and the tabletop became littered with dirt. When White squeezed drops of water to soften the matrix, he witnessed a close-up simulation of what happened during the rainy season at Aramis: the sediment expanded when wet, contracted when dry, and the back-and-forth destroyed fossils near the surface. In the lab, only chemical hardener kept them intact. By some stroke of luck, the discoverers happened upon this skeleton at an opportune time—close enough to the surface to be discovered, yet still buried just enough for the fragile bones to survive. In a few more rainy seasons, it might have been lost forever.

Even so, the bones had suffered more than 4 million years of geological assault. A prime example was the pelvis, which had been buried near the surface where the soil heaved most violently. No part of the anatomy would be more important in revealing how the animal moved, and no part suffered more destruction. In a foil pouch, White collected what looked like a fistful of rubble—shards that had flaked off the pelvic blades. "Extremely shattered," he noted. "Whether any of this will go back together again, we don't know." Expanding soils had exploded the hip socket. Black and white dots marked where calcite and manganese had crystallized onto bone. The sacrum turned out to be a bust; when White dug into the pedestal, he discovered most of the piece had disintegrated. Eventually, he ceased cleaning the pelvis due to the risk of further damage. "There was no room for error," observed Scott Simpson, "and Tim did not err."

By the end of the summer of 1995, White had rescued most of the bones from the recent excavation. Two more field seasons would be spent digging for more pieces. Two crucial parts, the pelvis and skull, would require years of extraordinary restoration. For the moment, however, enough had been revealed to embark on a new phase of inquiry: what did those bones reveal about human origins? As an old saying goes, the most important thing is not what you find but what you *find out*.

That summer, Lovejoy got his first glimpse of the beast. The investigators assembled in Addis Ababa to begin their analysis and hunched over a tabletop of newly cleaned fossils in the National Museum. One of the biggest surprises of the skeleton was its sheer existence.

"Are there carnivore holes in this thing anywhere?" asked Lovejoy.

"Nope," said White. "Doesn't look like carnivores got to it."

Years of reconstruction eventually would reveal the full animal. Ardi would have stood about 1.2 meters, or about four feet, tall (somewhat bigger than Lucy), with a grapefruit-size brain of about three hundred cubic centimeters, roughly one-quarter that of a modern human. In many respects, a layperson might mistake her for an ape, but a few key anatomical clues linked her to the human family.

Lovejoy fixed his attention on one particular feature of the pelvis poking up from the dirt pedestal: a curving bump on the front edge of hip bone called the anterior inferior iliac spine. Its anatomy signaled thigh musculature unique to bipeds. That trait alone, Lovejoy declared, "makes it absolutely, uniquely hominid."

"This is very Lucy-like," said Lovejoy. "You could take that and stick it on Lucy's pelvis."

And yet—that opposable toe, a clear signal of arboreality.

During the bitter locomotion debates of the 1980s, scholars had fiercely argued whether a creature could be a tree climber without a grasping big toe. Ardi turned that question inside out: a biped *with* an opposable toe?

It only got stranger. In humans, the big toe is much stouter than its neighbors because it carries the final force of pushing off the ground. Ardi had a robust *second* toe. With her grasping toe splayed off to the side, her second digit apparently did the work of toeing off. (Imagine your hand was Ardi's foot; with the thumb off to the side, your pointer figure would bear the load of the final push-off.) She seemed a congeries of anatomy: an opposable big toe like a chimp, a lateral foot like an early biped, big hands, and curved digits like an arboreal ape, a pelvis with hints of humanlike bipedal anatomy, and who the hell knew what else still hidden. In Greek mythology, a chimera was a creature with an incongruous assembly of parts from different animals—the head of a lion, body of a goat, and tail of a serpent. Ardi seemed like a simian chimera—a hodgepodge of parts in novel combination. In an interview with *National Geographic,* White hinted just enough to tantalize his scientific peers. "If you want to find something that walked like it did," he said, "you might try the bar in *Star Wars.*"

ONE DAY IN THAT SUMMER OF 1995, WHITE HOVERED OVER THE MUSEUM casting technician. Alemu Ademassu was the master entrusted with making duplicates of every human ancestor from Ethiopia. He had been nurtured by Berhane, trained at Berkeley and the Cleveland Museum of Natural History, and assisted many *farenjis,* none more demanding than the one looming over his shoulder.

At that moment, Alemu molded a copy of the pelvis—or what

little remained of it. The phone rang, and White briefly spoke to a journalist who had tracked him down in the casting lab and pressed him for details. "*Ramidus* remains very much a work in progress," he said.

White hung up. "Everybody wants to know about that pelvis that you're molding there, Alemu," he muttered. "The whole world wants to know about that pelvis."

The phone call had been prompted by big news of another discovery that closely followed on the heels of Ardi. In Kenya, the Leakey team had found another biped that they proclaimed to be a more likely ancestor of Lucy.

Photograph © 1995 David L. Brill

Casting technician Alemu Ademassu molds duplicates of *Ardipithecus* teeth in the summer of 1995. His work is scrutinized by his most demanding customer, Tim White.

THE NEW DISCOVERY HAD APPEARED ABOUT NINE HUNDRED KILOMETERS TO the southwest down the Great Rift. The Omo River flows from Ethiopia into Kenya and dead-ends into Lake Turkana, which sits in the bottom of a vast valley with no outlet to the ocean. Like the Afar Depression, the Turkana Basin forms a huge sediment trap. Richard Leakey brought the area to international prominence in the 1960s and 1970s but eventually became exhausted by the bitterness of scientific disputes and Kenyan politics (a low point came when critics published a book called *Richard E. Leakey: Master of Deceit*). His wife, Meave Leakey, a zoologist, took over the family expedition and undertook the same quest as her colleagues in Ethiopia—finding the oldest traces of the human family. There had been some promising earlier discoveries around Lake Turkana from the late Miocene and early Pliocene epoch: In the 1960s, anthropologist Bryan Patterson of Harvard found a fossil humerus and a partial mandible, but the pieces were fragmentary and their ages uncertain. In 1994 and 1995, the Leakey team returned to the same areas and found an assortment of jawbones, teeth, and a bipedal shinbone of a human ancestor older than Lucy. Meave had visited the Ethiopian museum, compared her new finds to the newly discovered *ramidus* teeth (it was a rare instance where the Ardi team showed unpublished fossils, because each team had something relevant to the other), and she and White both agreed they had two distinct species.

The Leakeys had long rejected the family tree drafted by White and Johanson that depicted Lucy's species as the trunk lineage of humanity between 3 and 4 million years ago. "Everyone assumes *afarensis* is the common ancestor of everything that comes afterwards," Meave Leakey said. "I've always said this is complete nonsense." Setting out to find something different from *afarensis*, she only wound up making Lucy's lineage a bit longer.

In August 1995, Meave Leakey and her team announced the new species *Australopithecus anamensis*—a 4-million-year-old member of the human family that claimed the title of the oldest member of the genus *Australopithecus* and oldest biped in the human family. Over the next several years, more fossils accumulated, and the scientific consensus held *anamensis* as the likely ancestor of *afarensis*, Lucy's species.

In *National Geographic*, Meave Leakey asserted her new fossils "may represent Lucy's ancestor and that *Ardipithecus* may belong on another branch of the hominid tree. Many hominid species may have evolved in those early years." In light of the new Kenyan discovery, some voiced suspicion that the new Ethiopian species might be an ancient chimp or a dead-end branch. Ian Tattersall of the American Museum of Natural History told the press: "*ramidus*' claim to be a lineal ancestor to later humans is greatly weakened." Don Johanson suggested the ostensible "root" species might be "an ancestor to later apes rather than later hominids." Such doubts would have grown even more grave had the world known the big secret of Ardi—the apelike grasping toe.

But the Ardi team had another secret: they also had found the same *anamensis* species that the Leakey team had just announced—in younger geologic strata right above the skeleton.

Their own discovery followed a familiar pattern: Alemayehu Asfaw wandered off and found something interesting. Early in the 1994 field season, Alemayehu spotted two pieces of maxilla, or the upper jaw, at another locality near Aramis. The fossil was duly collected, and soon the team became distracted by the jackpot Ardi skeleton. Only later did they realize Alemayehu's maxilla belonged to the very species the Leakey team had just found in the Turkana Basin. Alemayehu's jawbone appeared about eighty meters above Ardi in the geological strata and was roughly two hundred thousand years younger. Over the next few years, the Middle Awash crew picked up more *anamensis* fossils (including thirty pieces of at least eight *anamensis* individuals from another locality about ten kilometers west called Asa Issie). The younger creature differed from Ardi because it had smaller canines and more robust chewing teeth typical of the genus *Australopithecus*. The Leakey team called *Au. anamensis* a second lineage separate from *Ar. ramidus*, but the Ardi team saw no such diversity. In the Middle Awash, the older species abruptly gave way to the younger one—consistent with a single lineage evolving over time.

Did one evolve into the other? Ideally, fossils from layers

between the two hominid species would provide the answer. Unfortunately, a large lake flooded that section of the Awash Valley during the intervening period. The fossils in that layer were fish.

The Ardi team considered two possible explanations. One envisioned a continuous lineage: the species of Ardi begat *Australopithecus anamensis* and all subsequent members of the genus. The second scenario posited a splitting event: some earlier population of the *Ardipithecus* genus, or a close relative, branched off and evolved into *Australopithecus*. Critics later added a third option not entertained by the Ardi team: *Ardipithecus* was a dead-end lineage unrelated to *Australopithecus* or modern humans.

But the Ardi team considered that third option unlikely. In their judgment, the diamond-shaped canine teeth and hints of humanlike anatomy in the lower body testified that Ardi and her species belonged in the human family—if not direct ancestors, then at least on our side of the chimp-human split. In early comments to the press, White suggested all three species formed a single lineage evolving through time. ("It's just another great example of evolution being reflected in the fossil record," White told one interviewer in 1995. "*Anamensis* seems like a perfect intermediate between *ramidus* and *afarensis*.") Later, the team softened the single-lineage theory and allowed the possibility *ramidus* might have been an offshoot—but one within the human family.

THE FOSSIL SEQUENCE SHINED LIGHT ON A MYSTERY THAT HAD LONG TORmented all of the expeditions searching for the earliest fossils of the human lineage: why were human ancestors so rare before 4 million years ago and increasingly plentiful afterward? The Middle Awash team suggested an answer—ecological breakout.

Their theory went like this: Early members of the human family remained confined to a narrow habitat. Creatures like Ardi could come down from trees, but not leave them entirely. Sometime around 4 million years ago, *Australopithecus* expanded into a broader niche, abandoned the grasping toe, and became the first fully bipedal members of the human family. Granted, *Australopithecus* retained strong upper limbs and curved fingers and probably was more adept in the trees than later bipeds, but the overall thrust of natural selection shifted toward life on the ground. It sported big, thickly enameled chewing teeth, an adaptation to coarser foodstuffs of more open habitats. (Only with the origin of *Homo* did this big chewing apparatus subside, a shift often attributed to the advent of stone tools.) Lucy's kin represented our family's first "weed species"—one that popped up at multiple sites across Africa. *Australopithecus* became a pervasive genus that chewed its way into east, central, and southern Africa, endured roughly 3 million years, and encompassed multiple species. By about 2.5 million years ago, human ancestors split into at least two (and perhaps more) branches. One group expanded the chewing apparatus as exemplified by the nutcracker-jawed *Paranthropus*. Another more gracile group is widely accepted as the likely ancestor of *Homo*. To be sure, some scholars argue that far more splitting occurred and describe *Australopithecus* as a "wastebasket" genus—an overly broad, one-size-fits-all category that obscures the real biological diversity. Put aside that debate for the moment—the key point is that *something* enabled once-rare bipedal primates to multiply across Africa. The human family became—at least by ape standards—more cosmopolitan. By around 2 million years ago, *Homo* broke out of Africa and into Europe and Asia, and later colonized every continent on earth. The last surviving species, *Homo sapiens*, mastered not only every habitat it encountered but *altered* the entire global environment and explored beyond earth itself. Meanwhile, the other great apes remained trapped in narrow ecological niches and slowly edged toward extinction.

That grand drama began with primitive bipeds who stood upright beneath the trees, With her peculiar gait, Ardi sauntered into a raging debate about the entire human family tree.

Chapter 14
TREES AND BUSHES

The tree of life is both an enduring metaphor for evolution and a burning bush of controversy. Is the human family a sparse limb with few offshoots, as the Middle Awash team claimed? Or did our family tree sprout many species, most doomed to extinction? Historically, the debate has broken into two opposing camps: lumpers and splitters.

Much of the debate arises from the fact that science still hasn't quite figured out how to define *species*. Everybody agrees that the species constitutes the basic unit of zoology, but consensus breaks down on what that actually means. More than twenty definitions are in use. The most common is the "biological species concept" offered in 1942 by Ernst Mayr: species are breeding pools, or the full collection of populations that can potentially mate with one another. But even this trusty old container leaks at the edges. In some cases, what we call different species can mate and produce fertile or infertile offspring—zebras and donkeys beget zonkeys, and lions and tigers birth ligers. In North America, genetic analysis shows that the red wolf is actually a hybrid of gray wolves and coyotes. Darwin's finches, the famous examples of speciation, frequently breed across species. (Indeed, about 9 percent of bird species are known to hybridize in the wild, and some lineages remain capable of interbreeding for 10 million years.) Fossil species are even more difficult to classify because all the biological material has turned to stone and thus remains beyond reach of genetic analysis. Scientists can only guess what the former breeding pools and ancestry might have been. How does anybody know what constitutes a true species based on fossil fragments? How to recognize ancestors and descendants, or chop up evolving lineages into species? Those debates are about analytical approaches as much as fossils.

Biological classification was invented before there was any notion of evolution. The modern taxonomic system was pioneered by Swedish botanist Carl Linnaeus in the eighteenth century (he invented the binomial nomenclature of capitalized genus name followed by lowercase species name, e.g., *Homo sapiens*). To define species, Linnaeus borrowed two ideas from Greek philosophy: Plato's notion of ideal forms and Aristotle's notion of essences, or the traits that defined an entity. These notions persist in the type specimen—the one individual that exemplifies an entire species. Such a narrow definition is problematic for living species (what individual could embody all humanity?) and even more so for those that evolve over time.

The debate about "species" and the shape of the family tree has dogged paleoanthropology since it became a modern science in the mid-twentieth century. The year 1950 marked a watershed. By then, the theory of evolution had won wide acceptance in the scientific community, but consensus was slower to arrive about the mechanisms: *how* did evolution advance? (Darwin's specific theory of natural selection did not catch on as readily as his general notion of evolution.) Geneticists had begun to understand heredity, and the founders of population genetics used statistical techniques to demonstrate how advantageous genes could spread. These disciplines provided a mechanism for Darwinian natural selection, and the merger gave birth to a new paradigm called "the modern synthesis." The synthesis embraced Mayr's definition of the *species*—a mating pool of widely varying individuals.

At first, paleontology and anthropology remained stubbornly

indifferent to these new currents of thought: species were named willy-nilly without respect for current scientific practices. Many anthropologists practiced typology, an approach that viewed individual fossils as distinct classifications ("types") unto themselves. Typologists tended to classify each *specimen* as a separate *species*—thus the fossil record included Peking Man, Java Man, Broken Hill Man, Cro-Magnon Man, and so on. In contrast, a population-minded biologist would see those fossils as just variations within broadly inclusive taxa. By 1950, human taxonomy was an embarrassment of spurious names with more than one hundred species and about thirty genera—all when the fossil record was much sparser than it is now. Evolutionary biologist Theodosius Dobzhansky lamented that the "abuse" of species names was "making this fascinating field rather bewildering to other biologists."

In 1950, a group of eminent biologists and paleontologists gathered at the Cold Spring Harbor Laboratory on the north shore of Long Island to discuss the chaotic lineup of creatures then called "fossil men." The task of pruning the overgrown bush fell to Ernst Mayr, the bird biologist who had coined the modern definition of *species* and now sought to apply it to the human family. Mayr stuffed all human ancestors into three species of the genus *Homo* and declared, "In spite of much geographical variation, never more than one species of man existed on the earth at any one time." Later Mayr acknowledged his uber-lumping was an overzealous "temporary oversimplification" and recognized *Australopithecus* as a separate branch of the family that coexisted with a more gracile *Homo* lineage—basically endorsing the Y-shaped tree. But he admitted no more than two lineages, and this lumping mindset dominated anthropology for the next two decades. Why did human ancestors not speciate as exuberantly as other mammals? Mayr suggested that human ancestors "specialized in despecialization." In other words, they became *generalists* adaptable to many environments. In contrast, *specialists* occupied narrower ecological niches and tended to diversify more.

Recent history colored the discussion. In the wake of World War II and the horrors of Nazism, biologists recoiled from the old "racial anthropology" and its fixation with biological distinctions between ethnic groups and races. In the postwar era, biologists embraced a more inclusive view of humanity—in both present and past.

Other biologists at the Cold Spring Harbor conference also dragged paleoanthropology into the new synthesis. Adolph Schultz of Johns Hopkins University—an anatomist who devoted his career to cutting open apes and monkeys—demonstrated that modern *species* were big categories of varied individuals and geographic populations. Paleontologist George Gaylord Simpson extended these insights with an evolutionary perspective: over time, species evolved as *lineages*. Just as the species was the basic unit of zoology, the lineage was the basic thread of evolution. Simpson was an ornery paleontologist from the American West who amassed a magisterial knowledge of mammal evolution and became the preeminent big-picture systematist of his era. More than any other figure, he laid the foundation for modern paleontology, and even critics who later bashed him over the head actually stood on his shoulders to do so. Simpson distinguished three modes of evolution: *speciation* (splitting events), *phyletic evolution* (gradual change within a lineage), and *quantum evolution* (sudden "explosive" jumps). He considered phyletic evolution—slow changes in lineages—to be the main driver of evolution. Like the other architects of the synthesis, Simpson saw the human line as a spare branch in the tree of life—what he later described as "progressive advance in a rather unified group."

The lumping and linear mind-set became the new dogma. A muddle of names gave way to a tidy family tree with only two branches, *Australopithecus* and *Homo*—and debates narrowed to whether and how the former evolved into the latter. Human evolution was depicted as a gradual progression from ape to human along one main trunk with very few offshoots. The Cold Spring

Harbor conference dumped a bucket of frigid water on the old zeal for naming new species, and this conservative approach dominated the first decades of the bone rush in Africa. F. Clark Howell, the father of modern paleoanthropology, attended the conference as a young scientist and took to heart its lessons. His book *Early Man* included an illustration called the *March of Progress* from knuckle-walking ape to modern human, which became the iconic image of the linear view of our ancestry. Despite all his discoveries in Kenya, Richard Leakey never actually named a new species and just referred to his fossils by museum number and nickname. This conservative climate made the arrival of Lucy particularly dramatic: her species *Australopithecus afarensis* was the first new human ancestor named in fourteen years.

By the 1970s, the linear view of modern synthesis came under attack on two fronts. The first was a theoretical shift that emphasized splitting. The Simpson concept of lineages rested on the assumption that experts could identify ancestors and descendants, but skeptics began to raise questions: how could anybody really know what fossil ancestor begat what descendant just by the shapes of bones? Critics faulted the Simpson approach as "unscientific" because it rested on subjective judgment, not hypotheses that could be objectively tested.

Enter a German bug researcher. Entomologist Willi Hennig developed a more objective approach to reconstructing evolutionary relationships based on *shared derived characters,* or anatomical traits shared by descendants of a presumed common ancestor. His method sought to logically infer the relationships between species based on the appearance of distinctive novelties (such as the appearance of the spine in vertebrates, loss of a tail in apes, and so on through finer levels of anatomical detail) and parsimony (the simplest scenario was deemed most likely). In reality, paleontologists already had been using such approaches for a long time, but Hennig codified a more formal discipline—one that became very appealing to purists. This new method came to be known as *cladistics* and its followers became known as *cladists*. (The name derived from the word *clade*, a group of related taxa and their common ancestor.) Strict cladists sidestepped who-begat-whom debates—which they deemed unresolvable—by refusing to draw lines between specific ancestors and descendants. Instead, they only recognized *sister taxa*, or common descendants of hypothetical ancestors. Fish paleontologist Gareth Nelson of the American Museum of Natural History summed up their attitude: "Looking for ancestors in the fossil record seems to be like looking for honest men: in theory they must exist, but finding them in practice, alas, is another matter." Instead of one ancestor flowing into another in lineages, cladistics depicted evolution as a comblike diagram with every species splitting off into dead ends.

The second attack on the modern synthesis concerned the pace of evolution. Synthesis biologists assumed that most evolutionary change occurred gradually. But this idea seemed difficult to reconcile with a fossil record in which forms abruptly appeared and vanished. Why did lineages not show continuous change? Darwin had argued that the book of life was missing pages: "The geological record [is] extremely imperfect and this fact will to a large extent explain why we do not find interminable varieties, connecting together all the extinct and existing forms of life by the finest graduated steps." But in the early 1970s, a new school took the opposite view. Paleontologists Stephen J. Gould and Niles Eldredge asserted that the fossil record actually provided a faithful account of life on earth—evolution really did jump abruptly from one chapter to the next. Gould and Eldredge insisted gradual evolution was insufficient to explain the diversity of life. Raised a Marxist, Gould believed that western scientists were biased in favor of gradualism because it conformed to their bourgeois notions of progress. To Gould and Eldredge, evolution occurred by revolution—bursts of speciation followed by long plateaus of stasis. Their theory became known as "punctuated equilibrium."

A professor at Harvard, Gould was a gifted and prolific essayist who championed the idea of speciation and cited the human family as a prime example. In 1976, he famously declared, "We are merely the only surviving branch of a once luxuriant bush." Gould's metaphor spread like kudzu vine, splitting became became fashionable again, and the human family tree began to be redrawn as a bush.

As it turned out, *Ardipithecus ramidus* became the first of many new arrivals in the human family. By the turn of the millennium, other relatives cropped up beside Lucy between 3 and 4 million years ago, including *Kenyanthropus platyops* in Kenya, *Australopithecus bahrelghazali* in Chad, and *Australopithecus prometheus* in South Africa—and even more would come later. Some of the new offshoots on the human tree seemed otherworldly, such as the 2004 revelation of the so-called hobbit *Homo floresiensis*—a tiny species that coexisted with modern humans in Indonesia as recently as eighteen thousand years ago. As more oddities appeared, the human family began to look less like the *March of Progress* and more like a freak show. By 2002, Gould cast lumpers into a fringe group: "I don't think that any leading expert would now deny the theme of extensive hominid speciation as a central phenomenon of our phylogeny."

Actually one leading expert did dispute it—Tim White. Due to his success as a fossil hunter, he would name more human ancestors than any splitter, but in his heart he was a lumper—a lineal descendant of Mayr, Schultz, and Simpson. He remained haunted by what he called "the ghosts of Cold Spring Harbor." His default assumption was that any two hominids from the same time belonged to the same taxon until proven otherwise. His conception of the *species* was a big bucket that allowed wide variation (both at any moment in time and through time as the lineage evolved). As a field guy, White demanded a bedrock of stability—a classification scheme that wouldn't be overturned with every new discovery. He also happened to be the person who had named both the ostensible root and trunk of the family tree and defended the conservative interpretation that many discoveries elsewhere were just variations of the same taxa. He echoed the old rebukes of the Cold Spring

Harbor conference: splitters saw the normal variation *within* species and mistook it as *different* species; splitters carved up lineages into arbitrary segments, made up new names, and assumed separate lines of descent for each—"ghost lineages," he called them. He dismissed the bushy tree as an academic fad and accused colleagues of declaring invalid species to advance careers. After all, a new species in the human family made international news and top billing in major journals like *Science* or *Nature*, but another specimen of an already-known taxon got relegated to obscurity.

"Our family tree," White wrote, "still resembles a saguaro cactus more than a creosote bush."

To traditionalists like White and the Middle Awash team, the fossil record told a story of adaptation. It was a progression of *Ardipithecus* to *Australopithecus* to *Homo*. Splitters and cladists distrusted such narratives. They saw the fossil record as bursts of speciation—tangles of roots, each with its own particular history, no coherent theme, and no certainty that a fossil from one time period belonged in the same evolutionary story as another fossil from a different time.

IF TIM WHITE WAS THE TYPE SPECIMEN OF THE LUMPERS, IAN TATTERSALL was the same for the splitters. Tattersall was the curator of anthropology at the American Museum of Natural History in New York. He was a tall, bearded Englishman often dressed in a blazer and sweater vest who spoke with an air of erudition. In his office, mannequin heads of extinct members of the human family formed a lineup of big snouts and stout brow ridges. A row of skulls gazed toward a rack of wine bottles (he also was a connoisseur and, after writing more than twenty books on human evolution, he branched into the natural history of wine).

The American Museum of Natural History stands on the west side of Central Park in Manhattan and historically has been a citadel of the bushy tree camp. Tattersall arrived at the museum in

1971 when it was abuzz with cladistics and punctuated equilibrium. Educated at Cambridge and Yale, he looked back upon his schooling in the linear view of the synthesis like a man embarrassed by the naïveté of his youth. He studied lemurs, a group of primates isolated on Madagascar for 50 million years that diversified into a "riotous profusion of species" from mouse to gorilla size. Tattersall came to see diversity and speciation as the rule among mammals—including our family. To him, the human clade was like the trees outside the window of his paleoanthropology lab—a complex pattern of branches and dead ends.

He complained that "the dead hand of linear thinking still lies heavily on paleoanthropology"—nowhere more heavily than in the Middle Awash. "The Middle Awash people have this wonderful sequence of sediments that cover a very, very long time—and everything mysteriously seems to fit into a single lineage history," Tattersall observed with sarcasm. "*All* of human evolution apparently took place in the Middle Awash."

That comment came two decades after the discovery of Ardi and a flood of other fossils from Ethiopia. Over the years, Tattersall peppered his writing with acerbic references to White and his team. He sniped that they always seemed eager to draw lines between their discoveries and modern humans. In one book, he described White as an "enthusiastic unilinealist" relict of the single-species school of the University of Michigan. (This was an oversimplification because White did recognize multiple lineages within the human family, but demanded evidence of stark differences before doing so.)

Tattersall swiped with a foil, but White attacked with a battle ax. In 2002, White wrote a savage review of Tattersall's book *The Monkey in the Mirror* and called the author a master of "politically-correct paleoanthropological pontification" who spouted a "profoundly misleading" view of the family tree. White suggested readers would be better served reading science fiction such as Jean Auel's *The Clan of the Cave Bear.*

Tattersall and White represented two fundamentally different camps. White lauded the visionaries of Cold Spring Harbor; Tattersall condemned them—especially Mayr, whom he labeled a "traumatizing influence" for his hatchet job on the human family tree. The Middle Awash team depicted human evolution as a succession of evolutionary grades and adaptive plateaus; Tattersall dismissed those categories as "products of human minds, and not of Nature." To put it another way, imagine human evolution was a play. Traditionalists like the Ardi team focused on the overall plot progression, while cladists like Tattersall focused on the cast of characters and their relationships—the more the merrier.

Tattersall was hardly alone in his attack on the linear narrative of traditional paleontology. In 1999, *Nature* editor Henry Gee wrote: "To take a line of fossils and claim that they represent a lineage is not a scientific hypothesis that can be tested, but an assertion that carries the same validity as a bedtime story—amusing, perhaps even instructive, but not scientific." As more new members of the human family appeared, *Nature* often trumpeted the word "diversity" in headlines about new species.

Tattersall and White also exposed the gulf between scientists who focused on lab analysis versus those who pursued discoveries in the field. As a young paleontologist, Tattersall's study of lemurs was interrupted when he was expelled from Madagascar in 1974. His early attempts to find human ancestors in Africa, the Middle East, and Asia turned up no fossils. By his account, he abandoned one expedition in Djibouti after drunken Yugoslav sailors rampaged through his hotel. He became a museum curator with a mission: "every bit as important as finding new fossils was ensuring that fossils already on museum shelves had been properly interpreted." And interpret he did: his museum drummed diversity into the heads of millions of visitors per year. He was a go-to authority for the media and a clever quotemeister. His writings in academic journals, popular magazines, and books spread the gospel of a bushy human tree and kept up a steady acid drip of disdain on the traditional Simpsonian concept of lineages—and its modern champions in the Middle Awash. Tattersall also advocated for the growing ranks of lab scientists who demanded access to fossils—like the ones lying in the Ethiopian National Museum awaiting publication.

The species debate still rages today—but the choices are no longer so simple as everybody assumed years ago. In recent years, comparative genomics and fossil discoveries have affirmed that the human family once was far more diverse. Modern humans are a paltry remnant of our past variation, and one of the least genetically diverse primate species. For example, just one regional subpopulation of chimpanzees (*Pan troglodytes troglodytes*) has about twice as much genetic variation as all 8 billion humans worldwide. Yet our family was not always so homogeneous. At one point, at least four other archaic human ancestors around the world coexisted with *Homo sapiens*. But here's the catch: we only know about some of these extinct forms because they left imprints on our genome, meaning *they interbred with our ancestors*—and thus belonged to one big breeding pool and one species, by Mayr's classic definition. Our family tree might seem bushy, but the branches don't necessarily terminate in extinction. They split, diversify, rejoin, and rejoin again and again—which suggests the traditional metaphors of a family tree, bush, or cactus aren't such good comparisons at all. The debate has been trapped by semantics that simply failed to capture the true biological complexity. In fact, many geneticists have stopped referring to these archaic humans as *species* and instead speak of one big "metapopulation"—which faintly echoes the notions promulgated at Cold Spring Harbor in 1950, back before anybody had even heard of genomics. By the time Ardi finally appeared in public to shake the family tree, the very notion of that tree had begun to wither. But that's getting ahead of the story.

Chapter 15
VOYAGING

Ancient anatomists spoke of the body as "the lesser world." The history of its exploration spans two and a half millennia from the crude dissections of antiquity to modern molecular biology. The oldest writings come from ancient Greece in fifth and fourth centuries B.C.E. The science of anatomy was founded on cutting up bodies, with the name coming from the Greek *ana* (up) and *tome* (cutting). The Greeks left their stamp all over us. The shinbones are the *tibia* (flute) and *fibula* (needle); the midsection is the *thorax* (the Greek armored breastplate); and each finger and toe bone is a *phalanx* (the Greek military formation).

Even after the Romans became masters of the ancient world, Greeks continued to dominate the professional class of physicians. Eventually the Romans converted some terminology to their own language, and the Greek *kynodontes* (dog teeth) became Latinized as *dentes canini*—the canine teeth. The towering anatomist of the ancient world was Galen of Pergamum, a Greek doctor in the Roman Empire who spent his early career patching up wounded gladiators and rose to become the personal physician of Emperor Marcus Aurelius. Roman law forbade dissecting humans, but the emperor offered Galen a chance to expand his inquiries by joining a military campaign to Germany where he could freely slice up slain barbarians. The anatomist declined because he thought he could learn nothing new. Galen employed a retinue of scribes and dictated a huge corpus of work that established the anatomical canon for more than a thousand years. Unfortunately, much of Galen's research turned out to be based on dissections of other animals—an embarrassing fact not revealed for more than a thousand years and a cautionary tale of how the deadweight of prevailing authority can smother new learning.

During the European Renaissance, anatomical learning started to advance again. Beginning in the fifteenth century, Leonardo da Vinci explored "the cosmography of this lesser world" and dissected corpses to help him render the body more faithfully in his art. He discovered numerous errors in classical teaching and would have earned a place in medical history if he had realized his ambitions for a grand anatomical treatise; sadly, he did not live to complete it. The first great atlas of human anatomy was produced by Andreas Vesalius, who published *De Humani Corporis Fabrica* (On the Structure of the Human Body) in 1543. Vesalius finally revealed the errors in Galenic teaching (and, predictably, was vilified by academic peers for his heresy) yet faith in the canon remained so strong that one faithful Galenist expressed wonder at how much the body had changed since the days of the old master. Nevertheless, exploration of the human body began in earnest, and anatomy became something that we would recognize as a science. In 1555, Pierre Belon made a side-to-side comparison of human and bird skeletons and showed that both contained corresponding bones—what we now call homologous parts. Centuries passed before anatomists realized that these similarities stemmed from common ancestry.

In 1859, Charles Darwin published *On the Origin of Species* and introduced his theory of evolution—or what he called "descent with modification." Anatomy became a means to investigate evolution by comparing homologous body parts inherited from common ancestors—like those canine teeth, tibias, and phalanges. Since the oldest fossils contain no DNA, comparative anatomy remains the principal means to reconstruct evolutionary history, explain how one form evolved into another, and reconstruct what Darwin called the "great Tree of Life."

The next phase of the Ardi investigation involved a voyage to another unknown world populated by a strange assortment of creatures that have perplexed anatomists for centuries because they look so unlike anything alive on earth today. Appropriately, the voyage began aboard a vessel decorated with artwork from the Renaissance.

FOREIGN SCIENTISTS SOMETIMES REMARK ON AN UNUSUAL ASPECT OF human origins research in the USA: its reliance on philanthropy. Find a major fossil and the billionaires won't be far behind. The Ardi team had included one middle-aged woman who might be recognized by readers of the social pages and glossy lifestyle magazines. One fashion magazine described Ann Getty as "the queen of San Francisco society." She was married to Gordon Getty, an heir to the Getty oil fortune. A few years before the Ardi discovery, she took White's anthropology class, became smitten by fossils, and spent several seasons with the Middle Awash field team. (At the end of the Ardi excavation season, she bequeathed her clothes to a few penurious Afars who subsequently sported the finest ultraviolet-protection gear money could buy.) In labs, Getty often peeked over the shoulders of the scientists. She and her husband were generous benefactors of human origins research and put their private 727—known as "the Jetty"—at the disposal of the Ardi team.

In the summer of 1996, the anthropologists climbed aboard the Gettys' flying yacht, whose decor recently had been featured in *Architectural Digest*. The jet—Gordon's birthday present to his wife—had a private stateroom with a silk-sheeted bed, a gourmet kitchen, a wine bar, a bronze sink, and marble tabletops. One paleoanthropologist wisecracked, "I didn't know this much marble could fly." (The experience became even stranger after tabloids later revealed that Gordon Getty was a polygamist who had a secret second family with his mistress.) Leather parchment panels and imitation book spines lined the walls, and a recessed ceiling dome blazed with a gilded painting of the sun. Wherever they landed, limos whisked the party to the finest hotels. "Opulence beyond all comprehension," recalled Lovejoy.

Between the cabins of the plane ran a hallway lined with gold leaf panels and maps created by Dutch-German cartographer Andreas Cellarius in 1609. When the plane was refurbished, the interior designers suggested high-quality reproductions, but Ann told them to splurge and buy the originals—and there they hung in climate-controlled cases. They were fitting images as the anatomists explored the terra incognita of the past.

THEY WERE TRAVELING TO SEE THE RELICS OF A BYGONE PLANET OF THE Apes. The Miocene epoch was named by nineteenth-century geologist Charles Lyell; the word means "less recent," in contrast to *Pliocene*, which means "more recent." Ironically, this remoter epoch offers a much richer ape fossil record than the latter times. During the Miocene (roughly 23 million to 5.3 million years ago), warm, humid forests and woodlands covered the globe and scores of ape species flourished in Africa, Europe, and Asia. Of the early Miocene species, fourteen ape genera have been identified in Africa alone. About three dozen species of Miocene apes are known from fossils, ranging from the cat-size *Micropithecus* to polar bear-size *Gigantopithecus*. But the true abundance probably was much greater—more than one hundred species of apes by the estimates of some paleontologists. Somewhere among these varied forms lived the ancestors of humans and modern apes.

Then something happened to the apes. The Miocene climate cooled, forests shrank, eastern Africa turned more dry, and Europe turned temperate. In the Miocene, apes were abundant and monkeys rare. Somehow the situation reversed so monkeys proliferated and apes dwindled. The fossil record of Miocene apes vanishes after about 8 million years ago. Modern apes are a meager remnant of the former diversity. Unfortunately, it remains unclear which fossil species might be ancestral to modern ones (one exception is *Sivapithecus*, a plausible ancestor to the modern orangutan). To make matters worse, in the 1990s not a single fossil could be reliably attributed to the chimp or gorilla lineages (and today there is only a pittance more). In comparison, the human fossil record actually seemed a cornucopia.

Did any of these lost apes offer clues about Ardi? The team shuttled between museums in Barcelona, Florence, Nairobi, Addis Ababa, and Johannesburg. Many doors were opened thanks to the presence of Clark Howell, the elderly Berkeley anthropologist who had accumulated a lifetime of goodwill among colleagues around the world and possessed an equanimity and tact lacked by some of his younger successors. At each stop, the team commandeered museum workspace, and White organized assembly lines to examine Miocene fossil apes—*Proconsul* in Kenya, *Oreopithecus* in Italy, *Hispanopithecus* in Spain, and so on. Sometimes White caught the eye of teammates and wiggled his thumb open and closed, an inside joke about the secret of Ardi's opposable toe.

In Florence, the scientists walked the same cobblestone streets where Leonardo had conducted his anatomical investigations half a millennium before, gazed at Renaissance artworks such as Michelangelo's statue of David with his lifelike hands, and enjoyed a sumptuous feast at a Tuscan estate winery with the Gettys. In Nairobi, they were invited to dinner at the Leakey mansion, marking a period of detente in the old fossil cold war. At the University of the Witwatersrand in Johannesburg, Lovejoy ran into anthropologist Phillip Tobias, who was not exactly delighted to see him. Not long before, Tobias had published a paper about a 3-million-year-old fossil called "Little Foot" and described it as an arboreal human ancestor with a grasping toe. Lovejoy—who had privileged access to an indisputable grasping toe—told the press that interpretation

The Ardi team shares a laugh by opening and closing their thumbs—a secret signal about Ardi's opposable toe. This feature of the skeleton was kept secret for fiteen years. *(From left to right:)* Scott Simpson, Bruce Latimer, Berhane Asfaw, Owen Lovejoy, Tim White, and Gen Suwa.

was "patently absurd." The two men went into an office and companions heard loud voices behind the door. Lovejoy emerged looking shaken and mumbled, "He hates me." Meanwhile, Lee Berger, an ambitious young anthropologist at Witwatersrand University, was awestruck by the arrival of what he called the "Dream Team"—the famous scientists he had read about in the Lucy books. At that point, Berger was just beginning what would become a decades-long campaign to restore South Africa as the cradle of humankind and challenge "the East African hegemony of human evolution." Near the end of their stay, the visitors summoned Berger to face inquisition about a different set of claims that he had recently published—one that described a "chimpanzee-like" shinbone from Sterkfontein Cave and implied that South Africa fossils were more primitive and more similar to our ape relatives than *Australopithecus afarensis* from east Africa. "Tim just slaughtered him," recalled Meave Leakey, who tagged along on that leg of the voyage. After being forced to backtrack on some of his claims, Berger recounted that his interrogator (whom he nicknamed "the Great White Shark") pulled a pen from his pocket, slapped it atop a copy of his paper, and with a sly smile demanded, "Sign it and say you retract the title."

By the end of the trip, one thing seemed abundantly clear: Ardi looked like nothing ever discovered before. "At no stop in the tour did we see anything that might be confused for *Ardipithecus*," recounted White. Scott Simpson added: "There was no model in the zoo of the known." In that lost Planet of the Apes, however, they had glimpsed clues not fully appreciated until years later.

WE BELONG TO A FAMILY OF HANDY CREATURES. MORE THAN ANY OTHER feature, grasping hands distinguish primates. Most primates have opposable thumbs (a few species have lost their thumbs and climb with four fingers). In general, we great apes become unique on the end of our limbs. Ape shoulders are generally similar between species, but hands exhibit great specialization—witness chimp knuckle-walking hands, orangutan grappling-hook hands, and human nimble hands. In nonhuman primates, hands perform the double duty of manipulation and locomotion. In humans, hands were freed from the job of getting around and became specialized for dexterity.

Our hands can make tools, write symbols, paint *The Last Supper*, type seventy words per minute, or put spin on a 100 mph fastball. The celebrated examples of ape dexterity are clumsy in comparison. Chimps can throw, but not with much accuracy (thankfully, given their habit of hurling feces at zoo visitors). The human hand comprises less than 1 percent of our body weight, yet claims a disproportionately large portion of the brain (nearly one-third of the motor cortex, by some estimates) due to its intricate network of nerves for sensation and motor control.

In the fifth century B.C.E., the Greek philosopher Anaxoragas asserted that humans were the most intelligent animals because they had use of their hands. A century later, Aristotle took the opposite position and claimed nature endowed humans with dexterous hands because they *already* had smarts. Thus began a two-millennium chicken-versus-egg debate: which came first, human dexterity or intelligence?

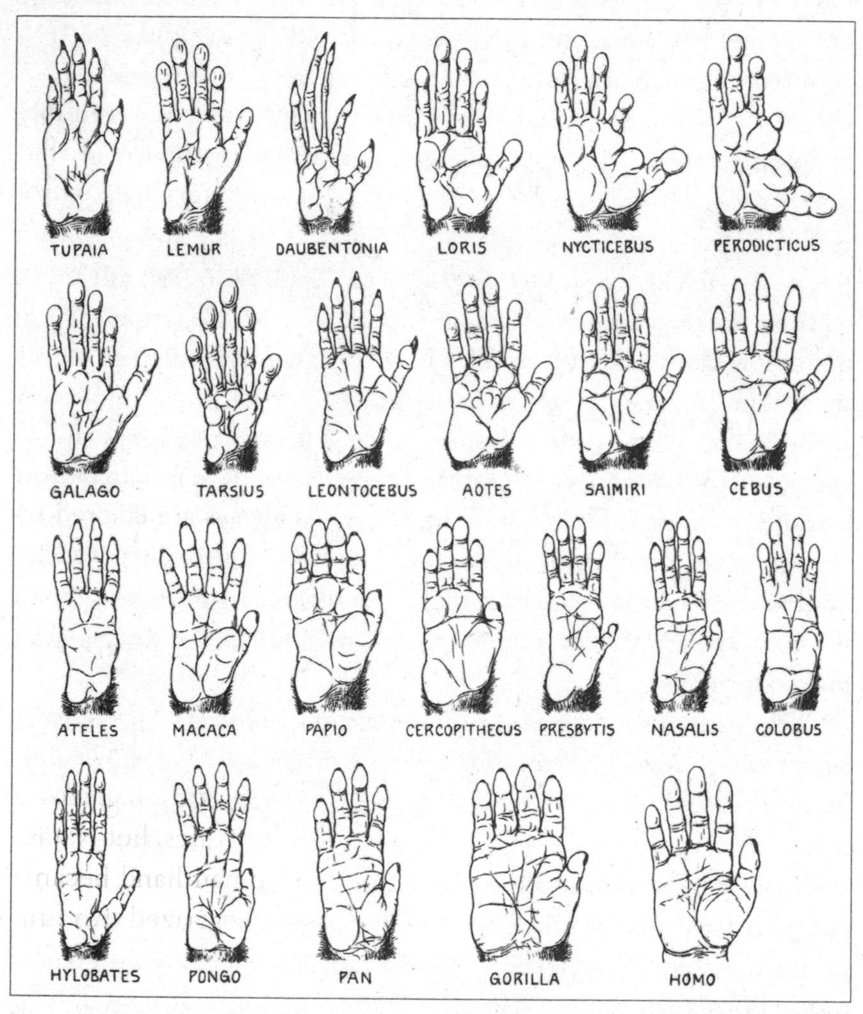

The study of evolution examines how species have adapted body parts inherited from common ancestors. These anatomical sketches by Adolph Schultz depict how primates have adapted hands for specialized evolutionary roles.

The first great treatise on the human hand was written in 1833 by the Scottish physician Sir Charles Bell, who celebrated it as an anatomical Swiss Army knife: "The hand supplies all instruments, and by its correspondence with the intellect gives him universal dominion." Bell recognized that its versatility was not so much anatomical as *neurological*. Around the same time, young Charles Darwin sailed around the world on the *Beagle*. Darwin cast hands as central characters in human evolution and asserted, "Man could not have attained his present dominant position in the world without the use of his hands, which are so admirably adapted to act in obedience of his will." Darwin theorized that dexterity, tools, big brains, reduced canines, and upright posture evolved as a package—an assertion falsified by Lucy and *Australopithecus afarensis*.

For a century after Darwin, the evolution of the hand remained a mystery due to the lack of a fossil record. That began to change in 1960, when Louis and Mary Leakey found a set of 1.75-million-year-old hand bones at Olduvai Gorge and named the species *Homo habilis*, or "handy man," because it was found near primitive stone implements. Interpretations always are colored by the prevalent ideas of the era, and in those days humanity was literally defined by tools—a paradigm framed by the 1949 book *Man the Tool-Maker* by Kenneth Oakley. The Leakeys put tools in the hands of the handyman.

The Leakeys assigned the analysis of the Olduvai hand to British physician and primatologist John Russell Napier. His mentor, Frederic Wood Jones, had characterized the human hand as a primitive organ that differed little from other primates, but Napier took the opposite tack and argued that the human hand became uniquely specialized through toolmaking. He categorized two fundamental modes of grasping: the power grip and the precision grip. In the power grip, objects are clasped by the palm and wrapped by the fingers and thumb as when swinging a sledgehammer or baseball bat. In the precision grip, objects are held with the fingertips as when wielding a paintbrush or scalpel. Napier asserted that the two grips marked key phases in human evolution. African apes cannot bring the tips of their stubby thumbs to meet the tips of their four long fingers, so they remain stuck with a power grip, which Napier took to represent the ancestral condition.

The Olduvai hand was woefully incomplete (mostly a few finger bones and a single phalanx from the tip of the thumb) and inferences about its proportions remained speculative. Assuming the opposable digit was short, Napier experimented by replicating primitive tools without using his thumb and concluded an apelike power grip could fashion the implements. Thus was established a model that prevailed for decades: human ancestors began with apelike hands incapable of fine manipulation. As our ancestors chipped stone tools, natural selection sculpted our hands. The thumb grew longer, the four other fingers shortened, and humans attained the precision grip—which Napier celebrated as the "ultimate refinement in prehensility." Such thinking lured some into a tailspin of circular reasoning: hand fossils capable of precision grip were deemed toolmakers, even if no tools were found nearby.

In the 1960s, revelations of the close genetic relationship between humans and African apes bolstered the idea that human hands evolved from long-fingered, short-thumbed, knuckle-walking ancestors like modern chimpanzees and gorillas. Sherwood Washburn called knuckle walking the "intermediate condition" between arboreal apes and terrestrial bipeds. Washburn could not muster much of an anatomical argument and cited football linemen leaning on their knuckles and the lack of hair on the backs of the fingers as evidence of such ancestry. Little fossil evidence existed to test this view. Lucy preserved only two hand bones, and other hand fossils of her species were isolated specimens. *Australopithecus afarensis* seemed to have a hand capable of manipulation, but the exact proportions remained unknown due to the uncertainty about which bones belonged to the same individuals. Ardi's skeleton, however, had a nearly complete hand.

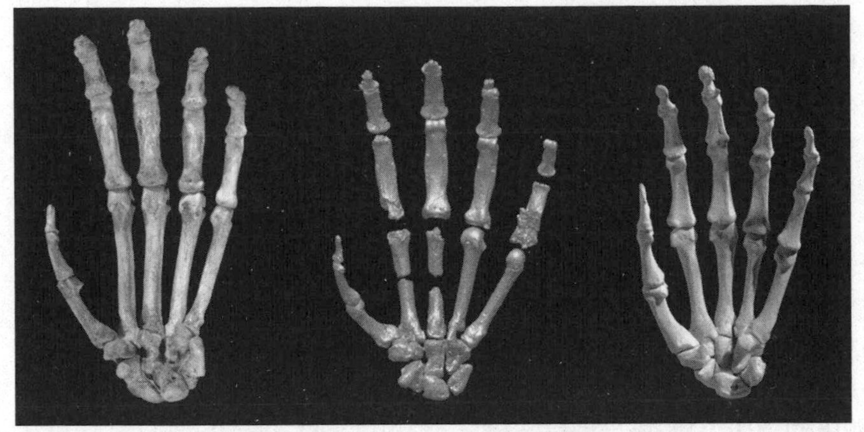

A comparison of hands of chimpanzee *(left)*, Ardi *(center)*, and human *(right)*.

Chimpanzee: long palms and fingers, and stiff wrist reinforced for knuckle walking and climbing.

Ardi: long, curved digits for climbing; palms shorter than chimp; thumb relatively longer than chimp but shorter than human; and flexible wrist. Human: relatively short palm and fingers, and long thumb able to rotate and join tips of all other fingers for precise manipulation.

In most sciences, investigators start with a hypothesis and collect data to test it. Those who study fossils operate in reverse: they begin with data—bones—then generate explanatory hypotheses. So it was with the hand of *Ardipithecus*. Other parts of the skeleton would require years of painstaking reconstruction, but the hand bones were small and hard enough to survive mostly intact. Investigators almost immediately could reassemble most of the hand skeleton—and it was unlike anything ever seen before.

The wrist, known as the carpus, is a snug masonry of irregularly shaped bones. In humans, the mobile wrist can bend backward, or dorsiflex, as it does when we crawl or do push-ups with palms flat on the ground. In chimps and gorillas, the wrist geometry limits dorsiflexion and sturdy ligaments stiffen the joint for knuckle walking.

Ardi bore few signs of any such stiff wrist. The finger bones appeared long and curved, as expected from a primitive ancestor, but the carpal bones resembled nothing the anthropologists had ever encountered. To the Ardi team, her flexible wrist suggested not knuckle walking but clambering on flat palms, a form of locomotion unknown in any living ape but common among modern monkeys and fossil apes from the Miocene.

As many had done before them, the Ardi team sought to explain how this fossil hand might have morphed into that of a tool user with a precision grip. Human dexterity relies on a relatively long thumb, four short fingers, and mobile joints that allow the fingertips to touch. By contrast, chimps have long palms and long fingers with dinky thumbs (they sometimes pinch objects between their thumb and the side of the index finger). And while chimps are much stronger than humans (estimates range from two to five times more powerful overall), they actually are weaker in key thumb muscles.

Ardi's thumb bore a robust attachment for a powerful flexor pollicis longus muscle, which would have provided firm grasping ability. This muscle is well developed in humans but absent or rudimentary in the other great apes. Although her thumb was short, it was a bit longer than that of a chimp relative to the rest of the hand. Despite her big climbing hand, Ardi probably had better manipulative capabilities than our chimp cousins.

This did not prove Ardi used tools, but suggested that a more grasping hand and powerful thumb muscles were *primitive*—and appeared millions of years before stone tools. Such a hand might have rendered Ardi less adept at climbing than a chimp, but more nimble in obtaining foods by fine manipulation. And that skill only grew: within 1 million years after Ardi, *Au. afarensis* had evolved hand proportions with more advanced manipulative abilities. The Ardi team later concluded, "Our earliest ancestors only had to slightly enlarge their thumbs and shorten their fingers to greatly improve their dexterity for tool-using."

In London, the Gettys put up the team in a swank hotel, where Simpson and Lovejoy holed up in a room and tried to make sense of the strange hand skeleton. How did Ardi evolve from an ape ancestor? And how might such a hand have led to humans? The anthropologists assembled casts of the Ardi hand on a tabletop beside comparative specimens of African apes. To give themselves room to work, they moved the table into the center of the room and struggled mightily with its weight. Being so engrossed in their mission to the past, they only belatedly discovered the source of the deadweight—the table housed a refrigerator.

A FEW MONTHS AFTER THE JET JUNKET, THE FIELD TEAM RETURNED TO AR-amis in December 1996 for the third and final year of excavation at the skeleton site. The first year had produced a bonanza, the second more bones from the front of the skull, but the last yielded nothing. White had exhausted the site as surely as he exhausted the crew. "Nobody, *nobody* other than Tim could have gotten all those bones—some as small as lentil grains," said Berhane Asfaw.

In that final season at Aramis, Gadi announced that he would give a baby camel to his friend *Doktor Tee*. It was a high compliment. A camel was the pickup truck of the Afar desert: a vehicle, transporter of goods, and symbol of prosperity. Better yet, it also provided frothy, fresh milk for a proper Afar breakfast. White explained that he could not take a camel on the airplane. Gadi knew nothing about aircraft or the faraway land of the *farenjis* who so desired the bones of the *Kada Daam Aatu*. A man could not take a camel to such a place? He questioned his American friend.

How many wives do you have?
None.

How many children do you have?
None.

Well, what do you have?
I have a cat.

Cat? In the Afar, wealth came in herds—camels, cattle, and goats. Herding cats made as little sense in Afarina as it did in English. Nonetheless, one morning, Gadi showed up at the excavation site with bundle tied to his rifle like a bindlestaff. He unrolled his *gabi*, the all-purpose cloth normally worn over shoulders or head, and revealed a kitten with hogtied legs. He dropped the cat in the dust at White's feet.

As a longtime bachelor, White had a soft spot for cats. The kitten scampered away as he and others gave chase. "Don't step on it!" White shouted. Another Afar friend rolled his *gabi* into a ball, bowled over the kitten, grabbed the animal by the scruff of its neck, and handed it to its new master. The cat promptly sank its needle-sharp teeth into White's thumb, hit bone, and drew blood.

White yowled, but it was love at first bite. He adopted the cat, leashed it, and nestled it on his lap at meals. The pet almost didn't survive the expedition: one night, a huge owl swooped over the dinner table with an ear-piercing screech and came within inches of grabbing the kitten off the ground. At the end of the field

The Afar wildcat: Tim White and Lubaka.

season, White secured permission to export the cat from Ethiopia, talked his way through airports, and took it home. It turned out not to be a domestic kitty but an African *wildcat*—a species known as *Felis silvestris libyca*, with long legs, bushy ringed tail, and feral temperament. "The males are solitary and extremely aggressive," explained White. (Louis Leakey once described the species: "I've never known one successfully tamed—those I've seen were spitfires to the end.") Back at Berkeley, the kitten grew into a hellion and beat the crap out of the neighborhood kitties even after White had it neutered. He named his pet Lubaka, after the Afar word for "lion." Once Lubaka charged into a neighbor's house and mauled another cat in its own home. Fights and vet bills eventually obliged White to keep the wildcat penned up inside. He built a solarium so Lubaka could spend his days in a sealed environment—an allegory, perhaps, for the scientific team's relationship with its profession.

Chapter 16

MISSION TO THE PLIOCENE

The voyage into the past entailed not only a broken skeleton but an ancient world—the soils Ardi walked on, the trees she climbed, the foods she ate, the animals she saw, and the carnivores who hunted her. For years, the team would relentlessly gather data to answer a question: what was the habitat of these early members of the human family? As a teen, White had been transfixed by television broadcasts of the NASA Apollo mission to the moon. He took to calling this one "the Mission to the Pliocene."

THE THEORY OF EVOLUTION RESTS ON THE NOTION THAT ENVIRONMENT shapes life. This idea slowly took root over several centuries. During the age of exploration, Europeans encountered exotic species in faraway places and realized the earth contained far more biological variety than previously imagined. In the eighteenth century, Swedish botanist Carl Linnaeus listed the locale of each species in his catalog of the diversity of life (he also popularized the words *fauna* and *flora*). Around the same time, French naturalist Georges-Louis Buffon suggested that animals adapted to local climates and proposed that species were not fixed but changeable.

In the mid-nineteenth century, Charles Darwin and Alfred Russel Wallace articulated modern evolutionary theory: species adapted to environments.

What habitat forged humans? For two centuries, theorists embraced the notion that humanity arose when primates stepped away from the trees—or when the trees vanished. This idea began to sprout even before any fully articulated theory of evolution. In his *Philosophie Zoologique* of 1809, Jean-Baptiste Lamarck speculated that "some race of quadrumanous animals" might "lose, by force of circumstances or some other cause, the habit of climbing trees." Darwin envisioned an ancestor who evolved "to live somewhat less on trees and more on the ground." In 1889, Wallace planted "ancestral man" on the grassy expanses of Eurasia: "It is more probable that he began his existence on the open plains or high plateaux of the temperate or sub-tropical zone." Similarly, in the 1920s, Henry Fairfield Osborn, a professor at Columbia University and the American Museum of Natural History, put the origin of humanity on the high plateaus of Asia. Forests were places of stagnation, he argued, while plains were "invigorating." After discovering the first member of the human family from Africa, Raymond Dart imagined *Australopithecus africanus* cavorting on the grassy veldt of South Africa. Despite being widely credited as its author, Dart did not invent the theory of grassland origins but simply imported the concept to Africa—where it later became associated with the common *savannas*. That term (which is not native to Africa and actually originated from a Carib language of the Americas) usually describes landscapes with continuous grass, and encompasses a huge variety of habitats and climates.

By the mid-twentieth century, savannas were commonly invoked to explain human distinctions such as upright posture, tool use, meat eating, hunting, brain expansion, body proportions, and more. But these assertions faced no real scrutiny. Scholarship examined not *whether* savanna theory was true but *why* it was true.

According to the classic evolution narrative, Miocene Africa, the ancestral home of the common ancestors of humans and apes, was a vast rain forest. By this theory, forests shrank, grasslands expanded, and human ancestors stepped onto the savannas—becoming upright, tool-using meat eaters in the process—while our ape cousins remained in shrinking pockets of trees. But the idea of a Miocene rain forest rested on scant data provided by a few sites with fossil wood that were assumed to typify all Africa. As it turned out, these assumptions were wrong—not only about the mythic coast-to-coast Miocene forests but also about the apes who lived in them.

In the 1970s, paleoclimate finally became a topic for serious scientific study. Deep-sea drilling of ancient geological strata began to reveal long-term patterns of ancient climate from carbon and oxygen isotopes. In 1981, South African scholar C. K. "Bob" Brain wrote a paper "The Evolution of Man in Africa—Was it a Consequence of Cenozoic Cooling?" He noted a chilling trend culminating in the late Miocene when polar icecaps expanded, sea levels dropped, the Straits of Gibraltar closed, and the Mediterranean dried up. He speculated that African forests shrank, savannas expanded, and new habitats opened for an upright ape. His paper kicked off a new wave of scholarship. If there was an environmental event, there was a theorist ready to speculate how it somehow made us human—the drying up of the Mediterranean, the deluge that refilled the Mediterranean, closure of seaways, uplift of mountain ranges, changes in global temperatures, volcanic eruptions, orbital cycles of the earth, and more. One recurring theme was the expansion of arid savannas—variously invoked at 5 million years, 2.5 million years, or 1 million years.

In the 1990s, the savanna theory faced its first major challenges—the discovery of early human ancestors such as *Ardipithecus ramidus* and *Australopithecus amanensis* that still lived among the trees. Fossils spoke first and loudest: *Ardipithecus* bones were found alongside fossil wood, leaf-eating kudu, and arboreal animals such as monkeys and squirrels—all described as evidence of a "closed, wooded" habitat. White had long expressed skepticism toward environmental determinism that depicted human evolution as a consequence of climate change. Likewise, Lovejoy doubted the grasslands-spawned-bipedality story; instead, he believed our ancestors became bipeds *under* the trees. The Mission to the Pliocene would put the savanna theory on trial with Ardi as the star witness.

The team was able to re-create Ardi's world with a massive haul of fossils and other data. But first, a major geological investigation provided the bedrock for the entire scientific mission and timeline for every scrap of bone. (The study of fossils depends on geology, so it is a cruel irony that the "rock stars" of human origins were never the people who actually specialized in rocks.) Big Science arrived at Aramis, and a team of geologists turned the badlands of the Aliseras into one of the best-documented stacks of ancient strata in the annals of paleoanthropology. The Central Awash Complex sprawled over nearly five hundred square kilometers riven with geologic faults. No single place preserved the entire 250-meter-thick cross section of stratigraphy. Instead, a few layers were exposed here, a few others there, so geologists trekked hither and yon to reconstruct what plate tectonics had ripped apart.

The Zipperman's ash collected in 1992 was just the beginning. Over five field seasons, geologists collected hundreds of ash and lava samples and identified twelve volcanic layers between 5.6 and 3.9 million years old. Paul Renne and his team at the Berkeley Geochronology Center correlated these dates with a second timeline: the magnetic orientation of rocks. Periodically, the earth's magnetic orientation reverses itself, and geologists found eight such turnarounds in the Central Awash Complex. (When *Ardipithecus ramidus* walked the earth, a compass needle would have pointed south, not north as it does today.) In addition, geochemical "fingerprinting" matched rocks between localities—and showed whether one fossil-rich exposure really belonged in the same time period as another. "Here, it's overkill," explained geologist Giday WoldeGabriel. "Nothing is left to chance." With all these lines of evidence, the age of any fossil collected in the area could be pinpointed within one hundred thousand years—remarkable accuracy for the deep past.

The stratigraphy revealed how this portion of the ancient valley changed as the rift slowly widened and sank. Geological layers testified to the evolution of the land: ancient lakes, fast-moving rivers, explosive volcanic eruptions, lava flows, and so on. By 4.4 million years ago, the landscape had settled into a floodplain where a river occasionally overflowed its banks and deposited reddish silts—the layer between the *Gàala* and *Daam Aatu* ashes so abundant in fossil bone. Radiometric dating showed that both layers were 4.4 million years old, meaning they bracketed a tightly framed snapshot of Pliocene Africa—probably just a few centuries or millennia.

That layer lay exposed in patches scattered around Aramis. At many fossil sites elsewhere, bones were deposited by moving water

Photograph © 1996 David L. Brill

The fossil team returned to the Aramis area over many years to look for new fossils that had eroded to the surface. Here a crew searches an Aramis locality in 1996.

and animals from different habitats got mixed together. At ancient Aramis, however, slow-moving floodwaters buried bones in fine silt but did not transport them; animals remained where they fell and could be traced to particular habitats with precision. For a decade, crews returned to pick up every scrap that had eroded to the surface with the last rains, eventually amassing nearly 150,000 specimens. Graduate student David DeGusta described "endless" crawls that picked the ground clean. "I was struck by how there were almost no fossils on the surface," he said of his last visit. "I remember joking with Tim that if other paleoanthropologists showed up here, they would think that we had intentionally published the wrong coordinates for the site."

Only about six thousand specimens of that massive haul could be identified by species or genus; the rest were too fragmentary. Long bones lacked ends because carnivores chewed off the spongy joints. Many fossils showed spiral fractures, the sign of bone broken when fresh. Some bones preserved toothmarks from rodent scavengers or boreholes from worms. Other fossils hinted at the agents of destruction: three hyena species, giant bears, big cats, and four species of pigs. Some fossils showed acid etching from passing through a carnivore digestive tract—including the molar of one *Ardipithecus.*

Only bones of the smallest creatures generally remained intact. Bats, shrews, and mice—known as micromammals—could be swallowed whole by predators, and they provided valuable information because certain species were known to live in specific habitats such as forests or grasslands. About one hundred meters from the Ardi skeleton, the crew excavated a "microfauna quarry" and recovered more than ten thousand mini fossils. Collectors scooped up dirt, rinsed it through screens in the river beside camp, and dried the remains in the sun. Back at the museum, specialists with microscopes and tweezers sorted tiny teeth, jaws, and bones so small they had to be mounted on pinheads. The bird specialist suspected the concentration marked the roost of an owl that disgorged or defecated the remains of its prey. They found the suspected hunter too: an ancient owl species called *Tyto*.

All around Aramis, the crew encountered thick slabs of calcium carbonate, a mineralized layer that forms in arid environments where water evaporates quickly near the surface. These rocks preserved the ground litter of Ardi's world—petrified wood, seeds, snails, dung beetle broodballs, millipedes, animal burrows, termite mounds, a monkey skull, and even a broken eggshell that looked as if it had cracked open yesterday.

Every bit of data became a pixel in an emerging picture of Ardi's lost ecosystem. It was defiantly slow science. The public gobbled up stories about evolutionary Gardens of Eden, but few cared about the fruit trees and serpents—nor all the other animals. Human ancestors comprised less than 2 percent of identifiable fossils, and ignoring all the other material would be expedient and cost-effective. It would also be abhorrent to the mission commanders. "People don't give a damn about the rest," White complained midway through the effort. "If I said, 'Fine, we'll play the game because, after all, we're funded by anthropology so I'll just do hominids and screw the birds, the mollusks, the pigs' . . . then it would be easy. But I'm not going to do that."

Each animal bore witness to the past. Among the most abundant were bovids, a family of cloven-hoofed animals that had diversified into many species of habitat specialists. They were among the most informative indicators: fast-running grazers such as wildebeests, gazelles, and oryx signaled open grasslands; waterbuck suggested lake or river margins; spiral-horned kudu browsed on leaves and indicated tree cover.

The expedition's bovid specialist was Elisabeth Vrba of Yale University (who had been a protégé of C. K. Brain early in her career). More than any figure from the late twentieth century, Vrba championed the idea that environment drove evolution, and she posited that recurring climate "pulses" forced bursts of speciation and extinction. She theorized that one such climate-induced pulse at the end of the Miocene epoch had produced the human lineage.

White didn't buy it, although he greatly respected Vrba as a scientist. Both agreed that the debate could be resolved only by fossils, so she joined the Middle Awash team as its bovid expert. One day in the 1990s, the two scientists sat over a collection of Aramis fossils at the Ethiopian museum.

Vrba picked up a bone. "That's an alcelaphine," she said, identifying a speedy runner of open grasslands.

"Bag it and let's put it over here," said White.

"Now, is this from the hominid layer?" she pressed.

"Yes," answered White.

She smiled at him. "Aha, so the alcelaphine is there. Isn't that nice? Nice open patches of grassland—probably next to the floodplain."

But that particular fossil turned out to be not so typical of Ardi's habitat. Leaf-eating tragelaphines (such as kudu) outnumbered grass-eating alcelaphines (such as wildebeests) by sixty to one. As they pored over the collections, Vrba and DeGusta developed a method to infer habitat preferences from bovid lower leg bones: forest animals had rugged bones adapted to darting between trees; grassland bovids had morphology for running straight and fast. DeGusta likened the body plan differences to Jeeps and drag racers—and those from Aramis were mostly like forest-adapted Jeeps.

The expedition had collected fossils from a series of exposures stretched out over about nine kilometers. Along three-quarters of this arc—the areas with the greatest concentration of *Ardipithecus* fossils—the evidence suggested a landscape at least partially wooded. Kudu represented nearly 40 percent of teeth in all the localities; colobus monkeys and baboons represented nearly 30 percent. The most common birds were woodland species such as peafowl, parrots, and peacocks. To the southeast, the evidence suggested a more open, grassy water margin where surveyors found a few aquatic animals such as hippos, crocs, and turtles—and no *Ardipithecus*.

More evidence came from ancient soils. Most plants use one of two forms of photosynthesis known as C_3 or C_4. Modern C_3 plants include trees, shrubs, herbs, vegetables, fruits, and some shade grasses. In contrast, C_4 plants such as tropical grasses and sedges thrive in hot climes. These two processes leave behind different ratios of carbon isotopes and have been widely used as indicators of wooded (C_3) or grassy (C_4) environments. Stan Ambrose, an isotope expert from the University of Illinois at Urbana-Champaign, dug trenches at Aramis to collect clean soil samples that hadn't been exposed to daylight for 4.4 million years. Back in the lab, his analysis showed that the areas where *Ardipithecus ramidus* dwelled had a stronger C_3 signature, suggesting Ardi's species preferred a more wooded habitat.

Tooth enamel preserves the isotopic signature of diet. Thus, a grass-grazing horse will have a C_4 signature while a fruit-eating chimpanzee will have a C_3 signature. Animals that eat both types of plants have a mix, and a carnivore will reflect its prey. White selected 177 mammal teeth, took enamel samples, and sent them to Ambrose initially identified only by number. When time came to match names to the samples, White intentionally misled Ambrose by mislabeling *Ardipithecus* as a pig to eliminate "any fear the blind test would be compromised." At the time, the analysis of dental isotopes still was a novel technique, and the ever-conservative White didn't yet trust it. "So yeah," admitted White. "I fooled him." When the results were complete, White instructed Ambrose to make a correction: change the "pig" to "*Ardipithecus*." The findings: Ardi's species ate mostly C_3 foods (a signature similar to woodland chimps) and mostly dined and died amid the trees.

With all the data in hand, the authors concluded that *Ardipithecus ramidus* lived in forests and woodlands—but *not* grasslands nor even *wooded* grasslands. In the final report issued at the end of the seventeen-year investigation, it was not hard to detect a prosecutorial tone: "grassland environment was not a major force driving evolution of the earliest hominids." In the media blitz following the seventeen-year investigation, White proclaimed the death of the savanna hypothesis.

Even some coauthors had qualms about the aggressive conclusions because the information gathered in the Mission to the Pliocene also could support a more nuanced view. The Middle Awash soils were abundant with phytoliths, the microscopic silica remains of plant cells—and most samples were half grasses and sedges. The paleobotanists interpreted the evidence as 40 to 60 percent grass coverage. "Tim White would not accept our results," said Raymonde Bonnefille, a French scientist who worked with her student Doris Barboni. Bonnefille added: "I think he was not enough objective. I had no prejudice. I just told him, 'Tim, you know we had fifty samples. It was not one indication. All the fifty samples contained grass phytoliths! There were grasses—and probably widespread grasses.'" But White and his colleagues put more faith in fossils than phytoliths; microscopic particles could blow all over the landscape (indeed, they were even found in seafloor drilling) but bones remained where the animals fell and were deemed more credible. After a long internal debate, the team leaders concluded phytoliths were unreliable—to some dismay by their French colleagues. The phytolith data were duly reported in published supplements but not emphasized in the main analysis. Critics later used the information from these supplements to depict Ardi's environment as more grassy and less wooded than what was described by the discovery team.

Ultimately, the debate became—as impasses often do—partially a matter of semantics. The term *savanna* has been used to describe everything from treeless grasslands to dense woodlands with up to 80 percent tree cover encompassing many types of vegetation and rainfall. About half of tropics or subtropics could be classified as savanna, so saying human ancestors evolved on them is barely more informative than saying they evolved on land. The United Nations Educational, Scientific and Cultural Organization (UNESCO) avoids the term due to its vagueness. But anthropologists have been unwilling to scrap the old favorite—and inevitably squabble about what they *think* it means. A woodland biped like

Ardi could both support savanna theory or refute it, depending on how one defined the term—and partisans tended to embrace whatever definition supported their view. Not surprisingly, when the Mission to the Pliocene became public years later, opponents argued past each other.

OVER THE LAST FORTY YEARS, PALEOCLIMATE STUDIES HAVE CREATED A paradox: as climate data becomes clearer, the effect on evolution becomes hazier. The boom in paleoclimate studies has generated detailed records of moisture, rainfall, temperature, and vegetation through geologic time in Africa—and the resulting story line is far from simple. The climate and botany of eastern Africa cycled back and forth in a very spiky fashion between wet periods when huge lakes formed in the rift valley, dry periods when dust collected on the coastal seafloor, and sometimes perplexingly both at the same time. Pollen studies show shifting proportions of grass, trees, and desert plants. A few long-term global trends stand out, such as an overall cooling since the Miocene, a drop in CO_2 levels, and a dramatic expansion of C_4 grasses. Some big events in eastern Africa seem intriguing, such as a retreat of tree coverage around 6.5 million years ago and another decline of forests coupled with expansion of C_4 grasslands around 2.5 million years ago (which some have linked to the appearance of the genus *Homo* and the development of stone tools). Nonetheless, the grand picture defies any simple rendering. Rather, the emerging portrait seems a mosaic of varied habitats, constant flux and disruption, and environmental drivers that may have been more local than global. The quest for a master narrative of climate-driven human evolution remains unfulfilled.

Contrary to past assumptions, even the spread of C_4 plants no longer appears to be a simple indicator of expanding grasslands and shrinking forests. In eastern Africa, dust accumulations in seafloor cores showed a puzzling contradiction: C_4 levels increased as grass pollens declined. Why would C_4—long used as a proxy for arid grasses—rise while grass pollens fell? Apparently the shift was not simply from trees to grass, but from one *kind* of grass to another—C_4 grasses displaced C_3 grasses. "The C_4 expansion happened without the grassland area expanding," explained Sarah Feakins, a professor of earth sciences at the University of Southern California. "That's quite a different story." But people initially missed this story because they assumed modern-day environments provided models of the deep past—but they don't. Just as with species, the present world doesn't always provide reliable analogies to lost worlds. "Today we don't have have C_3 grasslands in lowland Africa," Feakins added. "The Miocene was a long time ago, and you have to expect some of the ecosystems didn't look like today." If so, the notion of a C_3 forest versus a C_4 open grasslands may turn out to be a false dichotomy. The savanna paradigm, added Feakins, "should have been dead already."

"It's really strange," mused Feakins. "The data are right there on the page, yet it just refuses to die."

Modern paleoclimate data proves that Miocene Africa was not the lush forest of old imagination. Rather, it was a patchwork of evergreen and deciduous forests, woodlands, and yes—*grasslands*. Most fossil apes from the Miocene dwelled in partially open woodland savannas, not closed forests, and spent lots of time on the ground long before the human family forged off on its own path. So the origin of the human lineage cannot simply be ascribed to any sudden appearance of grass, nor disappearance of forests, nor abrupt descent of apes to the ground. The notion of grasslands progressively replacing forests is a myth, and the environmental context of our evolution far more complicated. The savanna saga serves as a reminder of a human foible—trying to describe an entire forest when we can see only a few trees.

Chapter 17

HARVEST OF BONE

The discovery of *Ardipithecus* and mother lode of Ardi had waylaid the team at Aramis years longer than planned. Even as the Mission to the Pliocene remained unfinished, the corps of discovery launched new explorations of other periods of the human past and pursued a grand agenda—the human saga from simian to *sapiens*.

The bone-strewn Middle Awash preserved 6 million years' worth of evolutionary history that White likened to snapshots in a family album. The team would fill Darwin's tree of life with bones—and reveal more members of the human family never seen before. The pace of the quest was exemplified in 1997 when the explorers made a series of major discoveries spanning 5 million years, including two new species of ancient humanity. That season provided a glimpse of the frenetic fieldwork, the team's impulse to place all its discoveries into a magnum opus, and the furor it provoked among some scientific peers.

The bonanza began with a catastrophe. For centuries, Ethiopian peasants have lived by a seasonal calendar. The *Kiremt* rainy season runs from June to September (the country's historic vulnerability to famine partly stems from its reliance on a short wet season). The *Bega* harvest season begins around October when the weather turns dry, farmers gather crops, and fossil hunters head to the Afar Depression to collect bones. In 1997, the normal rains did not arrive, and Ethiopia braced for the familiar scourges of drought and famine.

Then came the floods. Just when the dry season was supposed to begin in October, heavy rains began falling across eastern Africa. Two aberrant weather patterns, El Niño and the Indian Ocean Dipole, simultaneously hit and brought the worst storms in memory. The runoff from the rain-drenched highlands was so heavy that Lake Turkana and Lake Victoria both rose by nearly two meters. More than six thousand people died throughout East Africa. In Ethiopia, floods destroyed crops, interrupted harvests, and killed countless livestock. Only a quick response by the government and international aid agencies averted another famine.

In Addis Ababa, the Middle Awash team remained stranded at the museum, peering out fogged windows, and watching another field season slipping away. Anxious to find a way back into the field, Berhane made a reconnaissance trip and found the highway blocked by a flooded river. A few days later, the expedition descended the western escarpment of the rift and blazed a new route into wilderness where no cars had been before. They hacked through bush, forded swollen rivers, and endured swarms of mosquitoes in a four-day slog to the fossil fields. Along the new route, an Afar headman erected a roadblock and made death threats. He wanted to turn it into his personal toll road.

Rising floodwaters soon cut off all exit routes, including the way they had come. The expedition was marooned on its destination—an elevated peninsula called Bouri, which happened to be something of a treasure island. Measuring ten kilometers long and four kilometers wide, the formation contained fossil-rich sediments between 2.5 million and 100,000 years old, roughly coinciding with the lifespan of our genus *Homo*. How *Australopithecus* evolved into *Homo* remained murky—as did the appearance of the last species, *sapiens*.

On the first day of surveying, the team drove past an Afar village called Herto and found it eerily abandoned. With grass

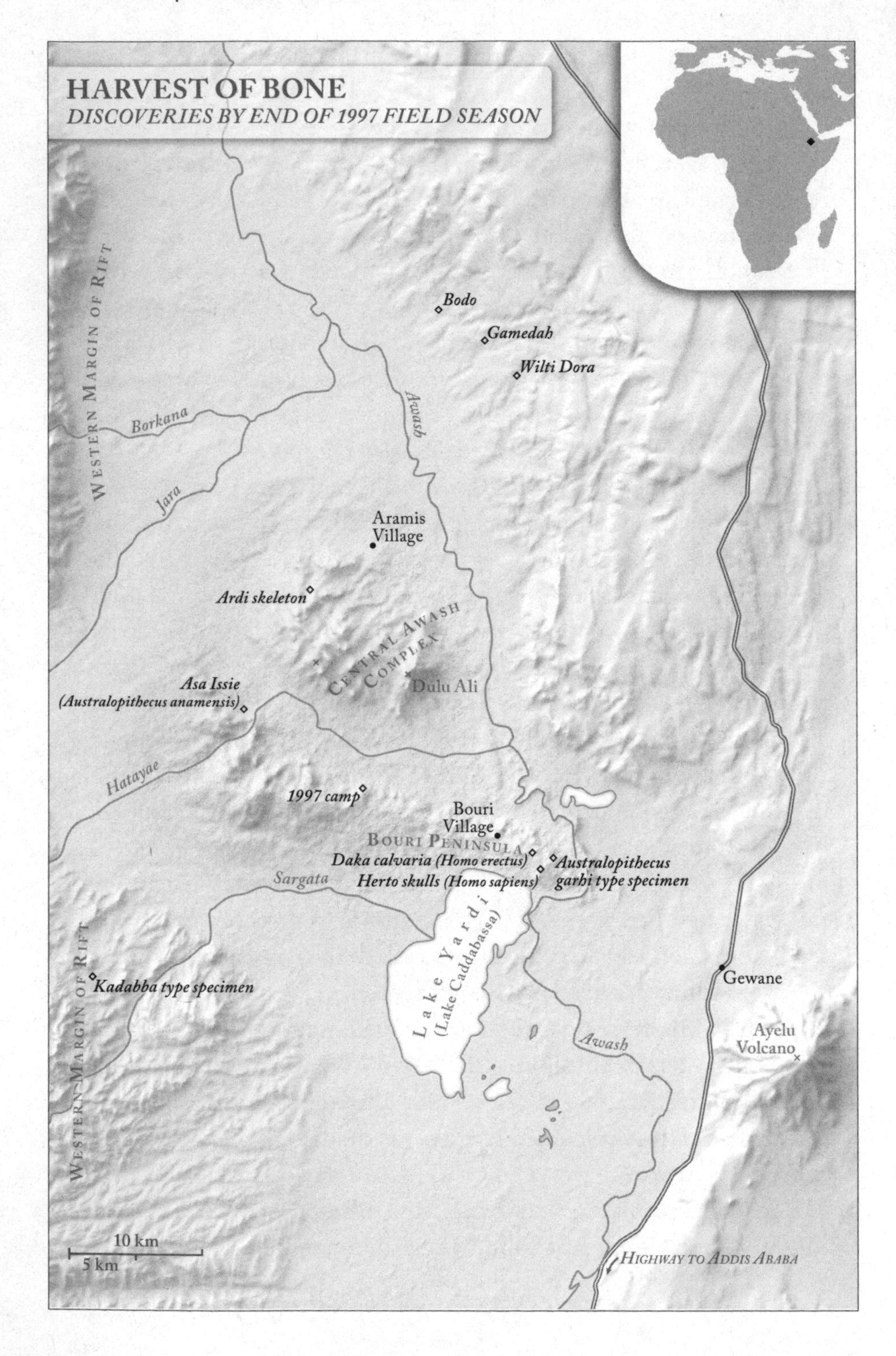

HARVEST OF BONE
DISCOVERIES BY END OF 1997 FIELD SEASON
Bodo
Gamedah
Wilti Dora
Western Margin of Rift
Borkana
Awash
Jara
Aramis Village
Ardi skeleton
Central Awash Complex
Dulu Ali
Asa Issie (Australopithecus anamensis)
Hatayae
1997 camp
Bouri Village
Bouri Peninsula
Daka calvaria (Homo erectus)
Australopithecus garhi type specimen
Herto skulls (Homo sapiens)
Sargata
Lake Yardi (Lake Caddabassa)
Kadabba type specimen
Western Margin of Rift
Gewane
Awash
Ayelu Volcano
10 km
5 km
Highway to Addis Ababa

Awash with Discoveries

SPECIES	AGE	MAJOR DISCOVERIES IN MIDDLE AWASH	LOCATION
Homo sapiens	160,00 years	Herto skulls, 1997 (Herto)	Bouri Peninsula
Homo erectus	1 million years	Daka calvaria, 1997 (Daka)	Bouri Peninsula
Australopithecus garhi	2.5 million years	Type specimen skull, 1997	Bouri Peninsula
Ardipithecus kadabba	5.5–8.5 million years	Type specimen, 1997	Western Margin

mats pulled off the frames, the empty huts stood bare as skeleton ribs. The storms had driven away the nomads, washed away the livestock dung and hoof-churned dust, and exposed fossils. White spotted a meter-long hippo skull emerging from the ground. On close inspection, the cranium bore a huge cleaver gash—the animal had been butchered. It was an enticing clue that the butchers themselves might be found nearby. Herto lay atop the youngest sediments on the peninsula (less than two hundred thousand years old), making this prime hunting grounds for finding early *Homo sapiens*. In 1987, biochemists at Berkeley had reported that all modern humans derive their mitochondrial DNA from a single maternal ancestor who lived in Africa within the last two hundred thousand years—a progenitor dubbed "Mitochondrial Eve." (This finding was based on only a small portion of the genome and led to the false expectation of a small ancestral population; newer studies of the full genome show that we descend from many population lineages.) At that point, the physical remains of those early *sapiens* remained scarce, and Herto offered a chance to find them.

The team later returned to the abandoned village and found pieces of three skulls of early *Homo sapiens*. Dated about 160,000 years old, the skulls briefly held the title as the oldest fossils attributed to our species, and the team described them as the "probable immediate ancestors of anatomically modern humans."* Yet something seemed odd: why no remains from below the neck? That mystery would grow more macabre in the lab—but first there was more searching to do.

---

* The Herto skulls became key evidence in a raging debate about the recency of our African origins. Early *Homo* left Africa by 2 million years ago and spread into Asia and Europe. The question was, did these archaic members of our family (including Neanderthals) persist in Eurasia and evolve into modern humans? Or did later waves of African *sapiens* replace them? One camp known as the "multi-regionalists" argued that modern *sapiens* evolved simultaneously in Africa, Asia, and Europe. Another camp known as the "recent African origins" school (Desmond Clark among its founding figures) argued that all modern humans descended from a population of sub-Saharan Africans much more recently and replaced their more archaic cousins. Early DNA studies swung the debate in favor of the recent African origins camp and the Herto skulls provided the first physical evidence of early *sapiens* with anatomy clearly different from the stouter Neanderthals. But research published since 2010 shows the truth is more complicated. First, some interbreeding occurred between *sapiens*, Neanderthals, and other archaic ancestors in Eurasia. Second, more comprehensive analysis of the full genome (which is far larger than the limited mitochondrial portion analyzed in the 1980s) reveal many ancestral populations, not a single lineage from any one locale. Finally, *sapiens* skulls from various sites in Africa and the Middle East exhibit a range of variation with no obvious "ancestral" stock to the exclusion of others. In short, modern humans descend from many populations that diversified, interbred, and coalesced into the composite of modern humanity. Currently, the oldest fossils attributed to *sapiens* come from Morocco, and many scholars have shifted to a new model of "African multiregionalism." For a detailed review, see Stringer, "The Origin and Evolution of *Homo Sapiens*" and Scerri et al., "Did our species evolve in subdivided populations across Africa, and why does it matter?"

Elsewhere in Bouri, they found a strange 2.5-million-year-old skull that was just the right age to illuminate the transition from *Australopithecus* to *Homo*—except it defied what anybody had predicted. Early *Homo* was expected to have smaller chewing teeth than *Australopithecus*, but this strange creature from Bouri had remarkably *large* molars and premolars, a prognathous snout, and a brain too small to classify as *Homo*. The discoverers named it *Australopithecus garhi* after the Afar word for "surprise." It was among at least three species in the human family that appeared around 2.5 million years ago. Despite its odd morphology, the discovery team asserted, "It is in the right place, at the right time, to be the ancestor of early *Homo*." (Some peers, however, later deemed *garhi* an unlikely progenitor of *Homo*.) The same time horizon also produced animal fossils with cut marks—an early hint of tool use. Earlier in 1997, the team just downstream at Gona announced finding 2.6-million-year-old stone tools, then the oldest in the world.

Four million years ago at Aramis, hominids were ravaged by carnivores. Two million years ago at Bouri, their descendants had begun doing the ravaging. These bipeds differed from Ardi or Lucy. *Homo* had bigger brains, greater intelligence, and technology—stone tools sharpened into blades for slicing flesh and shattering bone. "You go into this period with, in essence, bipedal big-toothed chimps," White said in 1999, "and come out with meat-eating, large-brained hominids." He likened them to primate hyenas—packs of carnivores who reduced huge carcasses to bone shards. They carved out a unique niche in the animal kingdom as tool-using bipeds and competitors of large carnivores. This new occupation boosted natural selection for intelligence, dexterity, technology, and social cooperation. In each successive strata, *Homo* left evidence of increasingly humanlike behaviors including more sophisticated tools*, culture, and finally self-destruction.

---

* Tools are the first visible evidence of culture, or the body of knowledge, beliefs,

When the floodwaters abated, the team encountered yet another surprise—fossils even older than Ardi and deeper roots of the human family.

THE LAST COMMON ANCESTOR OF HUMANS AND CHIMPANZEES IS SOMEtimes called the Holy Grail of paleoanthropology. Like the mythic chalice, it remains long sought, repeatedly claimed, repeatedly debunked, and never quite located. In the Middle Awash, the best place to search for this ancestor lay in the western margin of the rift, a series of stair-stepping hills whose oldest rocks roughly coincided with the time when the human and chimp families were believed to have diverged in the late Miocene epoch. In its 1996 funding request to the National Science Foundation, the Middle Awash team explicitly stated its hopes: "It is now possible to strongly predict that the remains of the last common ancestor will be found in sediments between 4.5 and 6.0" million years old. In the years before the 1997 floods, the team

---

customs, and skills shared by social groups. They also represent the first industrial revolution—a very slow one. "If these ancient people were talking to each other," Desmond Clark remarked drolly, "they were saying the same thing over and over again." The first recognizable tool style is called the Oldowan. In 1997, a team at Gona announced 2.6-million-year-old tools, the oldest Oldowan implements in the world (in 2015, another team reported 3.3-million-year-old tools in Kenya, but that claim remains controversial). Around 1.7 million years ago, toolmakers developed the more advanced Acheulean style known for the pear-shaped hand axe. The oldest identifiable weapon appeared only 400,000 years ago. Just as it did with bones, the Awash Valley revealed a saga of stones. Clark's archeology team had spent years collecting both Oldowan and Acheulean tools at sites on both sides of the river. A short distance to the north near an Afar Village called Aduma, another archeology team under John Yellen and Alison Brooks excavated more advanced implements from 80,000 to 100,000 years ago—finer-crafted projectile points, perforators, and blades from the Middle Stone Age. No tools have been discovered from the time of early human members of the human family such as Ardi. Apes are known to use sticks and hammerstones to obtain food, but such implements leave little trace in the archeological record.

launched a more intense reconnaissance led by geologist Giday WoldeGabriel and paleontologist Yohannes Haile-Selassie. Using space images to pick targets, they hiked along a thirty-kilometer frontier of hogback ridges, precipitous cliffs, and canyons so deep that sun touched the ground only at midday. Their reports were promising enough to divert the full corps of discovery on a side trip away from Bouri in the flood year.

One morning in December 1997, a caravan headed across the rift floor. When the terrain became too rugged for cars, the crew began walking. Yohannes forged ahead and called back on the radio.

"Tim, do you copy? I found a hominid. A hominid mandible."

"How do you know it's a hominid?"

"Because it's a hominid."

On the desert floor lay a fragmented lower jaw that became the type specimen of a new *Ardipithecus* species dated between 5.4 and 5.8 million years old. They had broken into the Miocene epoch and had found what they deemed an ancestor of Ardi. The son

Giday WoldeGabriel takes notes atop an outcrop on the western margin.

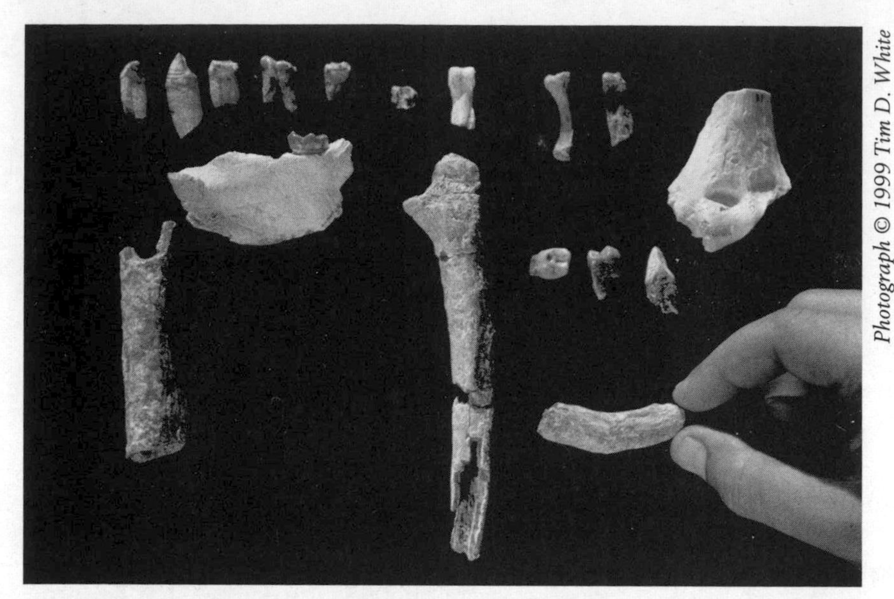

Photograph © 1999 Tim D. White

The collection of *Ardipithecus kadabba* fossils as they appeared in 1999. Years of intense searching for Miocene fossils of human ancestors yielded only a handful of bones and teeth. The type specimen mandible is at upper left.

of an Afar balabat suggested a name: *kadabba*, which means "big father" or "grandfather." Four years later, the team announced the new discovery "close to the common ancestor of chimpanzees and humans." (Initially, *kadabba* was deemed a subspecies of *Ardipithecus ramidus*, but subsequent discoveries elevated it to its own full-fledged species, *Ardipithecus kadabba*.) Yohannes exulted: "No one thought we would find any hominids there at all. Now everybody is hoping we will find a skeleton."

But there would be no grandfather skeleton. Nine years of relentless searching yielded a late Miocene menagerie of 2,800 fossils of carnivores, monkeys, pigs, hippos, giraffes, horses, and herds of elephants. This time the hominids amounted to about one-half a percent of the total, just fifteen pieces, including the mandible, assorted teeth, a broken finger, a toe, and two arm fragments. The deeper the past, the rarer the ancestor.

Those discoveries of older *Ardipithecus* joined a string of

Miocene finds elsewhere in Africa. In 2000 in Kenya, Martin Pickford and Brigitte Senut found a 6-million-year-old hominid they called *Orrorin tugenensis* ("original man" in the local language), including teeth and pieces of femur. They made an announcement within weeks of their discovery and submitted a paper within three months, so their species became the first member of the human family named in the Miocene. In 2002 in Chad, another team under French paleoanthropologist Michel Brunet reported the discovery of *Sahelanthropus tchadensis*, represented by a fine skull dated between 6 and 7 million years old. *Sahelanthropus* claimed the title of the oldest purported member of the human family, but it was a head without a body, so Ardi remained the oldest skeleton in the human family. Each team touted its Miocene fossils as early human ancestors, each claimed its species walked upright, and each emphasized its proximity to the last common ancestor of humans and chimpanzees. Even so, the three Miocene species shared few pieces in common for direct comparison.

The Ardi team included all these snapshots in the same family album. In their grand evolutionary scenario, all three Miocene species belonged to the same genus *Ardipithecus* and perhaps even a single lineage. They envisaged three evolutionary plateaus corresponding to the three genera of human ancestors—*Ardipithecus*, *Australopithecus*, and *Homo*. Each represented a grade of closely related species who shared similar physical form, diets, and locomotion. The first included *Ardipithecus*, *Sahelanthropus*, and *Orrorin*—the primitive bipeds who came after the chimp-human split. All were omnivores with reduced canines who foraged in the woodlands, sometimes climbed trees, and sometimes walked upright on the ground. The best exemplar of that group was Ardi—the skeleton lying back in the museum.

The second plateau began around 4 million years ago with *Australopithecus*—bipeds like Lucy with big chewing teeth who ventured farther from the trees, expanded over all Africa, and diversified into at least two lineages and multiple species—*garhi* being the most recent discovered. The third plateau began sometime after 3 million years with *Homo*—the tool-using predators who broke out of Africa and spread throughout the world. Finally rose *sapiens*, the superpredators who overran the world. By two hundred thousand years ago, the human body looked very much like our own, and the remaining drama of evolution became less about anatomy and more about brains and behavior.

"I can take you to one single place in the world where the whole record of human history . . . is represented," Berhane Asfaw declared at one 2006 conference, "and that is the Middle Awash of Ethiopia." At times, critics parodied the team for claiming *all* human evolution happened in the Awash Valley. The team didn't go that far but did emphasize a large swath of human history was *sampled* there—which was abundantly true. If all the far-flung sediments in the territory were combined into a single stack, it would stand one kilometer thick from late Miocene to present, a rare sequence of evolutionary history that they billed as "the continent's most important natural laboratory for the study of human origins and evolution."

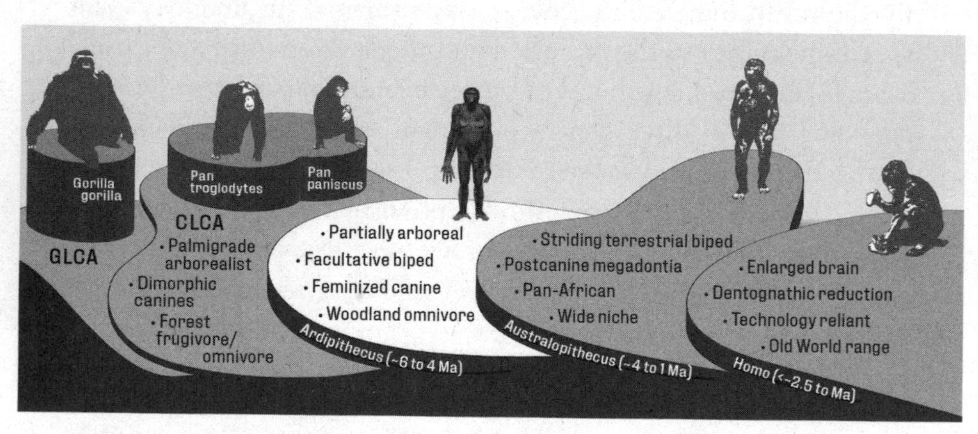

The Middle Awash team's depiction of three adaptive plateaus leading to modern humans. The plateaus correspond to the genera *Ardipithecus*, *Australopithecus*, and *Homo* (GLCA/CLCA = last common ancestors with gorillas/chimps). The 1997 field season produced new discoveries from all three, including two new species. Many more fossils would accumulate during decades of fieldwork.

But the simple story depicted by the Middle Awash family album was controversial. Some peers regarded it as retrograde for its steplike depiction of the human past—not to mention the team's eagerness to link its own discoveries to modern humans. Some suggested the Ethiopian team actually had uncovered evidence of more diverse lineages than they had recognized. As ever, the shape of the family tree remained one of the biggest points of contention in the science of human origins. One overwhelming factor thrust the Middle Awash team into the epicenter of that debate: they were finding more fossils than anybody else.

THE STORMS THAT WROUGHT HAVOC WITH ETHIOPIA'S FRAGILE FOOD SECUrity brought a record harvest of bone. In those weeks of the 1997 season of floods, the expedition had unearthed bookends to the human saga plus much in between. "Despite the obstacles and hardships," the team reported, "the season proved to be one of the most productive ever." One discovery after another added to the backlog of unpublished work as the fieldwork continued year after year. Eventually, the team would harvest more than four hundred fossil localities containing at least seven species of ancient humanity (plus far more other fauna) and more than three hundred archeology localities. Scientific peers clamored for quicker publication of all the fossils pouring out of the Middle Awash—especially the mysterious *Ardipithecus*. The National Science Foundation began to voice concern that the team had become *too* successful and its publications simply couldn't keep up with the flood of fossils. One NSF reviewer wrote: "The sheer volume of material coming out of the ground has this team swamped beyond being able to get out their brief reports to *Science* or *Nature*. . . . Getting the information published in a timely fashion is one of NSF's requirements for continued funding."

Back at the museum, Ardi—the skeleton from the ostensible root of humanity—remained locked in the safe. Soon events would put her bones off-limits to even her discoverers.

• • •

IN THE LABORATORY, THE BONE HUNTERS CAME FACE-TO-FACE WITH A darker side of humanity. At Herto, the searchers had recovered many pieces of skull, but all their scouring around the village turned up no bones from below the neck. Why? In the lab, White spotted gouges on the fragments. When the skulls were rebuilt, the scars aligned into long gashes—*cutmarks*. Near the dawn of our species, tool-using *Homo sapiens* not only butchered other animals. They also dismembered each other.

Nobody better understood the signs. In his first trip to Ethiopia in 1981, White intently examined the fossils found by his predecessors in the Middle Awash, particularly an archaic *Homo* skull called the Bodo Man, and he detected something that the Kalb team had missed—deep incisions in the eye sockets unlike typical carnivore damage. The gouges must have been made while the bone remained fresh and could not be blamed on clumsy lab workers. He also judged the skull had been improperly cleaned and imperfectly glued together. He secured permission from the Ethiopian government to export the skull to Berkeley for further restoration (the project served as part of Berhane's training). As he cleaned off remaining rock matrix, White uncovered more cutmarks around the eyes, face, and skull. The Bodo Man had been mutilated by another hominid.

When Ethiopia suspended field research, White poured his energies into studying the peculiar human habit of carving up other humans, a topic that for some reason struck a chord in him. In the literature, lurid claims of cannibalism were abundant but methodologies for recognizing it remained scarce. In order to develop a more scientific approach to postmortem mutilation, he borrowed a collection of human remains from an Anasazi pueblo in Colorado (anthropologist Christy G. Turner II previously had documented evidence of cannibalism at Anasazi sites in the 1960s and 1970s). White determined those humans had been dismembered like deer. Butchers hammered open long bones for fatty marrow, sliced meat off bones, and smashed skulls to harvest brains. Some bones shined with "pot polish" from being cooked, and others had been charred by fire. With colleagues, he studied more examples around the world, including cannibalized Neanderthals from caves in Croatia and France, and a mutilated jaw from South Africa. In 1992, White published *Prehistoric Cannibalism at Mancos 5MTUMR-2346*, which became a standard reference for cannibalism and postmortem modification in the osteological record. The newly found Herto skulls had fallen into the hands of the expert most qualified to interpret their peculiar markings.

But these *sapiens* skulls did not resemble typical cannibalism. Usually butchers of humans smashed open skulls to devour brains, which are rich in fat and protein. The Herto pieces had clean breaks from shattering *after* fossilization. Somebody decapitated the skulls, severed connecting tissues, and broke open the bases just enough to extract the brains—but took care to not smash the rest. Two skulls showed evidence of intentional scraping and polishing. Berhane reassembled a third, the braincase of a child about six years old, which shined with a distinctive patina, suggesting it had been handled repeatedly. It reminded White of ritual skulls from Papua New Guinea. Why would primitive *sapiens* carry around skulls? Trophies of slain enemies? A postmortem ritual? Memorials to loved ones? It was sensationally lurid stuff, but White refused to speculate. "We're like forensic scientists working at a very cold scene," he said at the time. "All we can say now is some kind of mortuary practice was exhibited by people who were our ancestors." This much stood clear: the desire to possess old bones was deeply rooted in our species.

His genteel elder colleague, Desmond Clark, took a more cynical view: such butchered skulls were prologues of inhumanity. "I think they were rather like ourselves," he opined. "We're dreadful, and I think we'll probably destroy ourselves unless we learn to sublimate it."

Chapter 18
BORDER WARS

In 1999, a fossil-hunting team at Gona, a territory downstream from the Middle Awash, began surveying some unexplored sediments more than 4 million years old and picked up fossils of a primitive human ancestor—which turned out to be more *Ardipithecus,* its first discovery outside the Middle Awash. One day that season, the sky began to roar, and the Gona team looked up and saw formations of fighter jets screaming north followed by waves of rumbling propeller-driven bombers of the Ethiopian air force. Ethiopia and its neighbor Eritrea had gone to war.

Eritrea was a former coastal province of Ethiopia and had gained independence by peaceful referendum after the civil war. Both governments grew out of regional liberation movements that had allied to overthrow the Derg. But the two erstwhile allies soon became enemies. The aerial attack marked the onset of another cycle of conflict that reverberated far beyond the northern battlefields. On the border between Ethiopia and Eritrea, half a million troops faced off along a six-hundred-mile frontier reminiscent of the trenches of World War I in what one correspondent called "the world's most senseless, inscrutable war." Afar herders fled south and competed for territory and water. Meanwhile, Issa continued to press into Afar territory from the east and south, and tribal warfare flared up anew in the bone fields. Year after year, the Middle Awash expedition leaders would hear that old Afar friends had died from violence. One victim was Neina Tahiro, the Afar politician who suggested the species name *kadabba*. Urging peace between the tribes, he was shot in 2000 after visiting an Issa settlement. His murder triggered revenge killings in the towns along the Afar highway. The tribes battled for territory, grazing land, water, and dominance over the highway shantytowns. In 2002, a band of Issa went to destroy a new Afar administration building and walked into an ambush that left thirteen dead and ten wounded—one of countless lethal conflicts that raged in those years.

One earlier loss struck close to the Ardi team. In 1998, Gadi crossed the Awash, got into a shootout with the Issa, and was wounded in the leg. After several days, his wounds became infected and, too late, the Afars sought medical care. "He was trying to get help by selling his gun to get treated," recalled Berhane Asfaw. "Nobody helped him. Finally he said, 'Please call Berhane or Tim.'" When the message reached the capital, the anthropologists jumped into cars and raced to help their Afar friend, but arrived too late. Gadi already had been buried. White grieved. Back at Berkeley, he placed a framed photo of Gadi on his desk so the scoliotic Afar, rifle yoked over his shoulder, kept watch over him just as he did during the excavation of Ardi.

AROUND THAT TIME, DON JOHANSON REGAINED PERMISSION TO WORK IN Ethiopia. The IHO had been put out of business in the country for about five years after the complaints levied by Berhane Asfaw and his Ethiopian colleagues had triggered the investigation by the Ethiopian attorney general. In 1999, Johanson and his partner, Bill Kimbel, recovered their permit to work at Hadar. "We were requested to spend three days with the attorney general," fumed Johanson. "We didn't know if we were going to go over and be put into jail. Bill Kimbel and I had to sit in this little room with our attorney and the attorney general and answer every single one of these accusations. At the end, they said, 'There's no cause here.'"

He blamed the Middle Awash team for his troubles. Johanson described the rivalry as "two groups that were at each other's throats." As he resumed work in Ethiopia, he found a sympathetic ear in the current antiquities administration—which likewise soon became embroiled in its own series of disputes with the Middle Awash team. A series of paleo quarrels entangled other research groups, the local press, diplomats, and government officials amid the already-turbulent politics in the country. At the time, Ethiopia had little tradition of open debate or dispute resolution; personal conflicts tended to fester until they exploded volcanically. "Compromise is not very easy among Ethiopians," reflected Zelalem Assefa, a former antiquities officer and American-trained Ph.D. scientist. "Grudges stay with you for a long time." Western fossil hunters were not exactly known as peacekeepers, either. The result was a combustive mix of turf wars over land, fossils, money, regulations, and all the elements required to write the biography of an egocentric species.

AN OLD AMHARIC ADAGE WARNS: ***HE WHO IS NOT VIGILANTLY SUSPICIOUS*** *will be displaced from this land*. The warning is well heeded in paleoanthropology. A fossil-rich territory does not guarantee success, and many amateurs fail simply because they lack expertise in the field, but a poor territory guarantees failure, even for the best operator. The Middle Awash project had nurtured the career of Yohannes Haile-Selassie. Nearing the end of his graduate studies at Berkeley, Yohannes launched his own field project in 1998 around Mount Galili, an area near the conflict-ridden Afar highway and the town of Gadamaitu, an Issa hamlet and haven for smugglers who ran camel trains to the Somali border. Aided by his mentors, Yohannes made initial surveys and found hominid teeth about 4 million years old—signaling that they had found another promising portal into the world before Lucy. Due to the danger of Issa bandits and smugglers, the explorers traveled with an escort of Ethiopian soldiers in an old Soviet army truck—"with a really big gun," recalled Giday WoldeGabriel.

But others coveted the same terrain. Horst Seidler was a famous anthropologist at the University of Vienna who specialized in using computed tomography (CT) scans for anatomical investigation. (He gained renown by employing CT to investigate the case of Otzi, the 5,300-year-old mummified body of the "Tyrolean Iceman" found in the Alps, and speculated that Otzi fell asleep and froze to death. Other researchers later discovered that the iceman's death had not been so peaceful: Otzi had an arrowhead lodged in his body, a fractured skull, and a brain injury from a blow to the head.) Eager to explore the deeper roots of humanity, Seidler went to Ethiopia and cultivated a warm relationship with Jara Haile Mariam, general director of the Authority for Research and Conservation of the Cultural Heritage (ARCCH), the Ethiopian agency that oversaw antiquities research. Seidler had no experience in Ethiopia and sought advice from those who did, including Jon Kalb, who was seeking to restart his career in Ethiopia. Somehow two teams wound up with permits for the same region.

In February 2000, Yohannes was working in the National Museum when a colleague informed him that the Austrian team had moved into his project area. Furious, he and Berhane made the long drive to Galili, found the rival team at the spot of his prior discoveries, and charged into the camp. "In hindsight, that was probably a bad move," recalled Yohannes. "They had armed guards with them and could have shot at our car." Yohannes presented the Seidler expedition with a government letter protecting his claim and departed. A few days later, the two sides met in Addis Ababa. Seidler apologized for the confusion and invited Yohannes to join his team, an offer that Yohannes indignantly refused. Both sides claimed valid permits but could not agree on boundaries or whose claim took priority. In the museum garden, Berhane and Yohannes also had a tense confrontation with Seidler's Italian colleagues who insisted they independently had discovered the fossil-rich Galili (which they had nicknamed "El Dorado") during their own explorations. The Middle Awash scientists alleged that somebody had leaked inside information to the rival team and the mixup quickly escalated into a bitter impasse. Later that year, Yohannes delivered a presentation at the annual meeting of the American Association of Physical Anthropologists and used the stage to publicly accuse Seidler's team of claim jumping before an audience of scientific peers.

Seidler and his colleagues insisted they had no idea that the other team already had begun work at the site. "On my word of honor, I never knew it was Yohannes's site," Seidler told one reporter. He maintained he had followed all proper procedures and obtained his permit "in a transparent and legal way" with the cooperation of antiquities authorities. Yohannes later produced photographs showing the rival team working beside the piles of dirt left over from his own sieving. But opponents had little sympathy for the Middle Awash group. "They had enough areas to keep them busy for two hundred years," said Kalb. "They sent in this graduate student to cover more territory which they didn't need at all."

In July 2000, the ARCCH issued a new directive: an Ethiopian student studying abroad could hold a permit for only three years—a provision that retroactively applied to Yohannes. Shortly thereafter, the authority awarded the permit to Seidler. "There was a big disagreement," recalled Yonas Beyene, an ARCCH employee (and Middle Awash team member). "At the end of the day, the Austrians got the area and I lost my job." Berhane Asfaw labeled the episode "the most outrageous claim jumping ever witnessed in the history of paleoanthropology."

No longer employed by the government, Berhane acted as a freelance watchdog and cried alarm when he perceived a threat to his science or country. Some foreigners had the mistaken impression that he still headed the museum (some journalists and even the journal *Nature* erroneously identified him as the director). "He

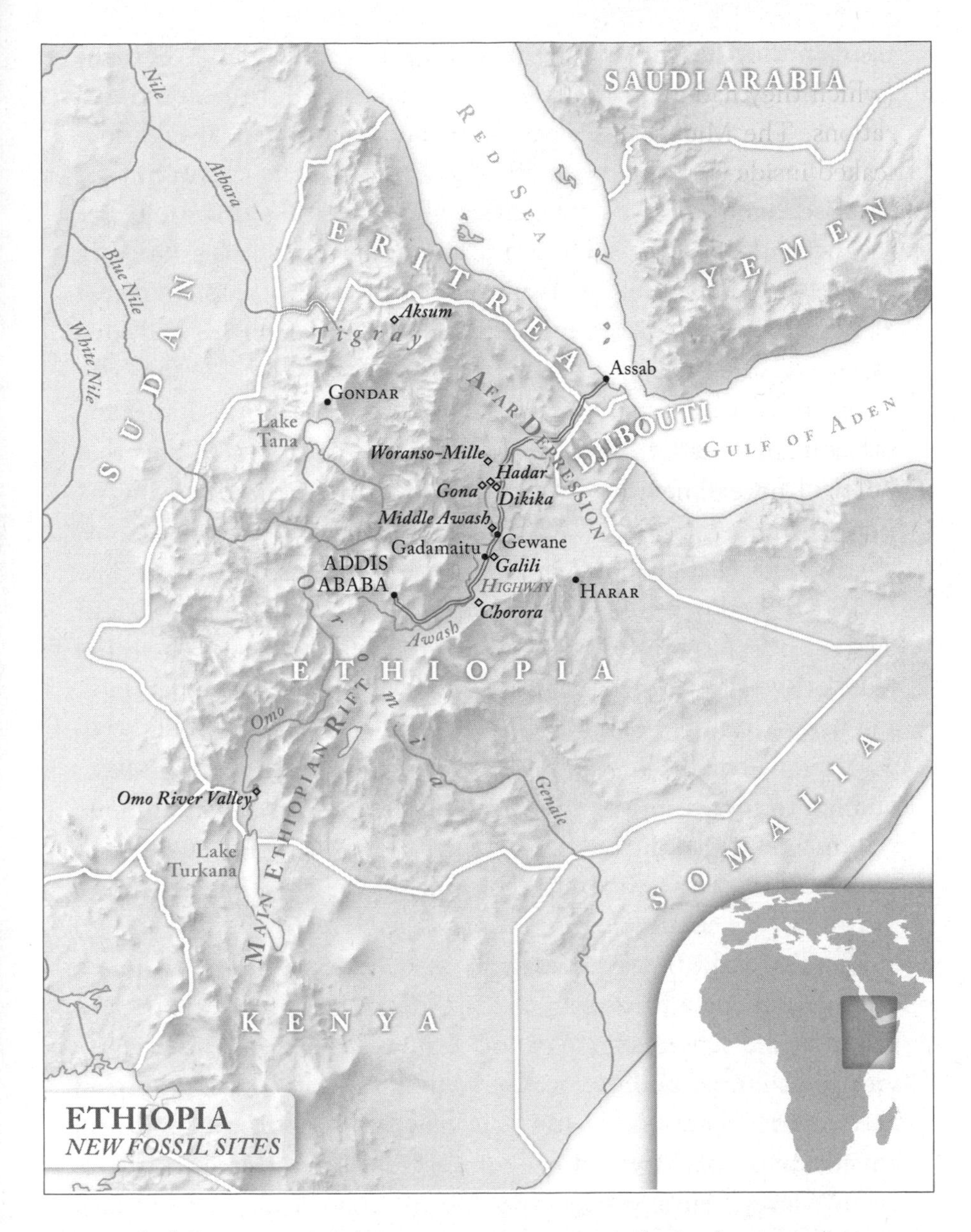

paraded for years as being a government employee because he was always on the museum grounds," seethed Johanson, a frequent target. At one point, paleoanthropologist Zeresenay "Zeray" Alemseged faulted Berhane for opposing one antiquities administrator after another: "Who ever is in charge is your enemy." Berhane made no apologies. In his eyes, politicians and bureaucrats didn't always understand the science, and foreigners didn't always concern themselves with the welfare of Ethiopia. His country had a sad history of seeing its antiquities plundered: an Aksum obelisk had been taken as a war prize and erected in a traffic circle in Rome to commemorate Mussolini's conquest of Ethiopia; ancient manuscripts sat in foreign museums; and countless fossils had been exported before Ethiopia developed its own fossil lab. Berhane had a personal mission: he had been trained to protect antiquities and represent Ethiopia before the international community, and that was what he would do—even when his own government didn't want to hear it. "Why is it," asked Berhane, "that such citizens blindly believe that foreigners are better than indigenous professionals?" (Another sore point: people kept dismissing the Ethiopian scientists as minions of Tim White.) Their team had invested years of sweat equity into the museum, organized the collections, and had strong opinions about how the science *should* be run. "No question, Tim and Berhane rub a lot of people the wrong way," said John Fleagle, an American scientist who worked in Ethiopia. "There's no question they feel they run the place—or should. At the same time, you walk into that museum and you can see who's put all the resources and money into that place."

Seidler and his team had embarked on a campaign to amass an electronic data repository of fossils from around the world and had brokered agreements to export Ethiopian specimens to Vienna for CT scanning. Berhane accused the ARCCH director of giving preferential treatment to the Austrian team in return for being wined, dined, and treated to "comfortable trips to cities like Vienna." Some scientific papers by the Austrian team credited Jara as a coauthor—but he was an *architect*. In turn, Jara dismissed Berhane as a conspiracy monger with a "history of spreading lies."

When the Vienna team published and marketed CDs of some of the fossil scans, Berhane accused his rivals of commercializing

Ethiopia's national treasures. At one point, Berhane and Seidler came face-to-face at a conference in Addis Ababa.

"When I opened the CD, I didn't see a single flag of the country," Berhane said heatedly. "I saw the Austrian government emblem."

"This is not true!" protested Seidler, who insisted all royalties would go to the Ethiopian museum.

Berhane pressed: "Our country has an emblem, the stamp of the country. *Your* country's emblem"—he jabbed his finger toward Seidler's chest—"was put in the CD! When you open it, you get the Austrian government's emblem!"

"Please have a look at the cover!" insisted Seidler.

Berhane wagged his finger reproachfully. Although the cover acknowledged the Ethiopian antiquities agency, he feared the actual data on the CD still could be sold for profit and foreigners would exploit his country yet again (Seidler insisted he would do no such a thing). "Are you disappointed because I belong here?" Berhane asked the Austrian. "I have a responsibility to react to anything that I think is unfair for the country. I have to!"

"Yes, but you also have to give us opportunity," chided Seidler. "And you have to be flexible enough to listen to all the arguments of others. You are always insisting on the same topic."

It took two years to reach a compromise. Later, a government minister brokered a settlement: the Seidler team kept Galili, and Yohannes picked a new site from unclaimed territories (which turned out to be a blessing in disguise when he made major discoveries there). Having at last secured the disputed land, Seidler prepared for a triumphant return and assembled an international expedition of scientists, journalists, and a film crew. A reporter from the German newspaper *Die Zeit* described Seidler at the Addis Hilton, passing time at the bar and pool, chain-smoking, and complaining virulently about Tim White with "insults that one is well-advised not to reproduce."

A few days later, Seidler led a large expedition back into the

Afar Depression. The reporter noted all the scientists were white and all the workers black. Training Africans, Seidler promised, would be part of future missions. A documentary crew filmed Seidler sitting with Issa leaders, chewing khat, and discussing development projects such as water wells, schools, and medical clinics. He entrusted a large sum of cash to the owner of a local bar.

The fossil hunting proved disappointing. Jon Kalb walked over the land and wondered aloud: Was it all a big trick? Had the Middle Awash people *pretended* Galili was a rich site, just to lure their rivals into the desert to make fools of themselves? He and his teammates blamed the Middle Awash people for all their troubles—the bureaucratic battles, the figures who spied on them from the distance, and their inability to find anything exciting on the ground.

On the fourth day, bandits raided the camp. The Issa guards proved mysteriously ineffective. "Twenty-five Somalis attacked our camp," recounted Kalb. "The wife of the chief geologist had the shit scared out of her—a guy came and cut open her tent with a knife." Seidler's team returned from fossil hunting and found their possessions gone. "I lost everything in my tent," said anthropologist Dean Falk. "We all did."

They broke camp and fled.

"Professor Seidler is convinced that Tim White's people have instigated the attack," reported *Die Zeit*. "Why else should they have targeted scientific material? The expedition is finished, premature and unsuccessful."

White responded to the allegation with derisive laughter. He was back at Berkeley when the raid occurred and offered a different explanation: a scientific vanity project ventured out of its depth. "They're completely incompetent," White said. "They don't know anything about finding fossils."

BUT GALILI WAS JUST THE BEGINNING. THAT CONFLICT PRESAGED AN EVEN bigger battle that threatened the very survival of the Middle Awash team and threatened to yank the Ardi skeleton from their hands.

In July 2000, the Ethiopian ARCCH issued new regulations that asserted the government's right to commandeer equipment and vehicles of scientific teams during the off-season and claim ownership of all property when research permits expired. The regulations also levied hefty daily fees for using the museum lab and reduced permit areas to one hundred square kilometers—a provision that could break up projects like the Middle Awash into literally hundreds of pieces. "We want more scientific competition," explained Jara. The new directives imposed more stringent deadlines: discoveries that remained unpublished after five years could become available to other researchers.

The Middle Awash team viewed the new regulations as an attack on science. "These corrupt individuals would like the areas under permit to be as small as possible," White vented in the heat of the dispute. "A postage stamp mosaic of them across the country will increase the opportunities for each of these corrupt government officials to solicit bribes from each of the research groups. . . . It's a simple matter of economics."

That fall, White and Desmond Clark sat down for an interview with an oral historian at Berkeley. Clark, the old lion of African archeology, was nearing the end of his life and recorded his memories for the sake of posterity, sometimes joined by his colleagues. White was scheduled to depart for Ethiopia in a few days, and the new regulations weighed heavily on his mind.

"They want to charge us $25 a day to go in and clean the goddamn fossils that we have found," White explained.

"What?" gasped Clark.

"And they want to charge our Ethiopian colleagues the equivalent of $100 a day for them on their salaries to go in and study the fossils that they have found."

"Oh no!" Clark exclaimed.

"Of course, these things will be waived on an individual basis [for] those individuals who kick a few bucks to the son of the antiquities director, or whatever he suggests," continued White. "This is how the corruption works."

"In the earlier days it was never like that. Never," observed Clark. "In the good old bad system of colonial times, it was a damn sight easier to get into the field, do the lab work, and so on. And much less expensive."

"But those days are gone," said White, "and they're not coming back."

White saw the matter—as he tended to see many things—as a battle for the soul of the science. He lived in a Manichean world of divisions: right versus wrong, scientists versus careerists, friends versus enemies, capable people versus incompetents. In his eyes, his Ethiopian colleagues were fighting to protect the integrity of science in their country—and their own independence. "If they lose this battle to the incompetent, greedy, selfish faux scientists that are supported by these colonialist institutions," he warned, "then you just turn the lights out."

The machinations of the Ethiopian government often remain inscrutable to outsiders. "It's a Byzantine process," observed former U.S. ambassador David Shinn. "The highland Ethiopians are very secretive, very suspicious of foreigners." Even so, some clues can be found in documents that leaked from the ruling party and were published by exile groups; they revealed an explicit strategy to coerce international organizations and the intelligentsia into compliance with the state vision of "revolutionary democracy." These papers outlined how foreign entities and their Ethiopian representatives would be "pressurized by legal instruments to toe the line." The state would "pile up the tax burden" on opponents and neutralize opposition by emptying "its belly and its pocket." For example, international aid organizations encountered restrictions when their missions did not align with the agenda of the ruling party. By the same token, the ruling party would reduce levies on loyal supporters. But this purist ideology was not always put in practice by frontline bureaucrats: party leaders complained about a deep-rooted culture of self-enrichment among government functionaries. Another factor was the bureaucratic culture. Regimes and ideologies came and went, but the bureaucracy endured and still bore vestiges of the emperor's feudal system. High officials faced little accountability, and the individual often lay at the mercy of the system. Low-level civil servants deferred to superiors and avoided risks that might jeopardize their jobs. Even simple transactions might pass through multiple offices, any one of which might halt the process because of some missing requirement or recalcitrant official. A permit application could turn into endless bureaucratic purgatory, or a word from above could expedite a stalled process—a potential instrument of punishment or reward.

Did any of those factors play roles in the antiquities disputes? Or was it just a low-level standoff between headstrong antiquities officials and headstrong civilian scientists? Those who really knew would not say.

IN THE FALL OF 2000, THE ETHIOPIAN PRIME MINISTER CAME TO THE UNITED States and met with a group of researchers at the Ethiopian embassy in Washington, D.C. Four Middle Awash scientists attended: Giday WoldeGabriel, Tim White, Clark Howell, and John Yellen (the NSF official who helped Desmond Clark launch the Middle Awash mission later joined the team as an archeologist). At the embassy door, the guards told Giday the reception was for scientists, not Ethiopians. He explained that he was both. The group entered and Giday spotted a familiar figure with a balding head, bushy eyebrows, and round spectacles. It was his old schoolmate Meles Zenawi.

When Giday had been at the university during the revolution, a group of other students from his home province, Tigray, drafted a political manifesto, slipped out of Addis Ababa, and headed to the northern mountains to form a guerrilla force against the Derg regime. One of the first to vanish was Seyoum Mesfin, who later led the rebels' foreign bureau. Soon the revolutionaries were joined by medical student Legesse Zenawi, who adopted the nom de guerre Meles, after a fallen comrade. The tenderfoot intellectuals trained themselves in military tactics, recruited peasants, captured weapons, and began hit-and-run raids on police stations, jails, and banks.

Over the next sixteen years, they grew into the Tigrayan People's Liberation Front (TPLF), one of the most powerful armies in the alliance that overthrew the Derg government in 1991. After the civil war, the thirty-six-year-old Meles became the prime minister of Ethiopia and youngest head of state in Africa. His old comrade Seyoum Mesfin became the Ethiopian foreign minister. Another TPLF veteran, Berhane Gebre Christos, became the ambassador to the United States. All three sat in attendance at the 2000 embassy meeting. Among all the pale American scientists, Prime Minister Meles espied his old friend. *Giday!* He greeted the scientist warmly and they embraced.

By all accounts, Meles was brilliant—a voracious reader and clever strategist who outmaneuvered rivals again and again. The *New York Times* once described him as "a man who could see around corners." Ambassador Shinn remembered the prime minister as "one of the smartest guys I've ever met." Another ambassador, Donald Yamamoto, cabled the U.S. State Department: "Whenever we raise concerns, he responds with highly nuanced and highly specific details to counter our arguments. . . . On numerous occasions we have observed Meles run circles around visitors." Giday had similar memories: "You could never corner him."

More than any other figure, Meles was the architect of modern Ethiopia—and reviled by political opponents for his party's viselike grip. Yet Meles did not fit the stereotype of an African strongman. He was soft-spoken, cerebral, and austere. Personal enrichment did not seem to interest him, but economic development did. (In his first press conference as head of state, Meles expressed the simple goal of allowing Ethiopians to eat three meals a day.) With that end in mind, his party kept firm control over the economy and politics of the country. The former Marxist assiduously courted allies in the west, which brought him to Washington to discuss the war with Eritrea, economic development, security in the volatile Horn of Africa—and to this embassy meeting to discuss research in his country.

When the meeting opened to questions, Meles heard a clear distress signal from the Middle Awash scientists about the antiquities bureaucracy. "To get a permit takes months and months—that's what we complained about," recalled Giday. "People spend more time in Addis instead of the field." The Middle Awash scientists warned that the country's new antiquities regulations ran contrary to the interests of science and Ethiopia. In the audience, Don Johanson watched the exchange and saw White hand over a packet of documents. ("He got up to give them to the prime minister and the bodyguards got really prickly," recalled Johanson. White dismissed Johanson's account as "simply fiction.") White recounted specifically pressing Meles about corruption. By his account, Meles initially bristled defensively but ultimately urged foreign researchers to resist any pressure for improper payment, adding forcefully, "Not one nickel!" Yellen recalled the prime minister's message: "He wanted to get rid of corruption and bribery. He wanted researchers to go along with that."

THE SCIENTISTS HAD NOT ONLY DARED TO CRITICIZE THE BUREAUCRATS who controlled their fate—they had done so in front of the Ethiopian prime minister, foreign minister, and ambassador to the USA. ("I'm sure that got back to Addis," observed Johanson.) The team soon felt the consequences.

One month later in October 2000, Jara demanded an inventory of all the Middle Awash project's equipment and keys to its vehicles—an ominous sign because the new regulations outlined how such possessions could be seized by the government. The team explained that everything was owned by the sponsoring institution, the University of California at Berkeley. Nonetheless, in February 2001, Jara cited the new directives and asserted the right to use the team's equipment outside the field season. When the team failed to comply, Jara suspended the Middle Awash research permit the following week because "you are not willing to respect the law of the land."

The scientists inquired whether the move was retribution for their critique of the new directives. For two months, they heard nothing. Meanwhile, an anonymous poison-pen critic launched an attack on the Middle Awash team on an Ethiopian news website. The free press had emerged as a new phenomenon since the change of government in Ethiopia, and factions sometimes pursued political agendas via character assassination and wild accusation. Berhane engaged his adversary in a back-and-forth that aired the dirty laundry of paleo politics for all the world to see. Berhane used the stage to repeat his complaints that Jara had helped Johanson regain his research permit. "It is unbelievable that a person empowered to safeguard Ethiopia's national treasures and regulate related research would decide to undermine his own institution and country," charged Berhane. "Rather, he stood on behalf of the IHO and its leader Donald Johanson, who had broken laws of the country for his personal gain." In April, Jara aired his own grievances in the newspaper *Addis Zemen:* "The Middle Awash project led by Tim White wants to monopolize the area and has been denying access to other researchers." He echoed the growing impatience about the Ardi skeleton—still unpublished seven years after its discovery. "These expansionist people," he complained, "are against the timely research and publication of *ramidus*."

Even as he faulted the Ardi team for failing to share its discoveries, Jara claimed that they had produced almost nothing of scientific value in the first place. "This group who claims to have undertaken field research in the Middle Awash for the last twenty years has only contributed defamatory allegations and never accomplishes research results important or worth mentioning," Jara declared, "and its contribution in promoting our country to the world is very minimal."

The scientists wrote to Woldemichael Chemu, the minister of information and culture—who oversaw the ARCCH and thus served as Jara's boss—and asked him to intervene. The minister urged them to apologize. The scientists refused. In late April 2001,

Jara wrote to say the team's permit was no longer suspended—it was *revoked*. The Middle Awash scientists were barred from the lab and field, and their vehicles and equipment were impounded at the museum.

Under the new regulations, the team had thirty days to appeal to the minister. Once the revocation became final, the government could assert the right to seize their possessions, carve up the project area, distribute territory to other research teams, and hand their fossils to rival scientists. The scientists hoped that the public dispute would oblige senior federal officials to intervene, but White lamented that their woes "keep falling below the radar of the folks who need to know in order to clean things up." A dustup over old bones ranked as a low priority for a government preoccupied with feeding its growing population, war with Eritrea, and restive opposition.

In addition, a power struggle was raging in the government at that moment. Meles had initiated a reexamination of the ruling party and had warned that corrupt elites threatened to turn the state into an instrument of personal enrichment. He ignited a ferocious pushback from opponents who nearly drove him from power. Yet again, Meles proved a wily tactician. Opponents were rounded up and arrested. The secretive nature of party politics made it difficult for outsiders to determine whether the purge was based on real misconduct or punishment for disloyalty. Meles warned of selling out to foreign interests—something that the Middle Awash scientists echoed in the realm of prehistory research.

The Middle Awash team fought for its life and sent letters to anybody who might help—the prime minister, the foreign minister, embassies, the National Science Foundation, the World Bank, and more. They found one ally who had a direct channel to the highest echelons of the Ethiopian government. American Paul Henze was a prominent and somewhat mysterious member of the class of scholars known as "Ethiopianists." His official biography identified him as a former diplomat, but in fact he was a veteran intelligence officer. In the twilight years of Haile Selassie's regime, he had served as CIA station chief in Ethiopia. Working under diplomatic cover, Henze became fascinated by Ethiopian history. He learned Amharic, explored the countryside in a Land Rover, slept in cheap hotels, camped in the open air, and trekked mountains with caravans of horses and mules. "I can't even remember a weekend he stayed in Addis Ababa," recalled his daughter, Libet Henze. "He was out in the countryside all the time." Henze left Ethiopia two years before the downfall of the emperor, and some of his old Ethiopian friends were liquidated by the Derg, leaving him with a deep disgust for the former Mengistu regime. During the Carter administration, he served on the staff of the U.S. National Security Council and pushed the Voice of America to begin broadcasts in Amharic. After retiring from the CIA, Henze continued writing books and journal articles about the country and during the civil war he met Meles in Washington (the savvy revolutionary recognized the old American spy as an influential voice on U.S. policy). Meles sat barefoot and chain-smoked, occasionally leaping to his feet when excited, and impressed Henze with his intellect and candor. With the overthrow of the Derg, that barefoot revolutionary had become the head of state, and the two men remained lifelong friends. Meles told Henze he was "always very welcome" in Ethiopia. Other scholars regarded Henze with some wariness due to his tendency to act as a cheerleader for the Meles government and turn a blind eye to its authoritarian impulses. Taking a realpolitik approach, Henze saw the current government as the best option for restoring stability, and repeatedly urged it to clean up petty corruption.

As a longtime observer of Ethiopia, Henze took keen interest in human origins research and befriended White. After learning of the fossil team's latest troubles, Henze e-mailed the scientist under the subject heading "Appalling situation!" He promised to help—and he knew where to pull the levers of power. Henze advised White that the "confrontation needs to be elevated to the highest level possible."

Over the next few months, Henze tapped his contacts in the Ethiopian government and pressed its embassy about the plight of the scientific team. Henze had enough rapport with the Ethiopian embassy to call upon Ambassador Berhane Gebre Christos at home and sit at the kitchen table to talk candidly about politics. Henze assured White that the Ethiopian diplomats "gave every impression of being deeply concerned about your problems" and sent "a strong communication to Addis Ababa." Ethiopia sought to bolster support in the USA to counteract dissident exiles who were denouncing political repression and human rights abuses (Washington, D.C., has the world's second-largest population of Ethiopians after only Addis Ababa). The embassy recognized that the paleo disputes undermined Ethiopia's prospects for NSF-funded research and its international reputation. Paleoanthropology brought Ethiopia headlines that were blessedly apolitical (the people involved had been dead for millions of years). Ethnic divisions were the powder keg of Ethiopian politics, but human origins brought a message of common ancestry for all humankind. In the arena of the deep past, Ethiopia could become a world leader just as it had in distance running. Case in point: two years earlier, a fossil skull from the Middle Awash had made the cover of *Time* magazine. "Foreign Minister Seyoum appreciated our work because he was seeing all kinds of publications—the cover of *Newsweek,* the cover of *Time* magazine, and all those kinds of things," recounted Berhane. "If the foreign minister knows about it, Meles definitely knows about it."

In the heat of the permit dispute, the scientific team brought more good press to Ethiopia. In July 2001, the team published *kadabba,* the fossils of the so-called grandfather of Ardi (it had been kept under wraps in the four years since its discovery). Typically such papers listed multiple authors, but this one in *Nature* credited Yohannes Haile-Selassie as sole author. The world thus witnessed an unfamiliar sight—an international discovery announced by an African scientist. Left unsaid: he remained locked out of the museum where those same fossils were stored.

The scientists did not publicly call attention to the irony of an Ethiopian celebrated around the world but blacklisted at home. Instead, they pursued a tacit strategy of letting discerning people recognize an injustice. White urged reporters to focus on the science and ignore the political disputes. In the back channels, Henze was more direct. "The position of the Culture Ministry in Addis Ababa is shockingly negative," he complained by e-mail to Brook Hailu, an Ethiopian embassy official. "Nevertheless it is disgraceful and contrary to Ethiopia's best interests. The announcement of these finds is the MOST POSITIVE PUBLIC RELATIONS DEVELOPMENT Ethiopia has experienced in many years—the government should capitalize on it as fully as possible." Henze urged the embassy to immediately bring the matter to the attention of the prime minister. As Henze astutely observed, the international press attention protected the scientific team, and any additional attacks by their bureaucratic adversaries would only appear "petty and obstructive."

On the day the *Nature* article went public, White visited Woldemichael Chemu, the minister of information and culture and the official who held the project's fate in his hands. Pointing to the news stories, White urged the minister to view *kadabba* as a point of national pride. He asked for a ruling on the team's appeal, but the minister only referred him back to Jara. Meanwhile, Jara's agency issued its own bizarre press release saying it had no knowledge of *kadabba* and no relationship with its discoverers, and it reminded the public that the Middle Awash permit had been cancelled.

But an appeal from the old-boy network could penetrate where outsiders could not. Having grown up a Tigrayan peasant, Giday WoldeGabriel shared a personal background with the leaders of the ruling party and intuitively understood its mission and how his science could serve it. They also shared ties from attending the country's elite national university during the cataclysmic revolution. Giday flew to Ethiopia and met with a senior Ethiopian dignitary (he declined to publicly identify which one) who had attended the embassy meeting. In an informal chat over coffee, the geologist explained the predicament. According to Giday, his contact assured him that top officials were aware of the problem and promised to remedy it. "He said, 'Don't worry,'" recalled Giday. "That was the end of it."

Later that summer, Chemu wrote again, but the researchers could not decipher the minister's letter, which seemed to both defend the permit revocation and hint that research could resume. As it turned out, the permit was restored, but too late to salvage the next field season. Not long afterward, the government appointed a new minister. The most controversial provisions of the new antiquities rules were never enforced. Jara remained in office but had discovered the limits of his power over the African wildcats of paleoanthropology. "They made his life hell," groaned Don Johanson. "Jara thought he was going to have a nervous breakdown." The Middle Awash team had lost most of a year and a season of fieldwork, but survived. "Many people don't like us in Ethiopia, especially at the ministry, for being vocal," said Giday. "It's a continuous fight, but people think twice before they cross us."

Chapter 19
BITING THE HAND

The Ardi team lived by an old dictum: science was not a popularity contest. One example concerned the hand. From the earliest days, the Ardi investigators had known that their skeleton showed no signs of knuckle walking. In 2000, the team was startled by a cover story in *Nature* under the headline "Evidence That Humans Evolved from a Knuckle-Walking Ancestor." The authors, Brian Richmond and David Strait of George Washington University, had employed "video based morphometrics" on two *Australopithecus* fossil casts and concluded that the measurements overlapped with modern knuckle-walking apes. "Between you and me," White complained to *Nature* editor Henry Gee, "publishing that thing as a cover was the worst thing that Nature has contributed to the field of hominids since the 'Flipperpithecus' dolphin rib-cum-hominoid clavicle." White expressed alarm that "flawed evidence and interpretation" would pass unchallenged into textbooks, and for months he hounded both authors and editors with questions, critiques about the study methodologies, and threats of exposure. Eventually, Gee begged off, handed off the problem to subeditors, and asked not to be copied on the e-mail correspondence. (The journal appointed a referee who supported the authors.) In military lingo, the clash was an asymmetric engagement: the Ardi team had privileged access to the oldest skeleton in the human family while their adversaries did not. In the heat of the dispute, Richmond and Strait noted that their prosecutor could resolve the matter by producing the best evidence available—the secret *Ardipithecus*. But the Ardi team still refused to reveal its hand.

AT THE TURN OF THE CENTURY, THE EDITORS OF THE *AMERICAN JOURNAL OF* *Physical Anthropology*, the flagship journal of the profession, asked a few prominent scientists to reflect on the state of their disciplines. Most made polite observations until White got his turn. "Something is wrong," he warned. "It is painfully obvious that most professional paleoanthropologists, like too many politicians, are unable to see it—or unwilling to change it."

He divided the field between *real* scientists and careerists. In his jaundiced eye, the rewards system had been distorted by publicity-hungry scientists and growth-seeking universities. He decried widespread ignorance of the core disciplines of paleoanthropology and questioned a large shovelful of published claims that he deemed dubious. "Such irresponsible proclamations momentarily seize the public's attention in popular news and go straight into textbooks," he complained. "The retractions rarely do."

He didn't have to list names. Among his many examples, peers easily could recognize colleagues—or themselves. "Armchair commentators abound," White wrote. "Actual producers of fossil data are increasingly rare." He saw anthropology replaying the classic tragedy of the commons with too many people competing for too few resources. The solution? White called for a halt to growth—no more herding students into graduate programs, flooding academia with unqualified people, and polluting scientific literature with shoddy work. Having written off the senior ranks of his profession, he appealed to the younger generation with a series of edicts that came to be known as "The Tim Commandments":

> In the field, do not think you're going to step out of the vehicle and pick up a hominid within 20 meters on the first day. Do not claim that you found the fossil if the other person did. The truth will eventually, some day, come out. Do not purchase fossils. Do not bribe officials. Do not steal another person's site, particularly when that person is a local scholar in a developing country . . .
>
> Do not let your ambition distort your ethics. If your career goal is to make a fortune, then go to medical school, become a knee surgeon, and practice on suburban soccer players. But get malpractice insurance; clinical performance will matter, and your errors will have real consequences for your future.

More than any single episode, his millennium manifesto signaled a rift between scientific camps. White lambasted colleagues who might be asked to peer-review his work, the journals that published it, and even the NSF—the agency that bankrolled his research. The scolding practically invited opponents to await any transgression, any error of fact or interpretation, that would show the self-righteous accuser wasn't so infallible himself. His vision of science would be embodied in Ardi, and enemies would respond in kind.

Chapter 20
IN SUSPENSE

Years passed. The Middle Awash team published several major discoveries, including *Ardipithecus kadabba, Australopithecus garhi,* and a comprehensive paper on the geology of Aramis but revealed nothing about what many people really wanted to know about—Ardi. Everybody knew they had the world's oldest skeleton. What the world didn't understand was that parts of the so-called Rosetta Stone still bore a disconcerting resemblance to a heap of rubble. One of the most basic tasks was determining the limb lengths and body proportions—all of which required painstaking detective work and caused an aftershock years after Ardi came out of the ground.

LIMB PROPORTIONS HAVE LONG FASCINATED ANATOMISTS—AND MISLED them. The Greeks and Romans idealized the anatomical ratios of the human body. Marcus Vitruvius, a Roman architect in the first century B.C.E., observed that the height of a human is equal to the span of the outstretched arms. In other words, when we stand in a T-position, our stature and wingspan supposedly form a perfect square. Vitruvius also observed that the outstretched arms and legs touched the circumference of a circle centered on the navel. Leonardo da Vinci immortalized this idea in his classic depiction of the

Vitruvian Man. The Renaissance Man was literally a well-rounded figure, or a perfect square, depending on how you spread him out.

Then a diligent investigator screwed up this enchanting story. In the mid-nineteenth century, English sculptor Joseph Bonomi measured eighty-four people and found that only six were perfect squares; in the vast majority of people, arm span exceeded height. In the twentieth century, the science of anthropometry (the measure of humans) boomed, and the myth of divine proportions was crushed by data. In reality, body proportions reflect evolutionary heritage—particularly adaptation to different styles of locomotion. Humans, apes, and monkeys share the same limb bones, and evolution has tinkered with these common elements to produce variations on the primate theme.

All other great apes are classified as "suspensory" primates because their long arms and long hands are interpreted as adaptations to swinging below branches. Orangutans are the most arboreal of the great apes and very well armed: their forelimbs are

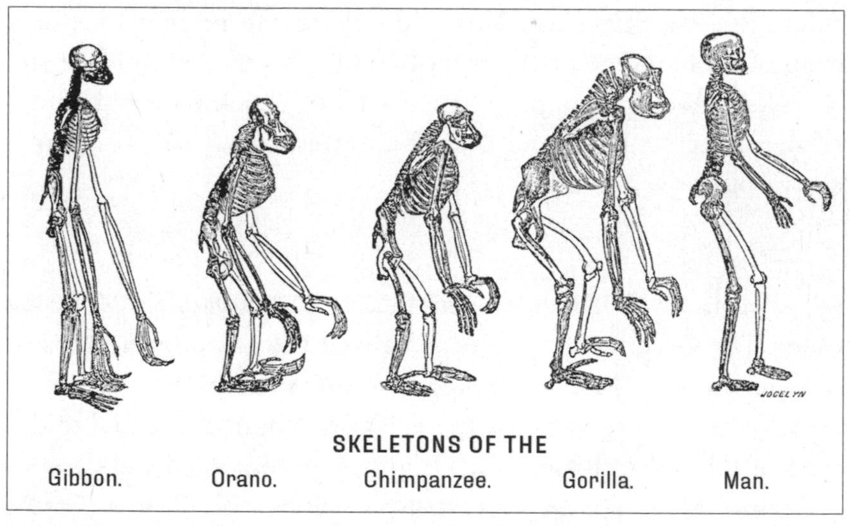

A classic illustration of ape skeletons from the 1863 book *Evidence as to Man's Place in Nature* by Thomas Henry Huxley. Note the long-armed suspensory gibbon on the left. All our cousin apes have longer forelimbs than hindlimbs. Humans are unusual because we have longer legs than arms. Despite being arboreal, Ardi also turned out to have legs slightly longer than arms.

about 40 percent longer than hindlimbs; when orangutans walk on the ground, their arms look like crutches. Even gorillas, the most terrestrial of our cousin apes, have arms about 16 percent longer than legs. Likewise, chimps have forelimbs slightly longer than their legs. In contrast, humans are leggy, with forelimbs only about 70 percent the length of our hindlimbs.

Scholars had long viewed the descent of humanity as a long-armed ape turning into a long-legged human. Lucy seemed to confirm this view because she had arms 85 percent as long as legs, proportions roughly midway between a chimp and a human. The Ardi investigators had every reason to expect their older skeleton would be closer to a long-armed ape. In the lab, White delighted in holding up Ardi's limb bones beside his own body. Despite standing only about four feet tall, Ardi had a forearm about as long as his own. Her shinbone, however, seemed much shorter than his. The early impression seemed consistent with the prediction: long arms and short legs.

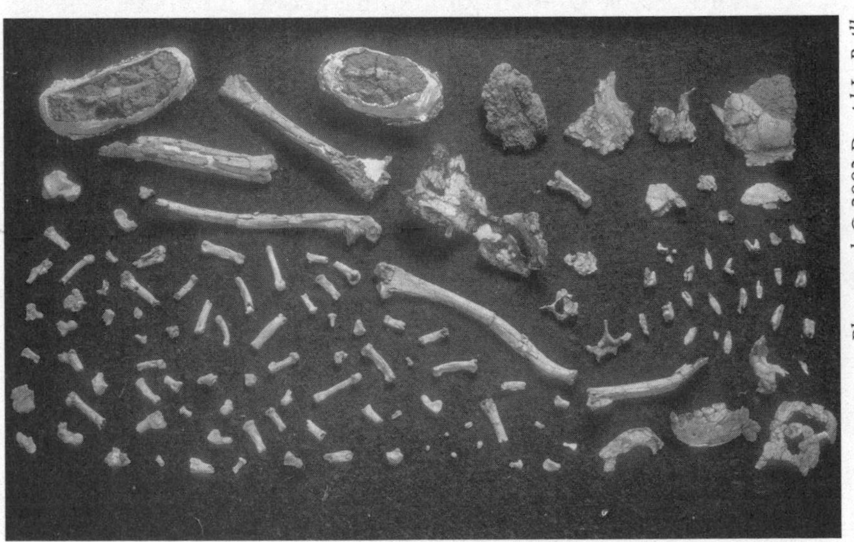

A puzzle awaiting assembly: the pieces of the Ardi skeleton in 2003.

ARDI DID NOT REVEAL HER SECRETS EASILY. THE SKELETON INCLUDED A mostly complete radius and ulna (two bones of the forearm) but the humerus (upper arm bone) remained missing. Luckily, the team recovered a humerus from another *Ar. ramidus* individual that could be scaled to Ardi's size.

In the leg, Ardi had only a portion of the femur (thighbone) and a long section of the tibia (shinbone) lacking its far end where it joined the ankle. Stuck with only partial bones, the investigators were forced to estimate the original lengths from other clues. Bones bear anatomical landmarks such as roughened surfaces for attachments of muscles and ligaments, which often are preserved in fine detail on fossils and provide hints about original lengths. For example, White spotted a ligament marking on Ardi's tibia. He laid Ardi's tibia beside the matching bone of another skeleton, AL-288, otherwise known as Lucy. When the two landmarks were aligned, it became obvious Ardi wasn't missing just a small piece of her shinbone but a *huge* chunk—its distal end portion, known as the medial malleolus, where it joins the ankle.

A more precise comparison required some numbers. The scientists collected measurements from hundreds of apes and fossil human ancestors and calculated regressions in order to estimate the lengths of Ardi's limbs, often taking bearings off anatomical landmarks to infer the sizes of missing sections. For the sake of accuracy, White, Lovejoy, and Suwa each made independent estimates. After months of crunching data, Ardi's arms turned out to be *shorter* than her legs. One basic comparison is the "intermembral index," the ratio of the long bones of the arm (humerus plus radius) divided by those of the leg (femur plus tibia). Ardi's forelimbs were only about 90 percent as long as her hindlimbs. Oddly, her limb proportions were very close to those of Lucy. Ardi was supposed to be getting closer to our common ancestor with chimps—but her proportions didn't look much like modern apes.

THAT DISCOVERY CHALLENGED SOME VERY OLD IDEAS. IN 1889, A YOUNG Scottish doctor named Arthur Keith boarded a ship bound for

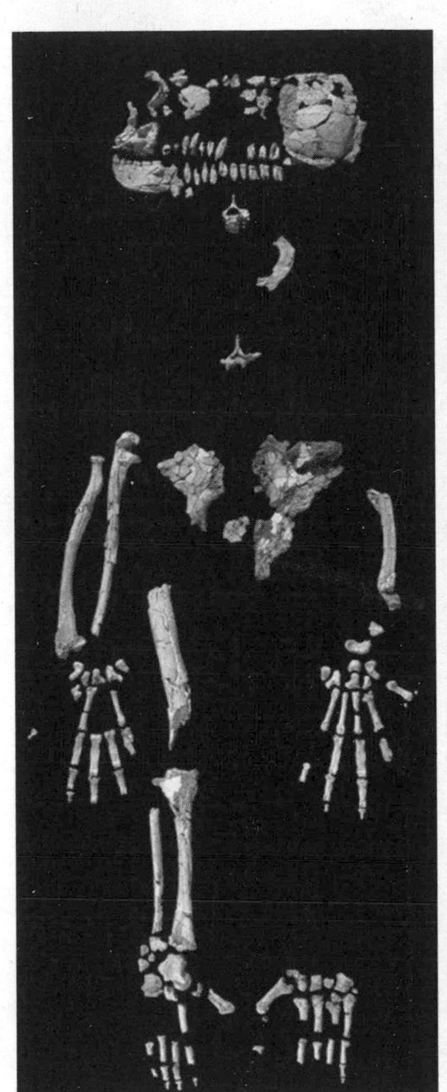

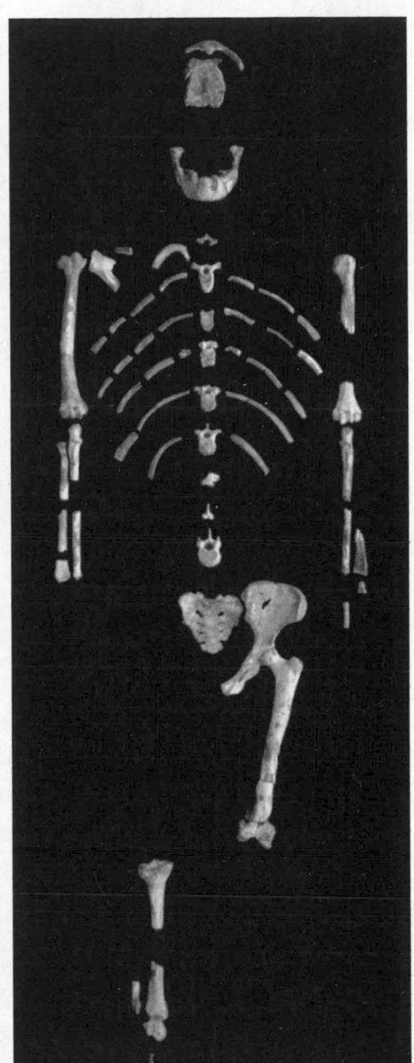

The Ardi (*left*) and Lucy (*right*) skeletons side by side.

Southeast Asia to take a job as a medical officer with a gold mining company in the Malay Peninsula of Siam (now Thailand). In his spare time at the mining camp, Keith hiked into the tropical jungle and marveled over all the exotic plants and animals, especially the agile monkeys scampering overhead. One creature entranced him:

a yellow primate that did not walk atop branches on all fours like monkeys and instead swung below the branches like a trapeze artist. Sometimes it strode upright atop branches while holding out long arms like a tightrope walker with a balance pole. Eventually, Keith realized these animals were not monkeys but gibbons—a species of so-called lesser apes.

Back at camp, malaria was killing an alarming number of mine workers, and the doctor began to wonder if the jungle animals also carried the disease. Keith shot a monkey and checked its spleen for infection, and the inquiry rekindled his interest in comparative anatomy. From bamboo, he built a laboratory raised on posts and spent days and nights at his dissection table cutting monkeys and gibbons. In medical school, Keith had been taught that the unique aspects of human anatomy were adaptations to upright walking. To his amazement, he discovered that gibbons actually possessed many of these same traits—despite being arboreal.

Keith returned to Britain, took an anatomy job at a hospital, dissected more primates from the London Zoological Gardens—and, like other evolutionists before him, saw that apes from Africa were even more similar to humans than those from Asia. In 1923, he gave a series of lectures at the Royal College of Surgeons and outlined a new theory of human origins: the radical shift to upright posture began as an adaptation to climbing—not walking. Old-world monkeys generally move with spines parallel to the ground in *pronograde* posture. By contrast, apes climb with upright trunks, or *orthograde* posture. In Keith's theory, the "great break" in human form occurred when pronograde monkeys became orthograde apes. In other words, erect posture began in the trees because our ancestors suspended below branches—just like those arm-swinging gibbons he had seen in Asia (gibbons' forelimbs are about 30 percent longer than their hindlimbs). In a series of publications, he theorized that humans descended from such a long-armed "brachiating" ancestor, an idea that persists in many evolutionary theories to this day. (In current classification, brachiation is a form of suspension.)

That notion helped inspire a major expedition to Asia. Harold Jefferson Coolidge was a Boston Brahmin who worked as curator of the Harvard Museum of Comparative Zoology. Between the two world wars, he organized several safaris to Africa and Asia in what we might view as the last huzzah of the great white hunters with khakis, pith helmets, and native gun bearers. In 1937, he organized the Asiatic Primate Expedition (cleverly abbreviated APE), which Coolidge touted as a quest for "our gibbonlike ancestors." The expedition provides a glimpse of the wholesale slaughter that produced comparative museum collections in the days before protections on now-endangered apes. In Siam and Borneo, hunters gunned down nearly three hundred gibbons plus many other primates. The carcasses were carried to the jungle dissecting hut of Professor Adolph Schultz, a legendary figure in anatomy. Speaking in a thick German accent, the anatomist worked with a cigarette dangling from his mouth and sometimes a glass of gin by his side as he oversaw a ghoulish assembly line where hundreds of animals were systematically cut into pieces. Schultz devoted his career to the single-minded collection of anatomical data—an undertaking that required huge numbers of specimens. To accomplish this, Schultz took dead apes from circuses, zoos, and pet dealers. Back in Baltimore, irate shipping clerks phoned his lab at Johns Hopkins University and demanded the anatomist immediately pick up putrid-smelling packages. During Prohibition, government agents stopped a suspicious courier at a railway station, opened the package before a crowd of onlookers, and were horrified to discover not booze but a dead monkey bound for Schultz's dissecting table.

Schultz preferred animals shot in the wild—the more the better. Like many anatomists in the days before computers, he made detailed sketches with the hand of an artist—an endeavor that required detailed observation—and his comparative anatomy drawings remain classic images to this day. In his jungle lab, he kept assistants busy day and night, including Sherwood Washburn, then just beginning his career. "Every day was a seminar with Schultz," Washburn recalled. But master and apprentice developed very different views of our family history. Based on his anatomical investigations, Schultz believed that humans separated from the other apes very deep in the past *before* our cousin species developed their long-armed specializations for suspending below branches—in other words, he didn't buy Keith's theory. Washburn, however, came to believe that our lineage branched off much more recently from an ancestor that *already* was a long-armed suspensory ape—one similar to living chimps and gorillas.

By the 1960s, Washburn had become the dean of American anthropology. Nonetheless, his views about a brachiating ancestor put him in a minority camp because most scholars had been convinced by the arguments of anatomists like Schultz. Even Arthur Keith, the original champion of the idea, abandoned his theory in old age. Washburn persisted as a forceful, but lonely, advocate for the notion that humans descended from arm-swinging apes only a few million years ago.

With the molecular revolution, the pendulum of opinion swung back in Washburn's favor. Many came to share the belief that humans descended from long-armed suspensory apes. Similarly, this school of thought held that our ancestors became upright to climb long before they walked erect. Ardi would test that theory—and provide the strongest challenge to it yet.

The question of whether or not the common ancestor of humans and African apes had a suspensory body plan is not just anatomical trivia; rather, it has far-reaching implications for understanding our evolution from head to toe. Similarly, the question of whether that ancestor had upright or horizontal posture is a debate between two radically different ancestral body plans. Ardi's limb proportions were reminiscent of some fossil apes from the Miocene—pronograde creatures with arms and legs close to equal length, and unlike any ape alive today.

• • •

THE WAIT FOR *ARDIPITHECUS* BEGAN TO STRAIN THE RELATIONSHIP BE-tween the Middle Awash team and the National Science Foundation, its main funding agency. Back in 1993, the Middle Awash proposal was rated "excellent" by all members of an NSF review panel and was ranked highest among fifty proposals. Three years later, reviewers again gave strong endorsement. One peer called it "a flagship project" and gushed: "This is how paleoanthropological field work should be done!" By the end of the decade, however, reviewers began to express impatience about the unpublished discoveries. In February 2000, Mark Weiss, the NSF program director for physical anthropology, informed White that the foundation would limit his grant to three years instead of five. He explained that scientific peers had become impatient. "The fossils are of very limited utility until they become accessible," he told White. White became angry and insisted the NSF panel had been misled. All but one reviewer rated the proposal as excellent, and the lowest rating was "good"—and White dismissed that last one as "uninformed, clearly hostile." His elderly colleague Clark Howell urged him to calm down and see the reviews in a more positive light because they were "overwhelmingly enthusiastic." With budgets already trimmed since his previous grant, White played brinksmanship and his proposals warned that reductions in funding would "jeopardize the entire research program." Lucy and *Australopithecus afarensis* were published in a suite of papers eight years after the discovery of the skeleton, but White insisted on comparing Ardi to different standards—including the comprehensive scholarly monograph for *Homo habilis*, which took twenty-seven years. "We will not publish before the research is ready," he vowed. "If we run out of money it will delay it further."

The dispute brought him to shouting, even years afterward. "He would call up on occasion, very upset," recalled Weiss. "I could understand that, but I had to pay attention to things other than

Tim's concerns." White urged the NSF to overrule its own peer reviewers. His insistence paid off: the NSF extended the Middle Awash grant in additional installments year after year—more than $630,000 over seven years.

Between 1982 and 2007, the NSF awarded White about $2 million for fieldwork in the Middle Awash plus another $2.5 million for a separate consortium with other research teams. Even as he expressed gratitude for the support, he publicly criticized the agency for investing too much on graduate training programs and too little on field research. "A few people go out and risk their necks and spend the time to get the fossils," White vented in 2000. "The people who sit in the labs—they're the 95 percent—they're doing the reviewing of grants. They don't understand the field work, but they do understand one thing—the sooner they can get their hands on the fossils that are found the greater the chance there is that they will be able to extract some analytical information out of them, which they can then publish and enhance their own career. And so they dedicate themselves to trying to take fossils out of the hands of those who find them."

ONE MORNING IN FEBRUARY 2002, BERHANE ARRIVED AT THE ETHIOPIAN National Museum and found Alemu Ademassu, the keeper of the hominid fossils, waiting anxiously with a letter in his hand. The letter authorized two foreign researchers to examine Ardi and other still-secret Middle Awash fossils. At that very moment, the researchers sat in the office of the museum director awaiting final clearance to see the fossils. Berhane gave the keeper firm instructions: "I told him no way."

Berhane hurried to the office and found two familiar faces: Ian Tattersall of the American Museum of Natural History in New York and Jeffrey Schwartz of the University of Pittsburgh. Tattersall and Schwartz were compiling a multivolume series on the human fossil record and had been traveling around the world to examine specimens. Two years earlier, Schwartz wrote White seeking permission to examine, describe, and photograph several already-published discoveries from the Middle Awash including *Ardipithecus*. It quickly became apparent that the two camps had different definitions of *published* enough to become available. In White's eyes, that meant comprehensive reporting, not initial announcements like the one that named the *ramidus* species back in 1994. He granted permission to see only fully described fossils—specimens discovered in the 1970s and 1980s. The correspondence ended politely, but privately Schwartz seethed: "I thought, what chutzpah!"

Tattersall and Schwartz decided to try again and bypass the discovery team. Don Johanson urged them to contact ARCCH director Jara Haile Mariam, who was certainly no friend of the Middle Awash team after all the recent bureaucratic battles. (Johanson also called their attention to the new antiquities laws that said discoveries that remained unpublished after five years could become open to other researchers.) Schwartz requested to see *Ardipithecus* and several other Middle Awash fossils, and Jara responded with a vaguely worded letter: "You can start your Laboratory study fulfilling the necessary requirements of the Authority." Schwartz and Tattersall took that as permission and flew to Ethiopia. At the museum, they presented their letter and asked to see the fossils inside the safe, including the mysterious Ardi. Alemu thought that sounded fishy and went to find Berhane.

To Berhane, such demands were offensive. To find the fossils, his team had *suffered*. And now these visitors presumed to harvest the fruits of the field team's labor? No, he would not abide. Berhane went to the office of museum director Mamitu Yilma (who occupied his former post) and argued that the visitors had no legal right to see the material. "Whatever they did behind closed doors, our permission was denied," said Schwartz. "Alemu wouldn't let us into the safe." The two visitors were furious. They had traveled halfway around the world and booked a long stay in the Addis

Hilton. "We spent a whole week twiddling our thumbs and blowing away our research money," fumed Schwartz, "just because of Tim!"

A reporter found Tattersall and Schwartz back at the Hilton sipping Chenin Blanc. They demanded to know why the Ardi team wouldn't produce evidence of the ancestor claimed to be at the root of the human lineage. The two scientists flew home and complained to anybody who would listen, including scientific peers, the National Science Foundation, and journalists such as Ann Gibbons, a reporter for *Science,* who wrote a long piece about fossil availability and cited the episode as the lead example of closed access.

For much of the twentieth century, paleoanthropologists formed a small coterie and negotiated access to fossils among themselves. In 1950, the United States produced only about two dozen anthropology and archeology Ph.D.s per year. By the 1970s, it produced more than four hundred Ph.D.s every year (and by 2013 more than six hundred). As these professionals flooded academia, competition increased for funding, fossils, and attention. Unfortunately, discoveries did not scale up at the same pace. Only a small fraction of anthropologists actually searched for specimens and even fewer found them. In the old days the science had been driven by discoveries in the field, but now it became driven increasingly by analysis in the lab. New specialties arose and people began to speak of "computer-assisted anthropology" and "virtual anthropology." The Internet age brought the expectation that data should be shared freely and widely. For example, the Human Genome Project—a global consortium transcribing our genetic code—put DNA sequences online even before scientific publication in order to maximize the benefit to society. Many aspired to bring this new ethos to old bones. Scholars cried for limits on the period of exclusive possession of fossil discoveries and for punishing alleged "hoarders" with denial of taxpayer funds, banishment from scientific journals, and ostracization from meetings. Gerard Weber of the University of Vienna called for a new era of fossil glasnost with a five-year limit on exclusive rights followed by open availability in a digital archive akin to that of the Human Genome Project. Weber wrote: "The paleoanthropologist walking with eyes fixed to the ground in the sunlight now has a respectable partner, who is to be found active in the computer lab gazing upon a phosphorescent screen."

Those who worked under the burning sun in the Middle Awash did not necessarily see technologists as *respectable* partners. White labeled them "computer jockeys." In his view, digital images were no substitute for originals and software was no replacement for expert judgment. No computer "modeling" could trump actual discovery. He rolled his eyes at the technological indulgence: fetishes of color graphics, bones rotating on computer screens, and beautiful presentations of mumbo jumbo. And he rejected the oft-used comparison to the Human Genome Project: fossils didn't constitute reliable data until they had been reconstructed, which often took years. Until then, a specimen was more like an unprocessed biological sample, not a sequenced genome. To him, the access campaigners suffered from a bad case of entitlement. *Glasnost?* Hardly. More like Napster, the file-sharing site notorious for copyright infringement.

In his eyes, the frame of the debate was all wrong. It was not a contest of greedy bone hoarders versus heroic data sharers. Rather, it was a battle over scarce resources. To White, the access cause really was a cloak for greedy careerists who wanted to swoop in and score quick payoffs. He felt compelled to defend his team—which included nearly fifty scientists from around the world—from being scooped by outsiders or treated like bone-fetching dogs.

Enraged, White fired off a fifty-five-page memo to Ann Gibbons. "Should the people who have never found or conducted primary work on hominid fossils be allowed to dictate what is best for the science?" he demanded. "Should they even be allowed to waste the time of the primary researchers by forcing them to respond to false accusations? A conspiracy theorist might cynically conclude that this is precisely their goal."

In this case, he was the conspiracy theorist. He viewed the museum episode as a setup—a sneak attack to embarrass the Middle Awash team, generate negative publicity, cut off funding, and ultimately kill the project. His suspicions were aroused because Tattersall and Schwartz had come to the museum for a scientific conference hosted by Horst Seidler and attended by numerous adversaries including Jara and Jon Kalb. "I know most of the 'collegial' sources of these allegations," he declared, "and I am familiar with all of their motives."

Such defiance—coupled with a growing collection of fossils—made the team lightning rods on the access issue. When they eventually published their volume in 2003, Schwartz and Tattersall (who examined fossils from sixty locales and managed to reach agreements with all but three teams) singled out the Middle Awash team for being among a small group of holdouts who kept discoveries "surrounded by a wall of curatorial protectionism that amounts almost to paranoia."

"What the describers are trying to hide," they wrote, "remains a matter for conjecture."

Shortly after the 2002 museum episode, White applied for another NSF grant with his emeritus colleague Clark Howell. The proposal sought $2.5 million for a collaborative investigation into early human evolution with several other research teams. The proposal was rejected, and one reviewer questioned the sincerity of White's "newly discovered spirit of inclusivity and eagerness to make new fossils available." Some noted that his proposal seemed oddly out of tune with his previously expressed opinions about "armchair theorists" and parasitic lab scientists. "Collaboration and the sharing of data do not require a grant," wrote another reviewer. "They require a decision by scholars about how to behave."

In 2003, White and Howell submitted another application to the NSF. This one proposed an international collaboration of thirty-three research teams—ranging from other fossil teams to primatologists who studied apes in the field—to investigate the pre-Ardi origins of the human family. Once again, some reviewers questioned his goodwill and voiced frustration over the continuing silence about *Ardipithecus*. One wrote: "Important hominid fossils, some recovered eleven years ago, are unavailable for study by bona fide researchers and their full publication is still awaited. To all intents and purposes the material collected, and the taxonomic and other hypotheses they generated, are not subject to normal scientific scrutiny. With this track record it is difficult to take the rhetoric of this proposal seriously."

Both the NSF and scientific journals faced a delicate challenge with the Middle Awash project because it was hard to find unbiased reviewers. White had pissed off a large portion of his profession—and vice versa—so his submissions typically included lists of prominent names who should *not* be allowed to judge his work (the list eventually grew to fill several pages). As before, White dogged the NSF officers, cajoled, and threatened exposure, and it paid off. In 2003, the NSF awarded $2.5 million to his consortium. Reviewers worried that the award would empower White to become a grant-distributing kingpin in his own right, so the NSF demanded more budget accountability and an advisory committee to oversee the project. The panel also expressed unease that data sharing would be left to the discretion of the investigators—a concern that White and Howell labeled as "unjustified, almost paranoid distrust."

WHILE THOSE CONFLICTS PLAYED OUT IN THE UNITED STATES, OTHER TENsions abated in Ethiopia. In June 2003, Berhane Asfaw stood before the international press corps, reached into a box lined with foam balls, and revealed a treasure that had been kept under wraps for six years—one of the Herto skulls. It was the debut of the then-oldest known anatomically modern *Homo sapiens*—the strangely tool-scarred skulls recovered during the floods six years earlier. "As a scientist I'd be happy if the skulls had been found anywhere," Berhane told the press, "but as a national I am proud to have found them here."

So was his government. The event—which coincided with a cover story in *Nature* and another wave of international media coverage—signaled a shift in the Middle Awash team's relationship with the host country. Beside Berhane sat the new minister of culture, Teshome Toga, a statistician and demographer who viewed science as a developmental tool, saw the fossil team as allies, and took great pride in the achievements of Ethiopian researchers. He expressed his honor at sharing the announcement of this ancient Ethiopian ancestor. The state Ministry of Information helped organize the press conference.

The government had taken a renewed interest in its national heritage. The state dusted off plans for a major expansion of the national museum. The old NSF-funded lab built in the previous decade could no longer hold all the fossils and artifacts pouring in from the country's many research sites, and in 2003 it was razed to make way for a six-story, $10 million structure. (The government capital expenditure dwarfed that of any single scientific team, but the state did not make plans to actually outfit the building, so researchers later stepped in to equip the labs and move the specimens.) Leftover funds from White's last NSF grant went toward outfitting the museum. The Middle Awash team had long allied themselves with the nation-building agenda and promised that developing Ethiopian personnel and facilities would take equal priority with the research itself. "When you are doing something on behalf of the country," explained Giday WoldeGabriel, "nobody can challenge you." His American partner also made clear his allegiance. In an address to a conference of Ethiopian government, business, and tourism leaders in early 2004, White warned about a certain class of foreigners bent on "hit-and-run projects aimed at collecting human ancestor fossils for publicity purposes" and those that sought to "exploit Ethiopia instead of building her capacity."

Those concerns played well in a state with a historic wariness of foreign exploitation and grand aspirations for its own development. (Prime Minister Meles envisioned an "Ethiopian Renaissance"—but not one beholden to the western neoliberal economic model.) White spoke to deep-rooted Ethiopian values: their national pride and sovereignty. Even as he alienated anthropologists back home, White demonstrated his fealty to the country that made his work possible and went so far as to suggest that Darwin himself would have been delighted by the emergence of Ethiopian scholars. "Ethiopia is now the leader in world paleoanthropology," he told the crowd, "and we will continue to bring this message to a world audience."

With peace on the bureaucratic frontier in Ethiopia, the Ardi team forged ahead with its odyssey into the natural history of the human body.

Chapter 21
UNDER THE RADAR

In December 2003, two men in business suits made their way through security at Bole International Airport in Addis Ababa. They closely guarded a blue suitcase and tried to look inconspicuous. They were transporting fossils out of the country.

The two men were Yonas Beyene and Gen Suwa. Their ordinary suitcase concealed an extraordinary cargo: for the first time in 4.4 million years, Ardi was leaving Ethiopia. With the antiquities battles mostly put to rest, the team secured permission from the government to export the *Ardipithecus* fossils to Gen Suwa's lab in Tokyo for further study. The trip posed a delicate matter—and not only because of the fragile bones. Berhane had complained about Seidler taking fossils to Vienna for scanning. (When the Ethiopian government sent the Lucy skeleton on a museum tour in the United States in 2007, the Ardi team and other scientists objected that the trip posed unnecessary risks to an irreplaceable specimen for little more than publicity.) In this instance, however, the fossil was their own discovery, the scans not for public distribution, the mission strictly for science and conducted discreetly. They traveled with no guards and followed a security strategy of simply blending with other business travelers as they dragged Ardi in a rolling suitcase and carefully stored her in the overhead bin of the airliner. "We were scared, both of us," said Yonas. "We took turns sleeping. Our eyes were always on the baggage compartment."

Upon landing, they took a train to the University Museum at the University of Tokyo. Before locking the fossils into the safe, they cracked open the suitcase. "There was not a single bone which was broken," said Yonas. "Everything was perfect as we packed it."

The trip marked a new phase of investigation. Traditional techniques had run into limits: the skull and pelvis were so fragile that White would not attempt further restoration. Instead, the investigators would illuminate the oldest skeleton with micro computed tomography (micro-CT), an imaging technology developed for industrial and medical purposes. Unlike traditional X-rays, micro-CT captured 3D images and allowed researchers to "electronically remove" rock matrix and reassemble pieces without risk. With it, the team also could examine inaccessible anatomical structures such as sinuses, bone cavities, tooth roots, dental enamel, and bony labyrinths of the inner ear. They could measure distances between anatomical landmarks, surface areas, and volumes.

Among scholars of human origins, Suwa stood out—or, more accurately, did *not* stand out—for his lack of visible hubris. He avoided press conferences because of the distasteful expectation to be hyperbolic. In reality, there would be less puffery about throwing out textbooks if people paid closer attention to the literature and fewer surprises if people carried fewer preconceptions. "Idea-renovating discoveries, even *ramidus,* is part of a cumulative process," Suwa explained. "The building blocks need to be put in place continuously and as accurately as possible. No one discovery is going to do it. You just have to just keep plugging it in."

With his painstaking attention to detail, Suwa was a walking database. "He has a photographic memory of every fossil he's ever seen," marveled Berhane Asfaw. "It's unbelievable. It's like he's transcribing a picture that he has in his brain." Suwa's contributions remained underappreciated by a large swath of his profession and highly valued by those who actually understood him. He didn't bludgeon, trap, or embarrass anybody. He just poured enough data into the barrel to drown the rat. "He has an extraordinary bullshit sensor," laughed Scott Simpson. "He *never* believes anything you tell him the first time!"

Suwa was the paleo version of the Japanese salaryman, a workaholic whose normal hours were double overtime. How many hours did he spend on the job? *Hmmm.* That question stumped Suwa because it was one bit of data he had neglected to quantify. "I don't do anything but work," he admitted. Even the taskmaster Tim White estimated that Suwa devoted twice as much lab time to Ardi as he did. When an important job needed doing, White would beg his overworked Japanese colleague to shoulder yet another task: *Please Gen, you're the only person I trust on this.* Suwa was the sort of exacting scientist who checked software in case it *might* have a bug—and actually found one. He won a $1 million grant from the Japan Society for the Promotion of Science for his CT machine, tested the software, discovered a flaw, and wrote new code to correct the problem.

In Tokyo, Suwa and his assistants scanned each tooth with hundreds of CT "slices" at a resolution finer than human hair. With the naked eye or microscope, Suwa could see only superficial features. With the CT machine, however, he could look *inside* teeth, measure enamel coating down to one-millionth of a meter, count growth rings to reconstruct life history, and scrutinize microscopic wear to infer what the animals ate.

The initial announcement of the species reported fewer than two dozen teeth. The Mission to the Pliocene increased the sample sixfold—145 teeth from at least twenty-one *Ar. ramidus* individuals, enough to analyze as a population. For example, the team recovered twenty-three canine teeth. All living and fossil apes bear interlocking canines. Daggerlike upper fangs sharpen themselves against the lower premolars, which is nature's way of keeping the weaponry sharp. In the human family, the daggers have been beaten into plowshares. Our canines are smaller and more diamond-shaped—a diagnostic trait used to identify early members of the human family who otherwise resemble apes in many other aspects of anatomy. Ardi's canine teeth fit the human family pattern. None of the *Ar. ramidus* upper canines or lower premolars showed any signs of honing. Instead, the canine teeth were blunter, diamond shaped, and smaller than any ape. They wore from the tips and became blunter over time. Wear served as indicator of age: older individuals had duller canine teeth.

More than any other tooth, canines reveal sexual differences because males have bigger fangs than females. Ardi's canine teeth fell in the bottom of her species range and affirmed the skeleton belonged to a female. Canine dimorphism is highest among species in which males compete violently for females. When the entire population was examined, statistical analyses showed that the range of teeth size differences remained low compared to other primates—an indication of low levels of aggression within her species.

None of the three Miocene species from the human family—*Ardipithecus*, *Orrorin*, and *Sahelanthropus*—displayed fangs like modern apes. *Sahelanthropus* showed that canines started shrinking as early as 6 million or 7 million years ago. By the time of Ardi, the fangs had waned even more, revealing that these early members of the human family had a social structure very unlike

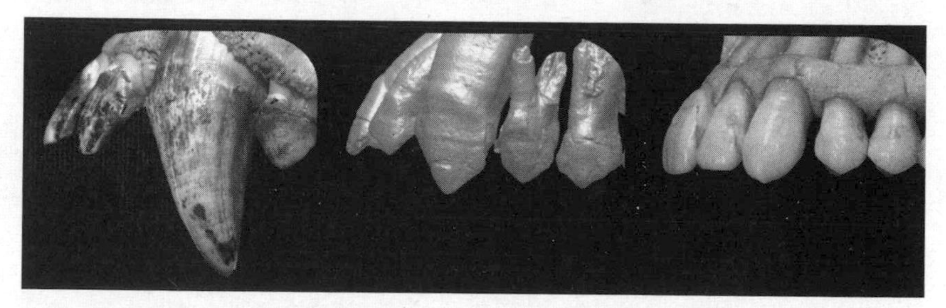

Canine reduction: Humans and apes differ profoundly in their canine teeth. A male chimpanzee (*left*) has huge fangs that constantly sharpen themselves by rubbing against the lower premolars. Humans (*right*) have more diamond-shaped canines that have shrunk since we diverged from the other apes. *Ardipithecus ramidus* (*center, a male individual*) showed that canine reduction was well underway by 4.4 million years ago and the honing canines had disappeared. Chimps exhibit substantial sex differences in fang size but canine dimorphism is low in members of the human family, including *Ar. ramidus*.

any of our cousin apes. In addition, Ardi showed canine reduction occurred *before* the appearance of the big chewing teeth of *Australopithecus*—a strike against earlier theories that suggested a link between shrinking fangs and expanding chewing teeth.

The teeth also testified about Ardi's place in the family tree. Animals that eat coarse foods tend to have thicker enamel; animals that eat soft foods tend to have thinner enamel (such as chimpanzees who dine heavily on fruits). Initially, the Ardi team described the *ramidus* enamel as thin—and many scholars seized upon this detail as confirmation of the long-expected chimp-like ancestor. Indeed, some even cited thin enamel to suggest her species was an ancestor of chimps, not humans.

But labeling enamel as thick or thin was a crude generalization—and much too imprecise for Suwa's taste. Even within a single tooth, enamel thickness varies. Much better to quantify thickness in *microns*, at specific landmarks, in individual teeth. Unfortunately, the literature lacked comprehensive data on primate tooth enamel. Undeterred, Suwa and his team embarked on the tedious task of building a new database with measurements from hundreds of apes, monkeys, and human ancestors.

With all this data, the Tokyo team depicted enamel thickness in images detailed with elevation lines and colors resembling topographic maps. By reading these swirls, Suwa reached a more profound observation: *Ardipithecus ramidus* bore a *distribution* of enamel unique to the human lineage—a nascent version of the pattern that distinguished *Australopithecus*, the genus of Lucy.

So was Ardi an ancestor of Lucy? Not given to cheerleading, Suwa deemed *ramidus* a viable *candidate*. He could not state with certainty whether *Australopithecus* descended from *ramidus*, the species of Ardi herself, or an earlier member of the genus *Ardipithecus*. He could, however, confidently assert that Ardi resided on the human branch of the family tree—as proven by the enamel pattern, canine reduction, and many other clues.

The teeth also provided more hints about Ardi's diet and

habitat. Sometime after Ardi, the genus *Australopithecus* moved into a wider ecological niche. *Australopithecus* bore larger, thickly enameled molars, big cheekbones (zygomatic arches, in anatomical lingo), powerful chewing muscles, and other signs that these creatures gnawed their way beyond the trees. For example, Lucy's species showed deep, parallel tooth scratches from abrasive foods typical of more open country.

Ardi had not made this transition. Her species had smaller teeth and shallow zygomatic arches, indicating weaker chewing muscles. Her microscopic tooth abrasions were finer and more random, indicating more variable diet. From this evidence, Ardi appeared to be an omnivore that fed in the trees and on the ground. Chemical traces also supported this view: Carbon isotopes of teeth showed the *ramidus* diet relied heavily on C_3 vegetation—again signaling it dined among trees. (No known human ancestors were heavy C_4 feeders until 3.8 million years ago.) Subsequent members of the human family displayed a broader range of isotopic values, suggesting that different populations and species learned to find meals in a variety of places. (Interestingly, one thing our ancestors don't share with us is bad teeth: tooth decay occurred in less than 2 percent of tool-using foragers of the Stone Age and was even rarer among early members of the human family. Cavities became more common only with the advent of cereal agriculture about ten thousand years ago and skyrocketed in the nineteenth and twentieth centuries with processed foods and sugars.)

So Ardi foraged on the ground but had not fully departed the trees. Her species occupied a different niche from Lucy and all subsequent human ancestors, yet the teeth testified that these woodland omnivores were closely related to the members of the human family who came later.

As Suwa peered into Ardi's mouth, his partner halfway around the world ran into problems on the other end of the body.

Chapter 22
TROUBLE AFOOT

One day, Lovejoy fingered a small foot bone called the cuboid. Why anyone would call such a thing cube-like was one of those maddening idiosyncrasies of anatomy. Nestled in the midfoot, the cuboid has irregular sides, not six square ones.

Then Lovejoy observed something else on the cast that didn't make sense. The outside edge of Ardi's cuboid showed a large, smooth surface known as a facet. The facet marked where the bone rubbed against another surface—a little sesamoid called the os peroneum. A sesamoid is a type of bone embedded within a tendon. Most sesamoids are small bones in the hands and feet; they function like little pulleys and pivots that reinforce tendons around turns. The biggest human sesamoid is the kneecap, but most are much smaller. The foot alone has a dozen sesamoids, mostly in the toes. Some are shaped like coffee beans, others like lentils or chickpeas. Sesamoids were so named because ancient anatomists thought they resembled sesame seeds.

Lovejoy puzzled over the cast. What did it mean that Ardi's cuboid had a facet for the os peroneum sesamoid? In humans, that sesamoid resides in the peroneus longus tendon, which curves around the outside of the foot and turns inward across the sole. In apes, the tendon helps close the grasping toe; in humans the tendon has been redeployed to propel toe-off and support the arch of the foot. Any change there must represent an important signal about the emergence of upright walking. But what?

Lovejoy checked his personal collection of chimp and gorilla bones. If our closest relatives shared a feature, he figured, it was *primitive*—and our common ancestor likely shared the trait, too.

Neither ape had facets for the sesamoid.

Strange. Ardi was a million years older than Lucy and thus should more resemble African apes. But unlike the apes, she bore a big, stinking facet.

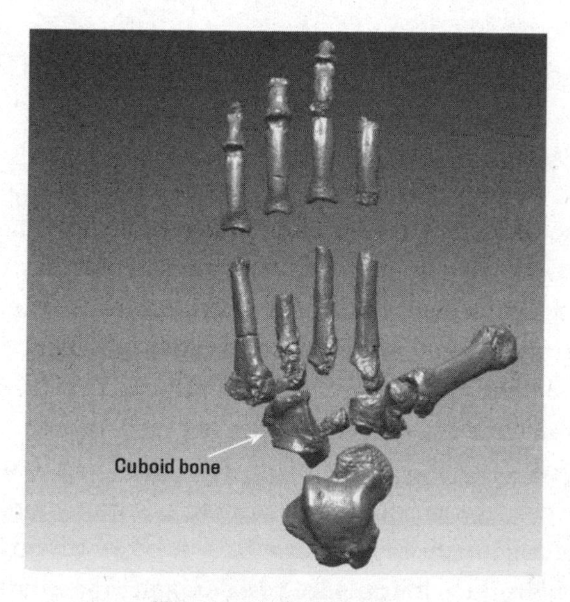

CT image of Ardi foot skeleton
(with cuboid marked) as seen from above.

Something didn't add up. Was this little facet and sesamoid just inconsequential minutiae? Or something more? Lovejoy had never pondered this structure before. Until Ardi, cuboids just didn't exist in the early fossil record. He sensed this little bone might tell him something new about bipedality, even if he didn't yet know just what.

• • •

BIOLOGISTS CALL US *HABITUAL STRIDING BIPEDS*. LONG BEFORE HUMAN ANcestors grew outsize brains or made stone tools, they stood on two feet and developed a bizarre form of locomotion otherwise unknown in the animal kingdom. No other living primate has abandoned the opposable toe. Of the millions of living species on earth, humans are the sole possessor of the arched foot, a contraption that absorbs our body weight on landing and instantly transforms itself into a springy lever that propels us forward. Some other animals move on two feet, but their gait is nothing like ours. Greek philosophers classified humans as a "featherless biped" because our two-legged stance puts us in the same category as birds. In fact, some birds are much better bipeds than we are. The nine-foot-tall ostrich can sprint faster than 40 mph with a bent-kneed gait that some biologists have described as "Groucho Running" (after the Marx Brothers comedian). The infamous T. rex was a biped, but nobody knows for sure how it moved, *Jurassic Park* notwithstanding. Kangaroos are bipeds but don't stride; they hop. Cockroaches and even some lizards can run on two legs. But none of these creatures moves with the peculiar stride of the human.

Yet lowly feet don't get their due. The brain and hand are widely celebrated as apogees of human distinctiveness while feet are treated like burros: dull, smelly beasts of burden. To an anatomist, however, the human foot is a delightful evolutionary innovation. Leonardo da Vinci was the first of many anatomists with a foot fetish, and his journals are filled with detailed drawings and descriptions of even the tiny *ossi petrosi* (stony bones)—the sesamoids. In 1944, the British anatomist Frederic Wood Jones added his own praise for the human foot: "It is the most distinctly human part . . . it is by his feet that he shall be known from all other members of the animal kingdom." Accordingly, anatomists wrinkle their noses when clueless colleagues characterize the foot as a clumsy version of the hand. The human hand is a minor renovation of the primate theme, while the foot is a complete remodel.

The appendages of our upper and lower limbs function very differently. The hand is *manipulative* while the foot is *supportive* and *propulsive*. The human hand is built for dexterity and can mold itself around objects. Muscles, tendons, bones, and nerves allow fingers to operate independently and precisely. In contrast, feet are weight-bearing supports whose muscles produce relatively little movement and whose main purpose is to stabilize, support, and push off. In the fingers, flexors and extensors act as classic opposing muscles; one fires as the other relaxes. In the feet, however, opposing muscles often both fire at the same time.

Ultimately, bipedality remodeled tree-climbing apes into mobile creatures who could trek long distances and use their hands to make tools, build fires, and populate the globe. It freed hands to become agents of human intelligence. But where did this bizarre foot come from? That mystery sent Lovejoy and his fellow investigators on an intellectual exploration that stretched more than a decade and tripped them up again and again.

Ardi offered the first glimpse of a foot older than *Australopithecus*. And what a strange one: she had a grasping big toe and curved digits like a climber, but her lateral foot bore joints shaped for pushing off the ground like a biped. She walked on a flat foot with no humanlike arch. Her skeleton had only a partial ankle, and its best-preserved part, the talus bone, suggested the flexible joint of a climber, and not the more robust, shock-absorbing heels of later bipeds. As with so many parts of the skeleton, her foot seemed a cobbled-up medley of parts. And now this cuboid posed yet another oddity.

WITH THE MYSTERY OF THE SESAMOID ON HIS MIND, LOVEJOY WALKED INTO the Cleveland Museum of Natural History and descended to a bunker-like room in the basement. The museum's Hamann-Todd skeleton collection, one of the largest in the world, contained the bones of more than 3,000 humans and 1,200 other primates. Two decades earlier, the Lucy scientists sat in the same rooms, pored over the *Au. afarensis* fossils from Ethiopia, and compared them to the collection. Now Lovejoy returned on another quest: what was Ardi telling them?

He walked past a human skeleton hanging on a stand and turned into the compactor. Despite a name that sounds like something that would crush a car, the compactor was a space-saving storage system that squeezed thousands of skeletons into floor-to-ceiling shelves that slide on motorized tracks. He turned into one of the aisles containing the museum's apes, opened the chimp and gorilla drawers, picked out the cuboids, and checked for facets. Unlike Ardi, the first one didn't have a facet. Neither did the second. Nor the third. He eventually checked about fifty chimps and fifty gorillas, and none had facets. He confirmed what he had seen in his personal collection: the African apes lacked this feature.

What about humans? He walked to another row and consulted some old friends whom he called the Boxcar Willies.

Anatomists rarely have much luck recruiting volunteers. In the early days of the science, Renaissance anatomists plundered bodies from the gallows. In eighteenth- and nineteenth-century England, bands of ruffians known as "resurrection men" snuck into cemeteries at night, dug up fresh graves, and sold the corpses at the back doors of anatomists and medical schools. Public outrage prompted Parliament to regulate the corpse trade by giving anatomists legal right to bodies of executed criminals and paupers. The bones in the shelves of the Cleveland museum represented a grim legacy of that tradition.

Cleveland became a mecca of comparative anatomy thanks to a vigorous Scotsman who took the down-on-their-luck, destitute, derelicts, and drunks of an industrial city and redeployed them into an A-team of human anatomy. T. Wingate Todd left an impression on everyone he encountered, living or dead. Born in 1885, Todd was educated at the University of Manchester in an era when the cutting edge of biology was literally the scalpel. In the early twentieth century, Western Reserve University in Cleveland aspired to become one of the top medical institutions in the United States, and the head of the medical school wrote to Arthur Keith (the preeminent anatomist of his era in the English-speaking world) seeking a recommendation for "the best young man in England." Keith knew just the candidate: T. Wingate Todd. A week before voyaging to America, Todd's obsession for his science was made clear when he arrived late to his own wedding due to an "engrossing dissection" back at the lab.

In 1912, Todd became chair of anatomy at Western Reserve. No serious anatomist could work without a comparative sample, the larger the better. Previously he had served on a classic archeological study of skeletal remains, the Nubian Archeological Survey, which excavated ancient cemeteries due to be flooded by the Aswan Dam on the Upper Nile. Archeologists dug up 7,500 skeletons spanning five millennia of human occupation and sent crateloads back to Manchester, where Todd cataloged and arranged them—a duty he remembered as "my greatest thrill." The survey pioneered the science of using human remains to reconstruct the history of health and disease in early civilization.

But the Egyptian bone assemblage bore one major shortcoming: the skeletons were anonymous individuals disinterred centuries after burial and their ages, races, occupations, and causes of death remained uncertain. Todd envisioned a collection with vital statistics. Building on a small collection begun by his predecessor, Carl August Hamann, Todd spent the rest of his life amassing bones. At the time, Cleveland was a burgeoning industrial city of factories, railroads, canals, and wharves and attracted laborers from around the world. The year before Todd's arrival, Ohio passed a law allowing medical schools to collect cadavers that otherwise would have been buried as indigents; bodies unclaimed by loved ones were appropriated by researchers like Todd. Each cadaver was dissected in anatomy classes, measured, and stripped down to bones that were boiled clean and shelved in surplus ammunition boxes. The skeleton room eventually came to resemble a munitions depot full of ammo crates. Over twenty-six years, Todd stockpiled 3,300 human skeletons plus a huge comparative sample of apes (by the 1930s nearly half of the world's skeletons of gorillas, chimpanzee, orangutans, and gibbons). "No collection like this had ever been assembled anywhere in the world," noted one colleague, "and probably never will be again."

The lab also collected other curios, including a mischievous chimp who delighted in untying people's shoes. Once Cleveland zookeepers shot a berserk elephant and dumped the carcass in the alley behind the medical school. Summoned by telegram from a meeting in Buffalo, Todd hurried home and found the disposed giant bloated by gas. His first knife cut unleashed an explosion and sprayed him with stomach contents. On the trolley ride home, other passengers cleared a wide circle around the anatomist.

Todd became a minor celebrity. He conducted a landmark study of how human skeletons change during the lifetime of the individual. Howard Carter, the discoverer of the tomb of King Tutankhamen, turned to Todd to determine the age of the mummified body. King Tut's skull and leg bones were sent to Cleveland, and Todd concluded that he died at about age eighteen and a half—an estimate confirmed by modern scientific techniques.

Nicknamed "the Chief," Todd was an imposing perfectionist who pushed himself from morning to night and whose blunt criticism sometimes drove employees to tears. At orchestra concerts, he whipped out his briefcase or simply fell asleep. He brought work home—quite literally: an Egyptian mummy called Senbi the Scribe stood by his fireplace. At age fifty-three, he died of a heart attack, leaving behind his wife, three children, and a monumental assembly of human remains. Within a few decades, however, his specialty became perceived as a dying science. As the frontiers of biology advanced deeper into the tissues at the microscopic and molecular levels, medical schools reduced anatomy instruction in favor of newer specialties. Eventually, the university rid itself of Todd's collection, and the skeletons went to the Cleveland Museum of Natural History. Todd brought classical anatomical sensibilities to the Midwest—a tradition venerated nearly a century later by the evolutionist picking through his charnel house.

LOVEJOY WENT TO THE AISLES OF HUMAN SKELETONS AND BEGAN OPENING drawers. Each individual occupied its own tray, with long bones stacked on one side, vertebrae and ribs on the other, pelvises in the center, and hand and foot bones in small boxes. Skulls, each sawn neatly in half, rested on separate shelves. The bones testified to the hard-knock lives of people who ended up unclaimed at the morgue. Some bones showed unhealed breaks, disfiguring pathologies, or ravages of syphilis. One fractured skull still had a shard of beer bottle stuck in the sphenoid bone. Some had gunshot holes or embedded bullets and knife blades.

Lovejoy started pulling out foot bones. Did humans have the sesamoid like Ardi? He surveyed the literature and confirmed that the os peroneum was as obscure as it sounds. Some modern studies reported that no more than 30 percent of humans had the sesamoid, often described as an evolutionary relic in the process of disappearing. That meant Lovejoy should find a facet on only one of every three human cuboids. The os peroneum itself usually wasn't preserved with the rest of the skeleton in the collections—much less recovered in fossils—so its facet on the cuboid had to serve as its proxy. That polished surface was the calling card of the little sesamoid.

The first human had a facet. So did the second. And the third. And dozens more. *Every* human cuboid had a facet—like Ardi.

He kept checking. Every species exhibits a range of anatomical variation, so dutiful investigators must examine large samples to appreciate the full spectrum of "normal." Most anthropologists, Lovejoy grumbled, remained oblivious to normal variation. Time and again, somebody would ballyhoo some fossil "discovery" because it exhibited a small difference from the "average" of modern humans. More than once, Lovejoy had reached into his collections and showed that the same traits fell within our normal variation. At one conference, an opponent claimed that Lucy's shoulder proved she was an arboreal climber. Lovejoy found a similar scapula exhumed from the modern human cemetery at Libben. People wisecracked, *Maybe it was the village's designated climber.*

Lovejoy checked hundreds of skeletons. The facet seemed universal in humans—which contradicted everything the literature had led him to expect. The sesamoid was just one little bone the size of a pebble, but it bothered Lovejoy like a rock in his shoe.

DECADES EARLIER, LOVEJOY'S GRADUATE STUDENT BRUCE LATIMER WROTE his doctoral dissertation on foot evolution. Latimer was a hale, blond, and garrulous young man who resembled one of the clean-cut early Beach Boys. Colleagues sometimes called him "the surfer"—never mind that he came from Ohio.

Latimer grew up with a grisly fascination for anatomy. In high school, he scavenged roadkill for a skull collection in his bedroom. Once he was driving a cute girl to a dinner date when they saw a fox hit by another car. Latimer couldn't believe his luck: *a red fox!* He desperately wanted the species for his collection—and providence finally had provided it. He pulled over, took a hatchet from the trunk, and decapitated the fox by the side of the road. He dropped the head in a plastic bag and climbed back in the car. His date said, "I think you better take me home."

Latimer had the good fortune to be in school when the historic discoveries from Hadar arrived in Cleveland. Don Johanson assigned him to lead the investigation of the *Au. afarensis* foot and instructed him to go down to Kent State and enroll in graduate school under Owen Lovejoy—a blessing appreciated more in hindsight than in the moment. Lovejoy trained his students the way ancient flint knappers sharpened stone tools: with repeated blows.

Latimer and his fellow student Bill Kimbel worked all day in the basement of the Cleveland museum, raced down to Kent in a rust bucket car, and often walked in late. Lovejoy interrupted his lecture and growled, "Well, look who's here. It's brains for shit and shit for brains." The two students never dared clarify who was who.

During the *afarensis* wars, Lovejoy decided it was no longer worth his time to attend conferences and argue with opponents whom he deemed unworthy of response. Latimer went as his proxy and endured a gauntlet of beatings by critics of the guy who beat him up back home. Those trials honed Latimer into a world-known expert on the foot. While teaching at the Case Western medical school, he had dissected hundreds of human feet and noted a bit of anatomical trivia about the very bone now bothering Lovejoy.

At the bottom of the cuboid lies a prominent groove. In old-world monkeys and apes, this groove forms a slot for the peroneus longus tendon. Most people assumed the human tendon ran through the slot just as it did in other primates. But Latimer noticed something peculiar: in humans, the tendon did *not* slide in the groove as described in anatomy textbooks and instead ran slightly lifted out of the groove. To help it glide over the cuboid, the tendon was embedded with the os peroneum sesamoid—right at the spot of the polished facet that had piqued Lovejoy's attention.

Lovejoy had filed this tidbit in the back of his mind as inconsequential trivia, useful only for impressing students that he knew more than the textbook. Suddenly it seemed consequential. Why in the name of Darwin would a tendon *not* run in a groove that seemed specially designed to hold it? He often voiced irritation at peers who cherry-picked anatomical trivia to support a favored story line. But he came to suspect *this* detail signaled a bigger truth—a redesign of the whole body plan.

THE FIRST ANATOMIST TO DESCRIBE THE APE FOOT WAS A SEVENTEENTH-century Englishman named Edward Tyson. From what little we know of him, he was a grave, serious bachelor who "devoted himself to celibacy" and whose chief diversion was cutting open dead animals and poring over their innards. "His studies were his chief delight," wrote one acquaintance, "only he took now and then a touch at fishing." In London, he became the go-to authority for biological oddities. Creatures strange to English eyes, brought by merchant ships, found their way to his dissecting table, including a porpoise, an ostrich, and a rattlesnake. In 1698, Tyson came face-to-face with another exotic creature that seemed neither man nor beast but something in between. Most incredible of all, its legs ended with another set of *hands*.

It had arrived in London aboard a ship returning from Angola in western Africa. Standing two feet tall and covered with dark hair, it behaved in an eerily humanlike manner. Tyson reported that it preferred the human company of sailors and eschewed monkeys aboard the ship "as if nothing a-kin to them." Lacking a better name, Tyson called the young animal a "Pygmie" after a legendary diminutive race described by ancient Greeks. At other times, he deemed it an orangutan (from the Malay word for "man of the forest"), the only ape known to European science at the time. In fact, it was an immature chimpanzee—a species that would not be recognized for another half century.

Shortly after arriving in England, the chimp fell ill. When it perished, it took its turn at the anatomist's dissecting table. Tyson described it as "more resembling a Man, than any other Animal" and classified it in the human genus as *Homo sylvestris*, or "man of the forest." But one feature was decidedly *not* human—the chimpanzee foot. It had a thumb-like great toe, four finger-like lateral toes, and a palm-like sole that could wrap around objects. Tyson described the appendage as "liker a Hand than a Foot." He could not classify this animal with the quadrupeds, or four-legged creatures, so he coined a new term: *quadrumanus*, or "four-handed."

In 1699, Tyson published the results of his investigation—a major scientific milestone as the first anatomical description of an ape. Much of Tyson's report was taken up with reconciling the anatomy of his "Pygmie" with the myths of wild men passed down from earlier generations. Those legendary creatures, Tyson concluded, probably stemmed from exaggerated accounts of apes and monkeys. In retrospect, we can see this pioneering anatomist wrestling with a timeless pitfall: trying to reconcile new evidence with old ideas.

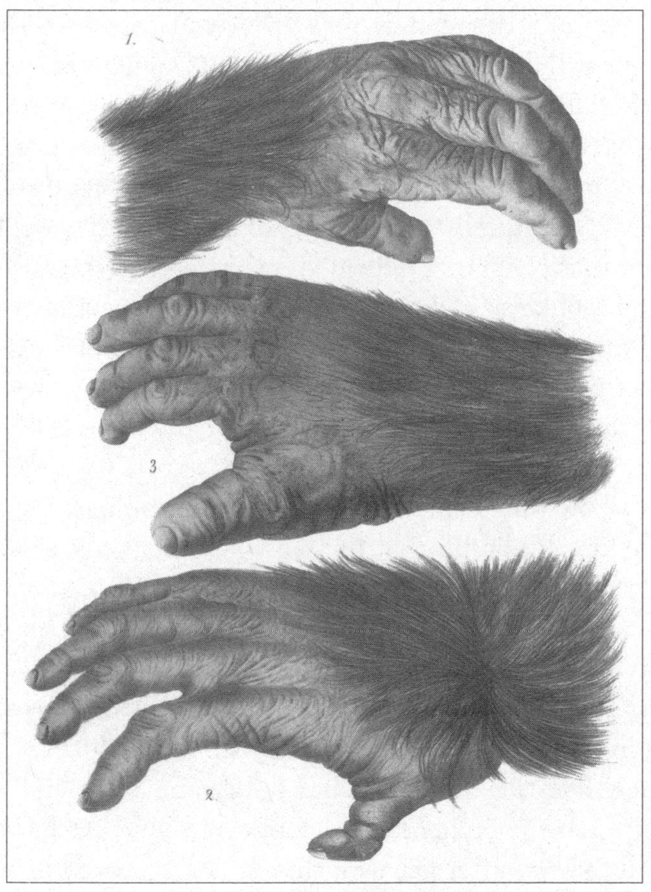

Quadrumanus: the chimpanzee hands (*figures 1 and 2*) and feet (*figure 3*).

During the Enlightenment, scholars had little notion of evolution (or its proto version of "transformism"). Instead, naturalists focused on classifying the diversity of life. The prime example is the *Systema Naturae* by Carl Linnaeus, a groundbreaking work that categorized species by physical appearance. Linnaeus did not question divine creation, yet anatomical similarities compelled him to lump humans with other primates. When a Lutheran archbishop accused him of "impiety," Linnaeus protested that the evidence left him no alternative. "It is not pleasing to me that I must place humans among the primates," he wrote to one correspondent in 1747, ". . . but I desperately seek from you and from the whole world a general difference between men and simians from the principles of Natural History. I certainly know of none. If only someone might tell me one!" One authoritative naturalist later did just that by borrowing Tyson's term for four-handed creatures. In his *Manual of Natural History* (1779), Johann Friedrich Blumenbach proposed that primates be divided into two groups: *Quadrumana* (apes and monkeys) and *Bimana* (humans). Manipulative feet became the trait that put a comfortable distance between humans and other primates. This distinction was adopted by Georges Cuvier, the most influential naturalist of the early nineteenth century, and held fast until the age of Darwin.

By the late nineteenth century, however, comparative anatomy convinced evolutionists that humans were not only anatomically similar to apes but *related* to them as well. In *The Descent of Man*, Darwin wrote, "If man had not been his own classifier, he never would have thought of founding a separate order for his own reception." Darwin called the apes the "nearest allies of man, and . . . the best representatives of our early progenitors."

A new question arose that has preoccupied evolutionists ever since: how did the prehensile ape foot evolve into the human foot? For a full century after Darwin, this debate occurred in a vacuum because few foot fossils were discovered before the 1960s. Instead, scholars looked to living primates for models of the ancestral foot. The most influential foot expert of the mid-twentieth century was Dudley J. Morton, an anatomist at Yale, Columbia, and the American Museum of Natural History. (Morton's toe, the second digit longer than the first, is named for him.) Morton speculated that the "prehuman foot" was halfway between gorilla and human. (Indeed, many comparative anatomists saw the gorilla foot as most similar to the human one.) To one degree or another, many anatomists continued a similar approach even as the fossil record improved. New discoveries were portrayed as waypoints in a progression from African ape to human. The molecular revolution inspired experts like Latimer to narrow the image of the ancestral foot even more—the chimp foot became the proxy. And that was why Lovejoy and Latimer found themselves deep in the compactor, trying to discern clues from the feet of the four-handed ones.

LATIMER SPENT HIS CAREER TRYING TO EXPLAIN HOW THE GRASPING APE foot turned into the springy human foot by connecting three dots: chimps, *Au. afarensis,* and human. The story went like this: the midfoot got longer, the toes got shorter, and the big toe turned from an opposable digit to one lined up with the rest of the toes.

Now Ardi had inserted a fourth dot between African apes and Lucy—but it didn't exactly line up as they had expected. Why would Ardi's midfoot be longer than a chimp's while keeping that opposable toe? Why would her toes be *shorter* if she still climbed in the trees?

"There's a procedure in anthropology," said Lovejoy, "when in doubt, measure!"

And so they measured. They recorded bone lengths, circumferences, articular surfaces, and degree of opposability in the big toe. They measured . . . and measured . . . and made little measurable progress. After every meeting, Lovejoy wrote summaries of their interpretations but wasn't convinced by what he had written, and his computer became cluttered with false takes. *Ardipithecus*—initially described as the most chimp-like ancestor ever found—wasn't slipping into its assigned role.

"We were trying to force it into that shoe," said Latimer, "and the shoe wasn't fitting."

Yet another thing didn't fit—that damn sesamoid. A

4.4-million-year-old foot should increasingly resemble an African ape, but it didn't. Which was primitive, the presence of the os peroneum or its absence? Neither the comparative collection nor modern literature could decisively answer the question. So Lovejoy turned to the anatomists he most trusted—the dead ones.

Often the best anatomical data came from papers published fifty or one hundred years ago. Lovejoy viewed the old masters like the Beethovens and Bachs of his science, practitioners from a golden age in anatomy that had never been equalled. They learned the old-fashioned way—by cutting human cadaver after cadaver, and ape after ape. "It was just the way anatomy was done back then—on the basis of logic and function," Lovejoy said. "Then the computer came in. Everybody thinks that if you go measure a bone, plug the information into a computer, and draw some beautiful graphics in thirteen different colors, you've got something close to truth. Ugh"—he grunted dismissively—"you've got something close to *chaos*."

Long after the field team ceased excavating, Lovejoy dug through the dusty archives. In some cases, old articles could not be found in any electronic database, so he ordered a musty paper volume from a remote warehouse. Even by the dry standards of anatomical literature, cuboids and sesamoids were not exactly sexy topics. In 1908, an anatomist at the University of Cambridge named T. Manners-Smith conducted a detailed study of the cuboid and os peroneum among 550 skeletons from Egypt (the ones curated by T. Wingate Todd before he came to America). The paper described the cuboid with the kind of precision that only an anatomist could appreciate—or endure. Then Lovejoy found this description of the human cuboid: "The most interesting point in connection with it is the presence of the facet for the sesamoid element in the tendon of the peroneus longus." Translation: humans had the same cuboid facet as Ardi—the very feature that first aroused his interest. (Similarly, another paper from 1928 reported the facet was present in 93 percent of humans.)

Manners-Smith also examined dozens of gorillas, chimps, and

orangutans—and reported that the apes lacked the same facet for the sesamoid. The old papers confirmed Lovejoy's observations that the os peroneum and its facet on the cuboid bone were near universal in modern humans and absent in apes. The modern literature was wrong—in this one trait, Ardi was like a human and not like a modern ape.

A perplexing question remained: which was more primitive, the modern African apes or the human family?

It was time to emulate the old anatomists, pick up the scalpel, and do some cutting.

A ZOO DIRECTOR COULD COUNT ON TWO THINGS ABOUT DEAD APES. FIRST, the zoo staff had to move fast before the carcass started to rot. Second, an anatomist would be calling soon.

Today great apes are endangered, protected by laws and international treaties, and rarely become available for dissection—unless they die in captivity. When an ape perished at the Cleveland Metroparks Zoo, a call from Latimer would follow as surely as rigor mortis. The anatomy professor would phone his friend the zoo director and say something like "I want to express my deep condolences for the loss of the chimpanzee" while trying to sound sincere. The director told Latimer that he was a sick bastard—and, yes, he could have the body.

So, a frozen chimp leg found its way into the freezer of the lab at the Cleveland Museum of Natural History. Latimer let it thaw overnight like a leg of mutton. In the morning, the staff arrived to find a bloody, hairy ape leg in a tray surrounded by two scientists in white lab coats. Curious colleagues peeked over their shoulders and recoiled. It could have been worse. Ape cadavers were so hard to obtain that the anatomists took whatever came their way, no matter how putrid. Sometimes they dabbed Vick's VapoRub under their nostrils to block the stench. Once the walk-in freezer at Kent State broke with two 350-pound male gorillas inside. The staff spent several days trying to preserve the bodies with dry ice, but it was a losing battle: decomposition generated heat just like steaming heaps of compost, and after a few days, the gorillas began to ooze foul fluid. Lovejoy and his colleagues donned rubber boots and gloves and began packing seven hundred pounds of gorilla goo into plastic bags. "Supposedly, the custodial people were going to help us," he recalled. "Let me tell you how fast they disappeared—they were *gone*." Luckily, this thawing chimp leg smelled just fine. The anatomists didn't even have to set up the fume hood.

Latimer cut down to where the peroneus longus tendon passed around the cuboid. They tugged on the tendon, and it slid through the groove like a rope through a pulley, making the dead chimp's big toe eerily open and close.

They cut into the tendon. No sesamoid. Latimer sliced deeper and exposed the cuboid bone. Sure enough, there was no facet, either. The dissection confirmed what they had observed in the bone collections: chimps lacked an os peroneum and facet.

Back at the medical school, the Ardi investigators already had dissected human cadavers and confirmed that every human had both the sesamoid and facet. They rolled cadavers into an X-ray machine and took radiographs. Sometimes the sesamoid showed up on the film and sometimes it didn't, because the os peroneum existed in humans but wasn't always *calcified* like a regular bone; sometimes it was cartilage and thus failed to appear on X-rays. The more reliable indicator was the facet, like a footprint that the sesamoid left on the cuboid—and just about every human had one.

The recent literature was wrong. Modern authors went astray because they relied on medical imaging technology and secondary literature instead of hands-on dissection and firsthand observation.

Another riddle: did the common ancestor of humans and chimps possess the little sesamoid or not?

The 1908 paper yielded one other curious detail: the sesamoid was much more common in so-called lower primates. Nearly all old-world monkeys bore the os peroneum, suggesting it was very primitive. Monkeys and humans last shared a common ancestor about 30 million years ago, long before our lineage split from the African apes.

The seed of a radical idea planted in their minds: what if the human lineage never lost that trait in the first place? What if they never had feet quite like gorillas and chimps?

The two men looked at each other over the bloody ape leg. Lovejoy spat, "Get a baboon!"

BABOONS ARE BARE-ASSED, BIG-FANGED AFRICAN MONKEYS WITH CLOSE-set, beady eyes, long muzzles, and nostrils like double-barrel shotgun bores. The olive baboon is named *Papio anubis* after the Egyptian jackal-headed god of embalming and mummification who supposedly bore the Pharaohs into eternity. Its long-snouted face will seem familiar to anyone who has ever perused an exhibit from the tomb of King Tut.

Unlike endangered apes, baboons breed like rats and have diversified into several species. Large troops wander throughout Ethiopia, and Afar herders sometimes shoot the fierce *Daam Aatu* to protect livestock from mauling. Males are bold enough to menace lions. Latimer had encountered baboons many times in Africa. During a stint as director of the Cleveland museum, he led major donors on safaris and repeatedly warned people not to put down water bottles or valuables. Inevitably, the tourists became careless and a baboon would swoop in, steal a camera or water bottle, and flee into a tree. "In their own way, they're laughing," Latimer said. In the Middle Awash expeditions, they fouled the wells, and Latimer picked out baboon shit before filling the water containers. In his mind, baboons were a subject of annoyance, not scientific study—until now.

Latimer and Lovejoy hurried back into the compactor. This time they did not follow the usual path into the ape or human aisles and went to the last row of the collection, where the museum kept a token collection of baboon skeletons. In truth, nobody paid much attention to those drawers, and their remoteness underscored the perceived lack of importance: monkeys were too distant to reveal much about human origins. Or so they had assumed.

The two anatomists pulled open the baboon drawer and picked out the cuboids.

Facets!

Baboons had an os peroneum—like humans and like Ardi. The scientists grabbed a so-called ligamentous cadaver, a baboon foot that had been stripped of the flesh but preserved with intact tendons and ligaments. Its sesamoid was *huge,* like a horse pill stuck to the side of the foot.

They started pulling out the oldest primates in the entire collection, Miocene apes from Pakistan and Kenya that lived 15 or 20 million years ago. *More facets!* In this feature, the Miocene apes resembled the baboons and humans. That os peroneum sesamoid was primitive, *really* primitive. The radical idea took root: *Could it be that modern African apes are not such good models for the feet of our ancestors?*

"All of a sudden we rocked back," said Latimer. "Shit, humans are the primitive ones!'"

SOMETIMES REVELATION DAWNS SLOWLY: THE SCIENTISTS FINALLY SEE over the maze where they have spent their lives. Perspective shifts from rat in the labyrinth to the *sapiens* above who marvels at how long it took that poor, benighted creature to figure it out. That sorry creature is the former you. You see all the wrong turns you took, all the dead ends you banged your head against, and behold the true path you *wished* you had seen long ago.

For decades, Lovejoy and Latimer had beaten a path into the same aisles of the compactor, which might as well have been the dead ends of a maze. They had accepted modern African apes as models of primitive human ancestors. Got a question? Check the chimps and gorillas. They barely looked at Miocene apes or monkeys. "You grow up with those ideas and you hold them," says Latimer. "It takes a sledgehammer to the head before you say, 'Whoa, wait a minute! I screwed that up!'"

It took Lovejoy and Latimer years to fully recognize the most profound message of Ardi: the old image of the common ancestor of humans and apes was an illusion. They had been deceived by the zoo of the known, and failed to imagine the menagerie of the unknown. When the investigators laid out the foot skeletons, it became easier to connect the dots from Miocene apes to Ardi to humans. Suddenly the living African apes seemed outliers. Modern apes were not faithful relics of our common ancestors. Instead, they evolved into specialized creatures *after* their common ancestors with humans.

The investigators had been asking questions backward. The question shouldn't have been, how did human ancestors *lengthen* the midfoot? They should also have been asking the opposite: how and why did modern apes *shorten* theirs? Likewise, it wasn't simply a question of humans shortening their toes; the ancestors of chimps also *lengthened* theirs.

The Ardi investigators conceived a new theory: the modern African apes abandoned the stiffer feet of quadrupedal Miocene apes and evolved feet more specialized for grasping. The os peroneum probably disappeared in chimps and gorillas because it restricted movement of the tendon. By eliminating the sesamoid, apes gained greater power, more range of motion, and feet more like hands. They became *quadrumanes*, the four-handed ones.

But the human lineage took a different path. Human ancestors remodeled a grasping foot into one suited for an outlandish form of upright locomotion. The tendon that formerly closed the grasping toe assumed the new function of pushing off the ground.

Ardi finally solved a mystery that had puzzled Latimer for two decades. Why would that tendon move out of the cuboid groove in humans? That once-arcane bit of trivia somehow had to be part of the story. But how?

They argued for weeks before the answer hit them: *the tendon became an arch supporter!*

Ardi retained a flat foot, but the arched foot appeared by the time of Lucy's species—as shown by the Laetoli footprints and later confirmed by fossils. At some point, human ancestors lengthened their feet, and the change in geometry angled the tendon forward and shifted its alignment in its groove—the peculiarity that Latimer had noticed years earlier but never understood until Ardi.

The odd arrangement made no sense on its face. No rational creator would design a machine with a tendon that failed to properly align with its own channel. But natural selection does not

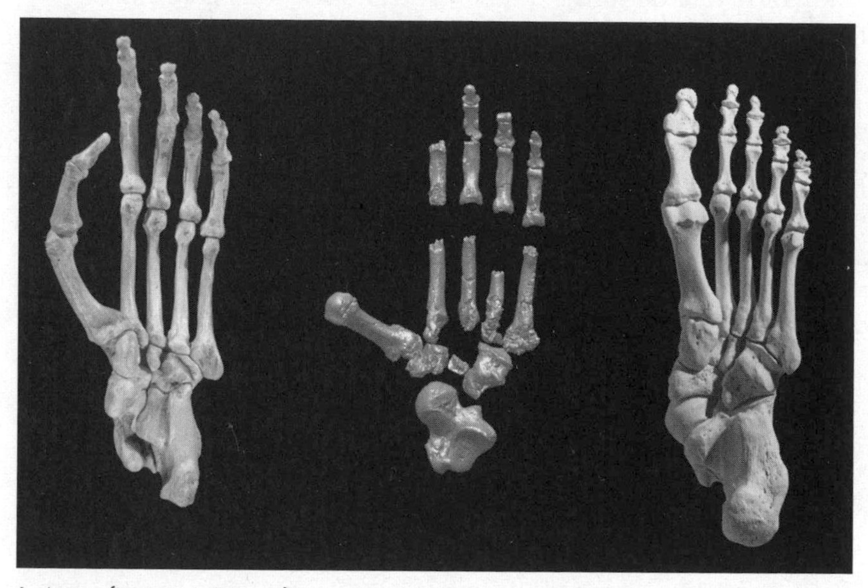

A plantar (bottom, looking up) view comparison of foot skeletons of chimpanzee (*left*), Ardi (*center*), and human (*right*). Chimpanzee: grasping toe; pliable midfoot; long, curved toes; and small calcaneus bone of heel.

Ardi: grasping toe—even more divergent than chimp big toe; bipedal but no foot arch; stiffer midfoot than chimp; joints at ball of foot adapted for pushing off four lateral toes when walking erect; and only fragments of the calcaneus were found, so that aspect of its locomotion remains unknown.

Human: robust, forward-oriented big toe; stiff midfoot with arch; metatarsal bones have domed heads, allowing toes to dorsiflex and push off the ground; and robust calcaneus bone of heel for shock absorption.

begin from scratch and must adapt whatever elements it has inherited. Consequently, anatomy is replete with odd compromises, old baggage, and less-than-ideal designs. As biologist Theodosius Dobzhansky observed, "Nothing in biology makes sense except in the light of evolution."

The os peroneum marks a pivot point in a tendon; it also turned out to be pivotal in the Ardi team's conception of her place in human evolution. The story was quintessentially Lovejoy, with his idiosyncratic eye for small clues, master body plans, and big adaptive packages. The rest of his profession moved inexorably toward objective systems, metrics, and software. Lovejoy believed that no analytical tool delivered better insights than the black box of the human brain. But not every brain: he believed that only a small cadre of people possessed the deep expertise to appreciate the revelations of Ardi. Not surprisingly, such attitudes ignited fury and derision among scientific peers. And that was fine with him.

Years passed before the Ardi investigators fully appreciated what they discovered the day of the bloody chimp leg. That little sesamoid ceased to be an anatomical anecdote and became the *synecdoche:* that which stands for the whole. More revelations from other parts of the skeleton would come in a cascade.

Chapter 23
TÊTE-À-TÊTE

The Institut de France stands on the left bank of the Seine in Paris, with its domed cupola arising over its semicircular wings like a head above two open arms. The Baroque building is a monument to French intellectual life, and among its five academies is the Académie Française, whose *Immortels* are entrusted with protecting the purity of the French language. Another is the Académie des Sciences, and in September 2004 it hosted a conference on human origins that united the discoverers of the three earliest members of the human family—*Sahelanthropus*, *Orrorin*, and *Ardipithecus*. In the wood-paneled *grande salle des séances*, statues of French intellectual giants gazed down coldly from their pedestals.

There a sparring match erupted between Tim White and Brigitte Senut, the codiscoverer of *Orrorin*, the 6-million-year-old species from Kenya, known by only a few teeth and two partial thighbones. Senut had declared *Ardipithecus* to be ancestor of a chimp and *Sahelanthropus* ancestor of a gorilla, making her discovery the only early human ancestor left standing. During the exchange, White tweaked Senut for espousing creationist rhetoric because she claimed *Orrorin* walked like a human but failed to provide evidence. *Créationniste? Moi?* Indignant, Senut challenged White to provide evidence of *his* unpublished fossil—the mysterious *Ardipithecus* skeleton, still secret ten years after its discovery.

Then White surprised everybody. He flipped open his computer and projected images of the Ardi skull on the screen—the first time they had ever been displayed in public. It was dramatic theater but didn't reveal much: the bones looked like a broken ceramic pot crushed by a steamroller. "This is the most fragile hominid skeleton ever found," White told the audience. "We are very sorry it's taken us this long to do, but I think you want the right answer instead of the quick answer."

TO A PALEONTOLOGIST, THE SKULL IS THE MOST VALUABLE PART OF THE skeleton because it contains so many elements under intense natural selection, including teeth, jaws, sensory organs, and brains. The brain arguably represents the most dramatic story line of all human evolution. Our brain is roughly three times larger than those of our closest ape relatives. If the rest of our body had grown at the same rate, we would stand as tall as giraffes. For most of human history—including the time of Ardi and Lucy—the brain expanded slowly or not at all. In the last 2 million years, it roughly doubled in size until attaining modern volume by around two hundred thousand years ago. Yet fully modern *behavior*—at least as expressed by sophisticated tools, art, and symbolism—appeared much later. Human intelligence was not simply a matter of brain size and probably involved changes in neurological circuitry and organization. A weak, slow primate became a force of nature—all because of a neurological drama inside the crucible of our skull.

Nature has designed the skull for durability. Skull pieces are the second most common fossils behind teeth. The human skull consists of twenty-eight bones: eight in the braincase, fourteen in the face, and six tiny auditory ossicles of the inner ear. (The tiny bones of the inner ear and paper-thin bones of the sinuses are rare in the fossil record.) The excavation team recovered most of Ardi's cranial vault, parts of the skull base, right face, the left mandible, and most of the teeth. The cranium had been crushed like an eggshell and compressed to about one inch thick. White speculated it had been squashed by a hippo or elephant as Ardi's carcass lay on the ancient floodplain. Some pieces were so delicate that he left them embedded in earthen matrix. The team made plaster casts of the pieces—which could be separated without any risk—and glued them into a crude prototype. They suspected they could do better, and placed their hopes in Gen Suwa and his computer.

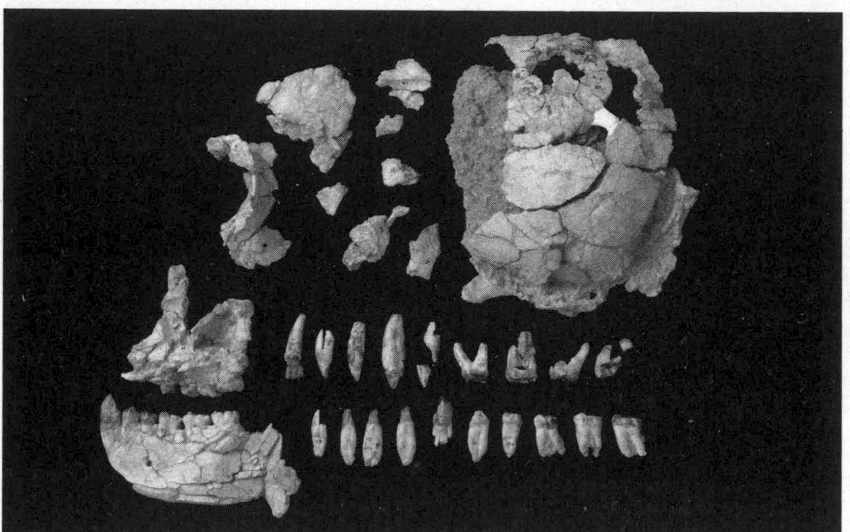

Photograph © 2001 David L. Brill

The pieces of the Ardi skull after cleaning. The pieces were molded to make physical casts and scanned to make images for digital reconstruction.

In 2007, Suwa devoted himself to the skull reconstruction to reverse the distortion of deep time. His team took more than five thousand CT scans of the skull fragments and created mirror images to replicate missing parts. Tiny landmarks provided hints, such as zigzag sutures, branch-like impressions left by blood vessels on the interior braincase, or holes where nerves and blood vessels once passed through the skull. Some scientists outsource their judgment to technology, plug in measurements, and trust the software to spit out answers. Suwa used his computer as a digital modeling studio where he rebuilt Ardi's skull millimeter by millimeter. Beside his computer he kept skulls of chimps and human ancestors for comparison. He cross-checked his placements against statistics from other primates. Each piece would be oriented as precisely as possible, checked against multiple lines of evidence, and adjusted again. The reconstruction became an exercise in anatomy, geometry, statistics, and the unquantifiable judgment of the expert.

Suwa was a slow, methodical scientist who submerged himself in work. Problem was, he kept drowning in *other* work. He managed the collections at the University of Tokyo museum, taught students, and supervised research. All day long, people peppered him with questions and broke his concentration. "You can't work like that on the reconstruction," he said. "You have to focus, *deeply* focus, and forget about everything else."

That left one option: moonlighting. After his colleagues went home, Suwa put in another eight-hour day. Sometimes he worked after the commuter trains stopped running, checked into a hotel near the museum, slept a few hours, and returned early in the morning to spend more time tinkering with his skull before his day job resumed. He paid for lodging out of his own pocket because no research grant would support someone sleeping in a hotel in his hometown. And he devoted his three-week summer vacation to his skull reconstruction—"literally a forced work period," he explained.

One important piece, the cranial base, remained missing. Fortunately, another ravaged *Ardipithecus* came to the rescue. Before the discovery of the Ardi skeleton, White picked up two broken chunks of occipital and temporal bone, which form part of the floor of the skull where it rests atop the spine. Gouges showed that poor creature had been mauled by carnivores who chewed the temporalis muscle down to the bone. Even so, the parietal and occipital bones remained fused, a testimony to the durability of the zigzag sutures that join the skull bones.

Both the ravaged cranial base and Ardi preserved a small portion of the same suture joint, which enabled Suwa to scale down the fragment to Ardi's size (92 percent of the original, to be exact) and add it to his reconstruction. The temporal bone also contains the opening of the ear canals and the delicate organs of hearing and balance (in life, these labyrinths are filled with fluid and tiny hairs that sense movement and body orientation). Suwa used high-resolution scans of the semicircular canals to fine-tune his placement. He made similar adjustments over hundreds of landmarks with increasing precision.

Anatomists have long recognized that the cranial base distinguishes humans from other apes. In humans, the floor beneath the brain is shorter (front to back) and wider than it is on apes. (Some scholars ascribe the unique human cranial base to upright locomotion, while others attribute it to an expanding brain.) This anatomical change at the base of the skull already had been documented in the human family back to *Australopithecus afarensis*—and it also was visible in *Ardipithecus*.

There was more. At the bottom of the skull lies a hole called

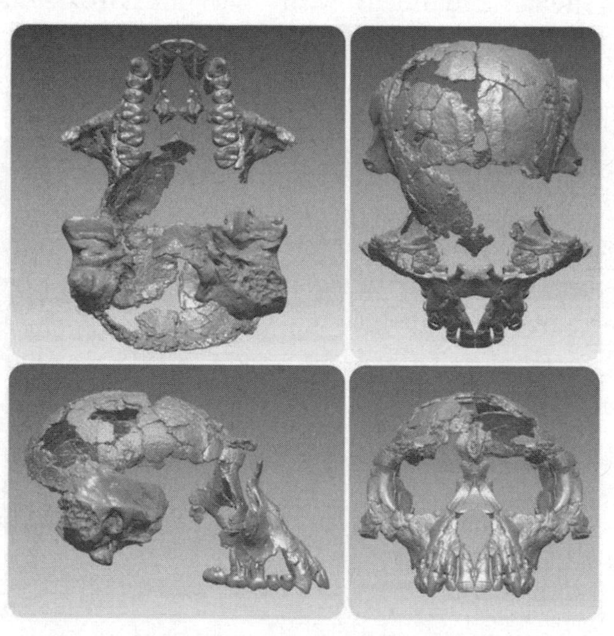

Gen Suwa's virtual reconstruction of the Ardi cranium.

the foramen magnum, where the spinal column enters the braincase. Just forward of the foramen magnum lies a portion of the cranial base known as the clivus, after the Latin word for "slope" because it angles steeply upward. In humans, the cranial base is "flexed" more toward a right angle; in chimps, this angle is more obtuse. In *Ardipithecus*, the clivus angled steeply upward, just as it did in other members of the human family. Despite its small brain, Ardi displayed what the investigators later described as "tantalizing evidence" of early neural reorganization.

Likewise, the *Sahelanthropus* skull from Chad, which was at least 1.5 million years older than Ardi, also had a cranial base that allied it with the human family. Even with ape-size brains, these earliest human ancestors displayed some kind of anatomical change at the base of the skull that distinguished them from modern African apes.

The skull rests atop the Atlas vertebrae, the topmost segment of the spine, named after the figure of Greek mythology who supported the globe on his shoulders. In chimps, this joint sits more to the rear and the spine enters from a more acute angle. In *Ardipithecus*, the joint sat almost as far forward as the ear opening, similar

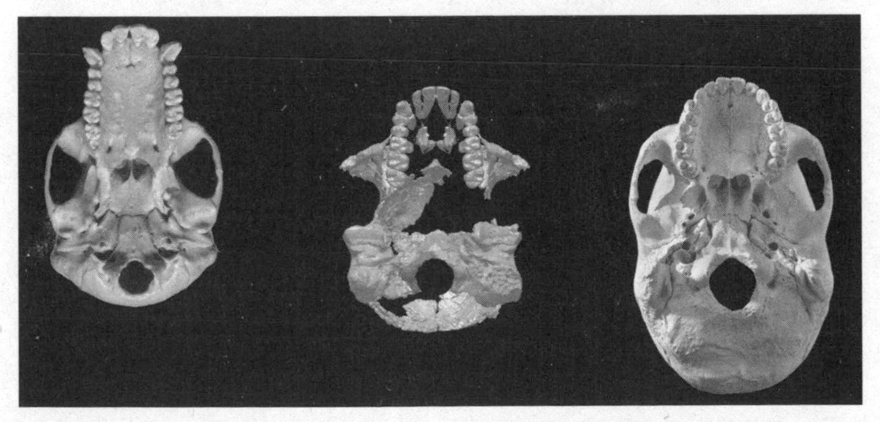

The underside of the crania of a chimpanzee (*left*), Ardi (*center*), and human (*right*). The skulls are aligned by the front of the foramen magnum, the hole for the spine. The comparison shows how chimps have evolved a more elongated cranial base and prognathous face. Ardi had a short cranial base, another clue that allied her species with the human clade.

to humans—another indication that the head rested on an upright spine.

When Ardi's snouted skull was laid beside the flat-faced, globular *sapiens* skull from Herto, the effect of 4 million years of evolution became dramatically apparent: it looked like a little melon beside a soccer ball. By filling the Herto man's skull with teff seeds (the grain used in Ethiopia's signature njera bread), the investigators estimated its brain volume to be 1,450 cubic centimeters, at the high end of the modern human range. Ardi had a brain less than one-fourth that size—another vivid example of evolutionary change documented in the Afar fossil fields.

The skull reconstruction took two years. Halfway through, Suwa took his prototype to Ethiopia, compared it to the original fossils, and returned to Tokyo to continue tweaking. He finished in January 2009, only a few months before the Ardi team submitted its final report for publication. He called the last version R10—the tenth iteration of the skull.

Meanwhile, the digital reconstruction faced another test—Tim White.

Back in Berkeley, he constructed another plaster model by hand the old-fashioned way. Even at the cost of duplicating effort, he believed in cross-checking. White had gone through a similar exercise back in the Lucy era. In 1979, he and Bill Kimbel reconstructed an *Au. afarensis* skull from many pieces. At that point, no complete skull of Lucy's species had been recovered, so the anthropologists created a composite from thirteen different individuals (including a partial cranium smashed into one hundred fragments). The scientists lined tabletops with rows of chimp and gorilla skulls and constantly consulted them as models for their bony puzzle. When they were done, the Lucy team was struck by how much the reconstruction resembled a modern African ape with its prognathous snout. "That's an incredibly protruding face," the young White said as he showed the reconstruction to a documentary film crew. "That's more apelike than any known fossil hominid." In the spirit of the times, the Lucy team predicted that older human ancestors would become increasingly difficult to distinguish from apes. As the *afarensis* skull neared completion, White absented himself from the Cleveland Museum of Natural History and spent several weeks sequestered in the basement of Don Johanson's house. He returned to the lab to put some finishing touches on the skull, and in a moment of inattention his precious plaster replica rolled off the table and smashed on the floor. Colleagues heard an agonized scream and burst of profanity. White was inconsolable and insisted

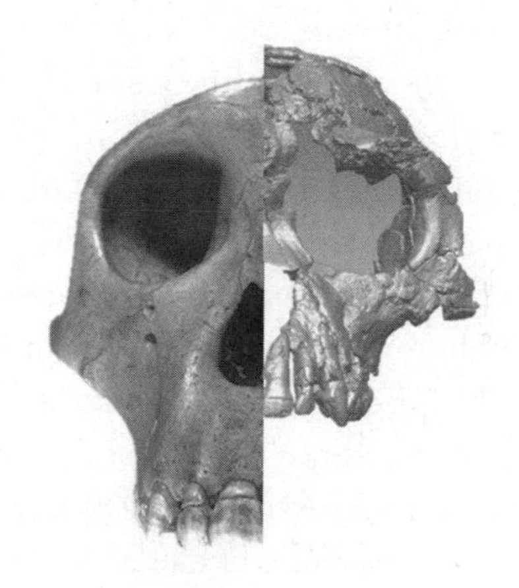

A side-by-side comparison of crania of a female chimpanzee (*left*) and Ardi (*right*). The chimp has more robust brow ridges and a deeper, more jutting snout with large canine teeth. Ardi has a smaller snout, reduced canines, and a higher-domed braincase. The images are scaled to equalize width of the eye socket and do not reflect true size differences.

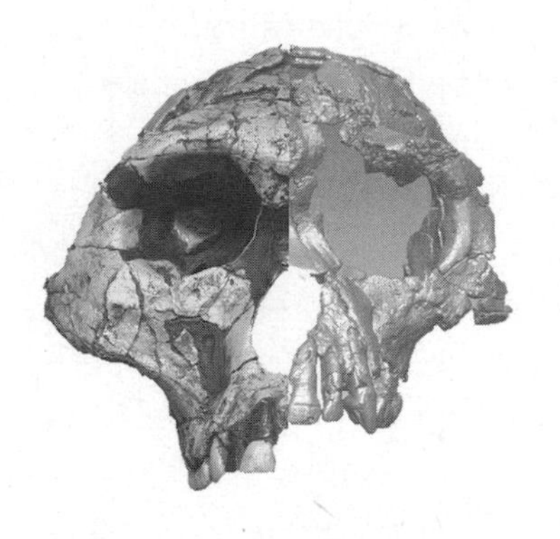

Another comparison shows the crania of *Australopithecus afarensis* (*left*) from 3.1 million years ago and *Ardipithecus ramidus* (*right*) from 4.4 million years ago. The *Au. afarensis* is another female of Lucy's species from Hadar. The *Au. afarensis* cranium has larger, more flaring cheek bones that anchor powerful chewing muscles, a distinguishing feature of that genus. The cheekbones of *Ardipithecus* are smaller and more typical of the primitive ape condition. Note the overall morphological similarities between the two skulls. Both include the higher-domed braincase and the abbreviated portion of the snout below the nasal opening, which houses the reduced roots of the canine teeth.

Illustration as well as chimpanzee and afarensis photographs by William Kimbel and used by permission.

he could not possibly go through the ordeal again, but eventually collected himself and rebuilt his model. He had learned a painful lesson. In his *Human Osteology* textbook, he advised using sandbags or cloth to prevent skulls from rolling off the surface.

Fast-forward thirty years. White was twice as old, his skull a million years older, and this time something else shattered—his expectations. In the first announcement of the species, his team had described the *ramidus* cranial bones with "strikingly chimpanzee-like morphology"—but that was based on scrappy pieces discovered before the skeleton. As the scientists rebuilt Ardi's skull in minute detail years later, they were struck by the opposite—features *unlike* our closest ape relatives. Chimpanzees have big snouts with jutting canines and broad incisors. Ardi's snout was shorter, shallower, and lacked self-sharpening canines and wide incisors. The older *Sahelanthropus* skull likewise bore small canines with no evidence of honing, a short cranial base and small brain. (Its skull was larger with more robust brow ridges, and the Ardi team also attributed those differences to sexual dimorphism because Ardi was a smaller female and *Sahelanthropus* a larger male.) The Ardi team suggested these two skulls belonged to the same genus and probably were close to the "ancestral state" of the last common ancestor of humans and African apes—yet looked surprisingly dissimilar to our living cousins. Their old notion of *primitive* had come unglued. A broken skull came together, and old notions fell to pieces.

"Those expectations of getting more chimplike as we go back, we never should have had them in the first place," White reflected years later. "You fill the void with analogy. That's what happened in the entire field."

WHILE THE FOSSIL TEAM WAS RETHINKING THE EVOLUTION OF THE HUMAN skull, neuroscientists elsewhere were reconsidering what happened within it. A framed picture above White's desk at Berkeley opened a window into that story.

It was a Victorian-era caricature of Thomas Henry Huxley, the British anatomist known as "Darwin's bulldog" for his combative defense of the theory of evolution. Huxley grew up in the dockside slums in a Dickensian London, and by his own account he had a "wild-cat element in me which, when roused, made up for my lack of weight." In the cartoon, Huxley had an oversize head with mutton chops and a lacerating gaze. In life, he wielded a tongue as sharp as his scalpel. "Cutting up monkeys was his forte," observed one writer in 1870, "and cutting up men was his foible."

Huxley left a century-long stamp on how evolutionists viewed the brain. Most famously, he fought a historic battle against Richard Owen, superintendent of natural history at the British Museum. Owen tried to draw a clear divide by creating a new taxonomic group for humans called *archencephala* (ruling brain) based on the false claim that only we had certain cerebral structures—specifically a posterior lobe, a posterior horn in their lateral ventricles, and a hippocampus minor. Huxley recognized that Owen's argument was bunk and spent three years dissecting primate brains to demonstrate that apes possessed all the same parts. This oft-told history established a model: human and ape brains shared common design and differed only in size, proportions, and the complexity of the folds. In short, the human brain was a reproportioned ape brain. Darwin called humans "big brained apes" who differed from higher animals only in "degree and not of kind." This view reigned for more than a century. Neuroscientists spoke of the "basic uniformity" of the mammalian brain and assumed that the basic cellular hardware was shared between humans and apes. Many scientists resisted the notion that the human brain was unique because the very thought seemed too anthropocentric and suggested that humans stood apart from the rest of nature.

Yet for most of the last century the study of brain evolution remained impoverished by lack of data. Soft tissues do not fossilize, so the record was limited to ancient skulls, endocasts (fossilized infill formed within the braincase), and evidence of behavior (such as archeology). Studying brain evolution with fossil skulls was like trying to reconstruct the advance of the computer by looking at the chassis. Changes in shape revealed little about internal circuitry. Since the oldest fossil skulls had brains of similar size to modern apes, chimps were widely studied as models of early human cognition.

All that began to change around the time Ardi was being reconstructed. Medical imaging technologies such as MRI and PET scans illuminated previously invisible structures in living brains. New techniques allowed researchers to label specific nerve cells and neurotransmitters and examine microscopic differences across species. Breakthroughs in molecular biology revealed how specific genes were expressed in the brain. With these tools, scientists documented a growing catalog of differences between human and chimp brains in cellular architecture, connectivity, morphology, gene expression, and biochemistry. Neuroscientist Todd Preuss of Emory University summed up: "Human brains are thus not simply scaled-up ape or monkey brains: they are rife with differences in virtually every system that has been examined in detail to date, and at virtually every level of organization, from the genome up." Humans were not just apes with enlarged brains, but *rewired* brains.

But it was hard to rethink old habits about thinking. For decades, chimps had been viewed as proxies for protohumans. In 2004, Daniel John Povinelli, a neuroscientist at the University of Louisiana at Lafayette, bemoaned that many experts had "radically anthropomorphized" chimpanzees and strained to find precursors of human behavior in our closest relatives. "After several decades of being fed a diet heavy on exaggerated claims of the degree of mental continuity between humans and apes," he wrote, "many scientists and laypersons alike now find it difficult to confront the existence of radical differences."

With the reconstructed Ardi skull, stark differences literally stared the investigators right in the face.

Chapter 24
ALL THAT REMAINS

For most of his life, Owen Lovejoy had been solving puzzles of bones. The ones exhumed from graves were the easy ones. More difficult were those butchered by other humans.

Every so often, the medical examiner called asking for help with murder cases. Putting aside the prehistoric, Lovejoy dutifully plunged into grim work: bones raked from furnace ashes, a decomposed corpse pulled from the Cleveland sewer, or dismembered bodies from lairs of serial killers. In one assignment, he reconstructed the remains of the first victim of "Milwaukee Cannibal" Jeffrey Dahmer and matched part of the vertebral column to a missing hitchhiker (Dahmer went on to murder seventeen people and eat some of them before he was caught). Another time Lovejoy drove home from the morgue and a maggot leapt from his pocket. "They can jump a foot and a half!" he exclaimed. Maggots and blowflies had their charms: they ate rotting tissues. Much worse were the bodies the maggots couldn't get to—especially the ones that killers left inside plastic bags. He never forgot the stench.

Some cases required creative investigation. In 1999, the Cuyahoga County coroner asked Lovejoy to help investigate the 1954 murder of Marilyn Reece Sheppard, the victim in the famous murder trial of her husband Sam Sheppard. (The case had been reopened after the couple's son had filed a wrongful imprisonment lawsuit against the state of Ohio.) The body was exhumed, and Lovejoy provided an expert opinion about the likely murder weapon. He obtained a cadaver skull, filled it with suet to simulate the fatty brain tissue, and bludgeoned it with a flashlight, fireplace poker, large pliers, and heavy wrench. First finding: the professor lacked killer instinct. "Not being practiced in this endeavor," he reported to the coroner, "I missed my primary target (the frontal bone) and delivered only a glancing blow . . . No damage to the cranium." No doubt, his friend Tim White would have smashed it on the first try.

In Ardi, Lovejoy took on the ultimate cold case.

CLASSICAL ANATOMISTS NAMED THE PELVIS AFTER THE LATIN WORD FOR "basin." Often called the keystone of locomotion, it would be the most informative part of the skeleton in deciphering how Ardi moved. It also would be the most difficult because no part suffered more damage.

The pelvis consists of three main pieces: two hip bones (one on the left and another on the right) and the sacrum (the triangular section of fused vertebrae at the bottom of the spine). Each hip bone actually develops as three separate bones that join into one by adulthood. These pieces are the *ilium* (the upper blade whose flaring edges can be felt on the sides and back of your waist), *ischium* (the lower protuberance upon which we sit), and the *pubis* (where the two halves of the pelvis meet in our private region). Ardi preserved pieces of all three.

But the punishment inflicted by geology was unlike anything done by a modern murderer. The left hip bone was mostly complete, but the pelvic blade was warped and lacking its upper edge. Portions of the right hip bone were badly damaged. The superior pubic ramus (a portion in the groin area) was broken into three sections. The center of the hip bone contains a cupped depression, known as the acetabulum (from the Latin for "vinegar cup"), where the head of the femur forms a ball-and-socket joint with the pelvis. It acts as a fulcrum that bears the weight of the body (and up to five times the body weight when running). Ardi's socket had been wedged apart by expanding matrix. Most of the sacrum and the upper pelvic blades remained missing.

The CT scans allowed the scientists to accomplish what hands and tools could not. Even so, digital reproductions could not surmount a fundamental limitation: there simply wasn't enough for a complete replica, and investigators could re-create some pieces only by inference. White insisted they not call it a proper reconstruction but a *sculpture*—a 3D hypothesis of what they *thought* the original looked like. Luckily, some of the most informative anatomy had survived to provide clues.

Two decades earlier, Lovejoy rebuilt Lucy's pelvis and demonstrated how much she differed from an ape. In chimps, the pelvic blades are much taller than in humans. Ape hip bones are oriented with their faces front to back. From the side they seem like thin paddles. Such orientation anchors large muscles for climbing. In humans, the hip bones are rotated more forward and laterally to form a bowl shape—hence the name bestowed by classical anatomists. As a consequence, the hip muscles are positioned more to the side of the body to balance on one leg. When a biped swings a leg off the ground, the hip muscles on the opposite side flex to keep the body upright. Lucy's pelvic blades were shorter and more flared laterally than a chimpanzee's. Muscle attachments on her pelvis and upper thighbone revealed a bipedal leg musculature. Likewise, her upper thighbone near the hip joint showed a bipedal pattern of spongy bone, an adaptation to the compressive forces of upright walking. In short, Lucy's pelvis already had adapted to balance on one foot. Upright walking, Lovejoy insisted, was "perfected" by the time of Lucy. He even deemed Lucy a *better* biped than modern humans because her pelvis had not yet been compromised by a birth canal for big-brained babies (*Au. afarensis* brains were only marginally bigger than those of chimps). Over the next 3 million years, brains expanded more than threefold and human ancestors ran into the "obstetric dilemma"—the conflict between locomotion (which favors a narrow pelvis) and birthing big-brained babies (which favors a wider pelvis). Subsequent pelvic changes were not *improvements* for locomotion, Lovejoy argued, but *sacrifices* for expanding brains. The ability to walk and run suffered—an evolutionary trade-off. Opponents savaged Lovejoy's reconstruction, but he waved them away and said they lacked understanding of biomechanics and anatomy. He remained steadfast that bipedality was so firmly established by Lucy's time that human ancestors must have been upright for a very long time before that.

In hindsight, Lucy had been easy—her pelvis was like fixing a broken teacup. Ardi was more like a teacup crushed in a grindstone. For help, Lovejoy enlisted a former student named Linda Spurlock, a forensic archeologist who taught crime scene technicians. She dressed pig carcasses in human clothes (sometimes sexy lingerie), shot, stabbed, or ran over the bodies, buried the remains in shallow graves, and assigned students to exhume and interpret the evidence. For seven years, Lovejoy and Spurlock sculpted iterations of Ardi's pelvis. "We'd get it to where we felt it was perfect, make a cast, and send it to Gen," said Lovejoy. "And he'd tell us where it wasn't perfect." Sending your idea to Suwa was like donating your body to an anatomist: you knew he would slice you apart. Back they'd go for another round. And another. "He's the most exacting person I've ever met, save one—that would be Tim White," said Lovejoy. "It's torture."

THE PELVIS IS AN ANATOMICAL CONGRESS WHERE COMPETING DEMANDS make compromises. Early in his career, Lovejoy and his fellow excavators at Libben unearthed a chilling example of a woman who died with a fetus stuck in her birth canal. Small-brained apes rarely suffer birth complications—nor, most likely, did *Ardipithecus*. Her compromises involved two forms of locomotion—walking on the ground and climbing.

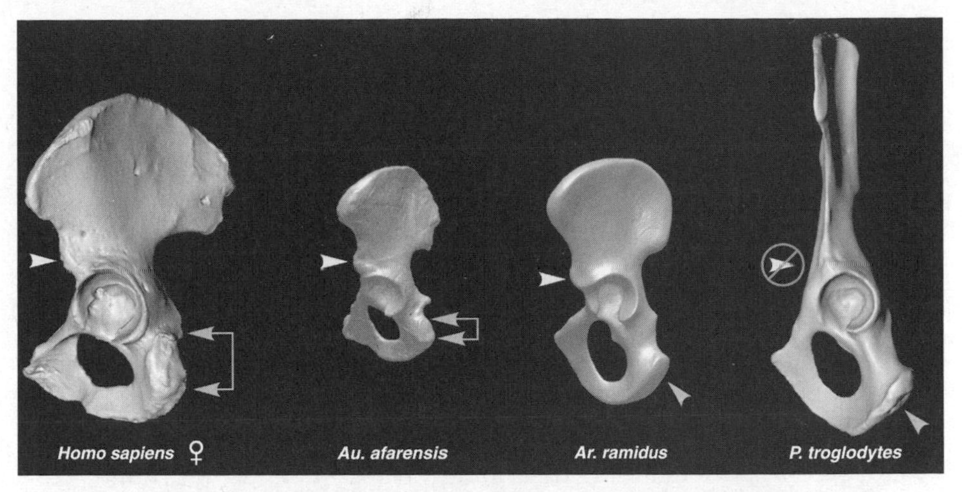

A comparison of hip bones. (*Left to right:*) human, Lucy, Ardi, and chimpanzee. In the three members of the human family (*at left*), the pelvic blades are short and wide but in chimps they are tall. The members of the human family also show bumps for the anterior inferior iliac spine (*single arrow to left of each specimen*) to anchor muscles for upright locomotion. In humans and Lucy, the ischial surface (*double arrows*) is angled to anchor the hamstring muscles for upright locomotion. In Ardi and chimps, the same structure (*single gray arrows lower right*) faces downward for climbing muscles. Ardi's pelvis combined the anatomy of both climbers and bipeds.

Despite their fondness for likening bones to familiar objects, early anatomists were stumped by the hip bone because it resembled nothing they recognized. It became the *innominate*, or the nameless bone. Ironically, the bone so elusive in description is very elucidative for locomotion. No part of the skeleton differs more between humans and chimpanzees, a distinction noted by Edward Tyson in his pioneering dissection of a chimp in 1699.

In their first glimpses of the pelvis, White and Lovejoy recognized a vital clue—a prominent bump of bone of the anterior inferior iliac spine. This portion of the pelvis anchors the rectus femoris (part of the quadriceps muscles of the thigh) and the iliofemoral ligament (the strongest ligament in the body). The structure is unique to bipeds of the human family—and visible on Ardi. (Some apes show small bumps but none so robust as ours.)

Then things became more complicated.

The lower part of the pelvis, the ischium, bears two prominences called the ischial tuberosities, commonly called the "sitting bones" because we rest on them when we plop on our behinds. In chimps, the ischium anchors powerful climbing muscles. In Lucy and other bipeds of the human family, this structure had shrunk compared to apes and signaled to Lovejoy that they had sacrificed climbing to enhance upright locomotion on the ground.

But Ardi had not yet made this shift. As he cleaned the fossil, White uncovered a long ischium—an apelike climbing adaptation never before seen in a member of the human family. In one area after another, the investigators had been surprised to find Ardi unlike chimps and gorillas. But in the lower pelvis, Ardi proved *similar* to arboreal apes. The pelvis, like the foot, seemed a hybrid adapted to both climbing trees and moving erect on the ground.

The pelvis reconstruction spanned fifteen years, three continents, and eleven iterations. Even after staring at it for so long, Lovejoy never outgrew his shock at the final result. "There was a divide, a continental divide, between the upper and lower pelvis," he said in wonderment. The humanlike upper half seemed designed for walking upright while the apelike lower half seemed built for climbing. Before Ardi, Lovejoy would have dismissed the notion of such a hybrid as a violation of evolutionary law. With his own hands, he had reconstructed the creature he once thought impossible.

ALONG WITH MANY OTHER FEATURES OF THE SKELETON, THAT STRANGE pelvis afflicted the Ardi team with uncomfortable questions. Could the hybrid pelvis of *Ardipithecus* have evolved into the bipedal pelvis of *Australopithecus* in only a few hundred thousand years? Ardi lived 4.4 million years ago, and the first known species of *Australopithecus* appeared soon afterward, by 4.2 million years ago. Moreover, could chimps and gorillas have evolved their pelvic similarities independently? To answer such questions, Lovejoy turned away from bones to molecules, fruit flies, sea urchins, and little fish.

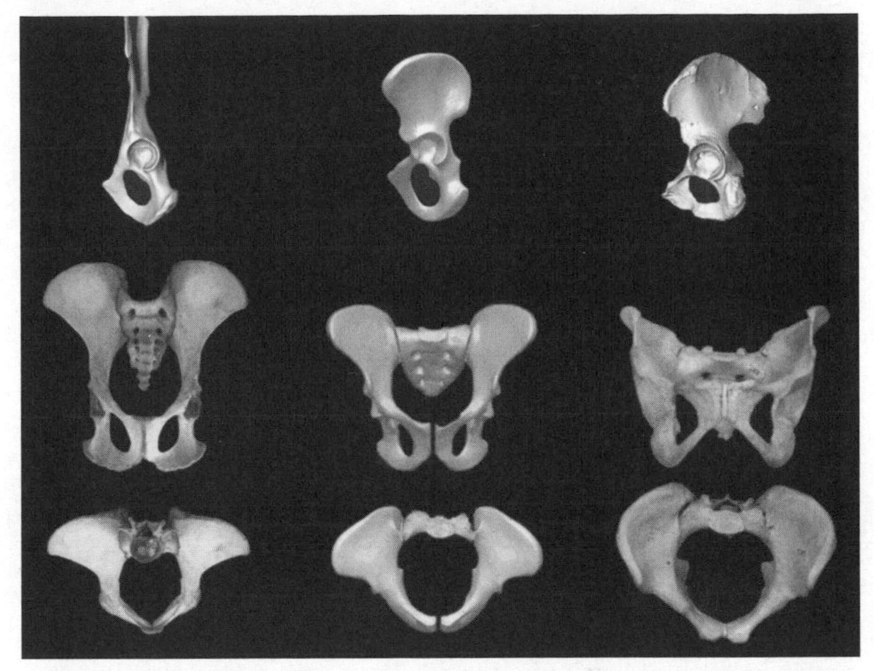

A comparison of the pelvis of chimpanzee (*left*), Ardi (*center*), and human (*right*). The chimp has tall pelvic blades with a narrow isthmus in the middle of the hip bone. In humans, the pelvic blades have rotated to form more of a bowl shape. The Ardi reconstruction combined elements of both a bipedal walker and an arboreal climber.

A few years prior to the discovery of *Ardipithecus,* Kent State University named Lovejoy a "university professor," a lifetime appointment that relieved him of teaching responsibilities and freed him to research and think. Thus liberated, he set out to write his masterwork on the evolution of human locomotion. Until that point, his work had focused on adult skeletons and biomechanics, and he'd never thought much about how bones formed in the first place. He began catching up on the literature about how bodies developed in utero. "I sat down to write the first chapter," he said, "and lost my mind."

Since the mid-twentieth century, evolutionary biology had operated under the paradigm of the "modern synthesis"—the model that took root in the 1940s and hit paleoanthropology at the Cold

Spring Harbor conference of 1950. The synthesis ascribed big evolutionary changes—even the emergence of new species—to the cumulative effects of microevolution (small incremental changes) over vast stretches of time. Yet the synthesis had one big shortcoming: it failed to explain *how* forms changed. When the synthesis was articulated in the 1940s and 1950s, the gene remained an abstract concept and scientists had little understanding of the mechanisms that translated genetic code into morphology. How did fertilized eggs develop into complex creatures? As noted by molecular biologist Sean B. Carroll, the synthesis treated embryology as a "black box"—an unknown process that somehow turned genetic code into fully formed animals. Was gradual variation really sufficient to produce entirely new species? Turn fish into four-legged animals, land mammals into whales, and arboreal apes into bipedal humans?

Scholars had long noted so-called *homologous* body parts—such as mammal forelimbs and bird wings—as evidence of common descent, but synthesis biologists doubted the underlying genes were homologous and assumed that the genetics of animals became increasingly dissimilar with the passage of time. That belief was squashed by the humble fruit fly.

In the 1980s and 1990s, a series of breakthroughs opened a new field of molecular biology known as evolutionary developmental biology, or evo-devo. Researchers discovered that animals as diverse as fruit flies, mice, and humans shared basic sets of "tool kit" genes that had changed little over hundreds of millions of years. Mouse eye genes could be put into *Drosophila* fruit flies and produce normal insect compound eyes (with many visual receptors). Similar limb-building genes made fly wings, fish fins, chicken wings, and human arms and legs. Genes that formed fly body segments were redeployed to form the segmented spines of vertebrates. Evo-devo brought a breakthrough in the fundamental problem identified by Darwin—the origin of species. Nature didn't rewrite every line of genetic code from scratch to spawn evolutionary novelty. Instead, it largely reused old code in new ways. The main driver of evolution was *regulatory*—genetic switches that turned ancient tool-kit genes on and off.

Genes include both functional protein-building elements and regulatory elements that control their deployment. In other words, there are tools for building things and instructions about when, where, and how those tools should be used. The tools—the structural genes—remain little changed even between distantly related species over hundreds of millions of years. But the instruction manuals—the regulatory genes—used the trusty old tools in novel ways to create diversity in the animal kingdom. Such complex and varied inputs meant that shifts in timing, spatial pattern, or vigor of gene expression allowed tool-kit genes to be *pleiotropic* (Greek for "many ways"), or to have multiple effects. In 1976, Emile Zuckerkandl, one of the pioneers of molecular biology, mused that human genes, with the right regulatory tweaks, theoretically could be used to make *any* primate.

Traditionally, most biologists had assumed that evolution advanced slowly by mutations in structural genes. Evo-devo blew apart such old notions of gradualism and showed that shifts in gene regulation could bring rapid changes in physical form. In the fossil record, where tens of thousands of years get compressed into thin strata, big transformations could appear out of nowhere. Parallel evolution (the independent acquisition of similar traits by different species) ceased to seem so unlikely because animals with similar tool kits are likely to make similar adaptations.

Evo-devo struck Lovejoy like a flash of light. He came to the painful realization that much of what he thought he knew about bones and evolution was simply wrong. His mentors had trained him to think biomechanically, but now he retrained himself to think developmentally. Instead of bony hardware, he began talking of genetic software. "Eventually the scales fall off your eyes," he said, "and you realize this mechanical stuff is a dead end."

Lovejoy's conversion coincided with the arrival of a pupil named

Marty Cohn. Cohn began as a traditional anthropology graduate student with little biological background. He had come to Kent State in 1991 to study the biomechanics of chimp knuckle walking (on the recommendation of Sherwood Washburn, who long believed it to be the precursor of bipedality). Lovejoy quickly steered his student in another direction. One night in the lab, Cohn remarked that anthropology seemed fixated on explaining the *why* of evolution but was oblivious to the *how*. Lovejoy asked how much Cohn knew about *Drosophila* and homeobox genes. *Huh?* "The answer was the same for both—nothing," recalled Cohn.

"Tabula rasa!" exclaimed Lovejoy. "You need to start reading development."

Lovejoy hired Cohn as a research assistant, and together they waded through the burgeoning literature. After a few months, the two men decided they needed to import new techniques into their lab, so Cohn found a summer job at University College London under developmental biologist Cheryll Tickle. (Down the hall was Lewis Wolpert, a pioneer in the field who developed two fundamental concepts, known as positional information and pattern formation, that governed embryonic organization.) Cohn and Lovejoy spoke by phone every week. Sometimes Cohn reported a new discovery, and Lovejoy responded by leaping ahead to some insight that the student did not grasp until much later. Lovejoy advised his student to forget about anthropology, remain at UCL for his Ph.D., and become a biologist: "For god's sake, stay," he advised. "It's a lot better career than working with old dead hominids."

Cohn studied molecular mechanisms of limb development, and one experiment provided a vivid lesson. Cohn applied a growth factor—a cell signaling protein—to a chicken embryo. That brief signal mimicked the effect of turning on a single gene switch and induced the formation of an extra limb with normal digits, skin, nerves, muscle, and tendons. "When I saw that, I realized that's one gene that sits at a top of hierarchy of thousands of downstream signals," said Cohn, now a prominent researcher at the University of Florida. "That gave me a real appreciation for how many downstream effects followed this single switch being thrown." One small genetic tweak and *boom*—a wholesale transformation across the organism.

Some theorists already had sensed the revolutionary implications of developmental biology. In 1979, Stephen J. Gould and Richard Lewontin wrote a famous essay warning that evolutionists had become infected by an "adaptationist" mind-set—the mistaken belief that *any* new trait signaled some vital adaptation forged by natural selection. In fact, Gould and Lewontin argued, some evolutionary changes were collateral effects of big cascades—nothing more than excess baggage with zero evolutionary value. They coined the term *spandrel*, named after the ornamental features on cathedrals that had no functional purpose. To evolutionists, spandrels were fool's gold. The fossil record, Lovejoy believed, was *full* of spandrels, and careers were made concocting "just so" stories about them.

Lovejoy began to view evolution in terms of package deals, often with lots of excess baggage and trade-offs. The key was to look for *patterns*—packages of traits that revealed the overall direction of natural selection. (Earlier theorists had made similar arguments about total morphological patterns, body plans, and so on.) Lovejoy never finished his book; the introductory chapters on bone development dispatched him on a detour from which he never returned. On one hand, he spoke like a morphologist from the bygone century and thus seemed archaic to some contemporaries. On the other hand, he spoke a language of developmental biology that most anthropologists did not comprehend or regarded as hocus pocus. "I have to be honest with you, I read that stuff and I can't even understand what he's saying," said Jack Stern of Stony Brook University, an old rival from the locomotion wars. "If he turns out to be right, it will just have gone over my head." Another Stony Brook anthropologist, Bill Jungers, scoffed: "Owen's armchair evo-devo has not swayed many people." He was right about that. In 1999, Lovejoy, Cohn, and White published a paper in the *Proceedings of the National Academy of Sciences* that sought to introduce developmental biology to the profession of old bones. "We tried to bring paleoanthropology into the developmental era," said Cohn. "It didn't work."

But it did rework the minds of the Ardi investigators. The eyes that analyzed Ardi saw matters very differently from their younger selves who had interpreted Lucy. Some skeptics proclaimed *Ardipithecus* too primitive to give rise to *Australopithecus* only a few hundred thousand years later. To a traditional gradualist who expected to see incremental change in the fossil record, the shifts seemed discordantly abrupt. To a biologist gobsmacked by evo-devo, however, such rapid changes seemed entirely possible. The abrupt appearance of such shifts was accentuated by the nature of the fossil record in which eons of evolutionary change were compressed into thin geological strata. A few tweaks in the developmental parameters of pelvis and spine and *boom*—a quadruped could become an upright biped. A few changes in gene expression and bone growth plates and *voila*—radically different body proportions.

Such notions were deeply unsettling to a science conditioned to expect a step-by-step progression in the fossil record. Indeed, some skeptics would dismiss such ideas as wishful thinking—the equivalent of waving a magic wand to explain why an unexpected fossil still could belong in the human lineage. And if that wasn't enough, the final element of the Ardi skeleton would plunge even deeper into the theoretical realm. The pelvis obliged Lovejoy to reconstruct a structure that was badly damaged, but the next part remained almost entirely missing.

Chapter 25

SPINE QUA NON

The spine requires no metaphor: it literally forms the backbone of the body. Its interlocking segments link vertebrates from head to tail. It serves the main axis of the skeleton, anchor of limbs and conduit for nerves. In human evolution, it was a place of radical transformation. At some point, our trunk redeployed from horizontal to vertical and assumed a form unique in the animal kingdom—the S-shaped lordotic spine. This recurved spine centers our weight over the hips and feet, so standing requires only slightly more energy than sitting. The lordotic spine is the sine qua non of bipedality, and without it we could not manage our strange manner of walking and running.

This makeover came with severe costs. Standing erect required anatomical compromises that rendered the human spine vulnerable to compressed nerves, slipped discs, osteoarthritis, strained muscles and ligaments, sacroiliac joint dysfunction, and more (low back problems afflict 70 percent to 85 percent of people at some point in their lives). Such troubles are old news: even fossil skeletons are rife with spinal pathologies. Lucy might have suffered from the postural deformity Scheuermann's kyphosis. The best-known skeleton of *Homo erectus*, the 1.5-million-year-old Nariokotome Boy, has been diagnosed with scoliosis. Numerous Neanderthals from caves in Europe and the Middle East suffered from osteoarthritis.

The great diversity of vertebrates bears testimony to the spine's adaptability in evolution. With regulatory tweaks during embryonic development, the basic elements of the spine—the vertebrae—have been repurposed to form long-necked giraffes, long-thorax snakes, prehensile-tailed monkeys, stiff-backed apes, or upright humans.

The spine is divided into regions: the cervical (neck), thoracic (chest, with rib-bearing vertebrae), lumbar (lower back), sacral (fused vertebrae that form part of the pelvis), and caudal (tailbone). Evolution has done its work largely by transferring segments (the individual vertebra, or plural vertebrae) between spinal regions and altering the functions of these regions. In utero, Hox (short for homeobox) genes determine whether each segment develops into, for example, a rib-bearing thoracic, a low-back lumbar, a fused sacral, and so on. The architecture of the spine fundamentally shapes the overall body design. Major anatomical changes may result from redeploying individual vertebra from one region to another or altering the biomechanical roles of those regions.

In human history, the center of action has been the lower back—the lumbar region, where the spine curves inward just above the waist. Our low back is prone to injury from lifting, twisting, or sitting too long, but great apes have avoided this liability by trapping the lowest lumbar vertebrae between the blades of the pelvis. Chimps and gorillas rarely suffer from spinal pathologies. The stiff back of apes enables powerful climbing but makes them ungainly on two feet. Chimps and gorillas stand by leaning forward and bending the hips and knees to balance their center of mass atop the feet. They can walk upright, but only with an awkward "bent hips, bent knees" gait that generations of scientists have scrutinized for clues to the origins of bipedality.

Vertebrae are uncommon in the early human fossil record because they are fragile and rarely survive long enough to fossilize. Lacking historical evidence, anthropologists turned to our primate cousins and made inferences about the ancestral spine. Humans have a "mobile" lower back with an average of five lumbar vertebrae. In contrast, our cousin great apes have short, stiff lumbar columns with three or four lumbar vertebrae, and many scientists deduced that our spine evolved from an ancestor with such a

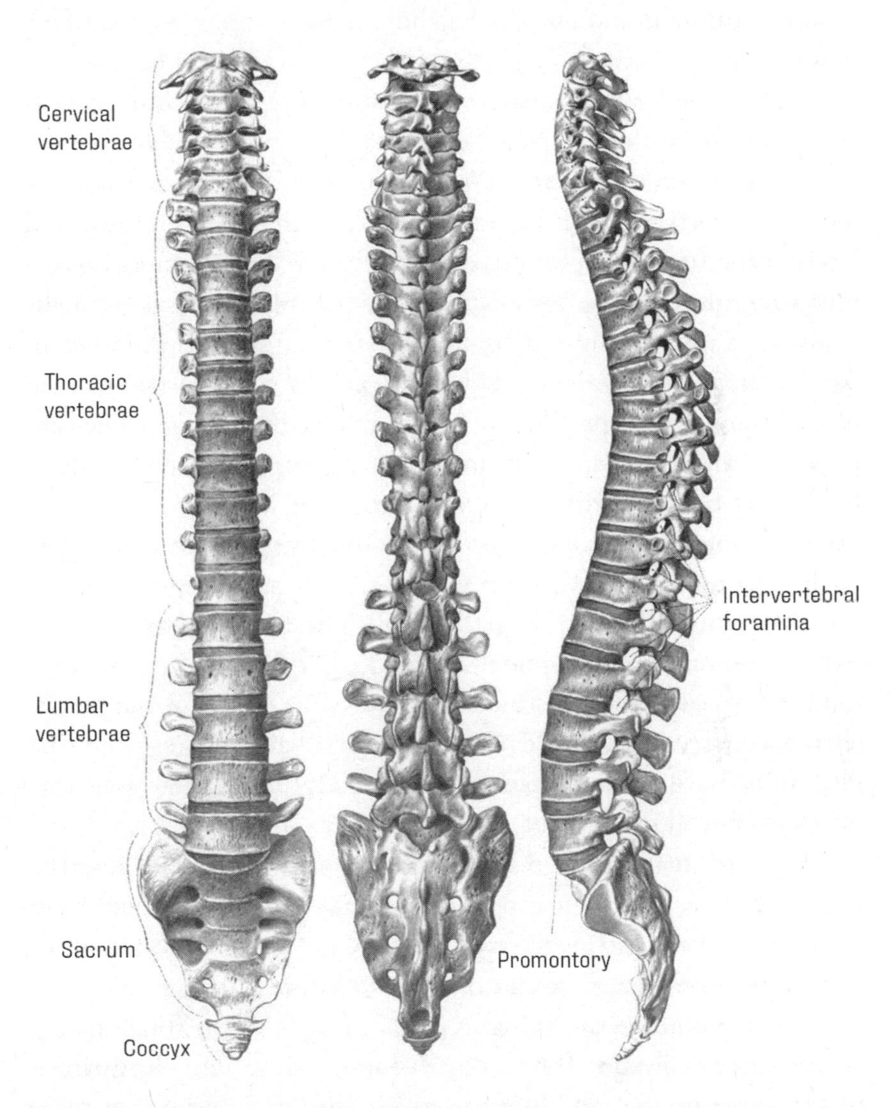

The sine qua non of upright locomotion: the S-shaped human spine shown in views from front (*left*), rear (*center*), and left side (*right*).

short, rigid lumbar column. This inference was based on comparative anatomy and logic: it seemed more likely that humans lost the stiff back rather than other living apes all gained stiff backs independently. For example, in 2004, David Pilbeam of Harvard published a lengthy analysis that concluded the last common ancestor of humans and apes had a short, rigid lumbar section like a modern ape.

Ardi made Lovejoy suspect otherwise. The field team had recovered only two identifiable vertebrae and a small bit of sacrum, so inferences about the *Ardipithecus* spine would have to be speculative. To Lovejoy, all the bipedal features of the skeleton suggested a pattern: Ardi must have had a mobile spine like a human—not a stiff back like our ape cousins. Lovejoy had remarried to an anatomist, Melanie McCollum, and together they parried theories. In particular, Lovejoy pondered the sacrum. In chimps, the sacrum is narrow. In humans, the sacrum forms a broader triangle and prevents the lower spine from becoming immobilized between the pelvic blades. Ardi was missing most of her sacrum, so that portion would have to be reconstructed from other clues. The Ardi team judged that a chimp-like narrow sacrum would leave too little room for Ardi's birth canal. The Lucy skeleton had a well-preserved sacrum—and it was wide, like a human's. The team made a replica of Lucy's, scaled it up to Ardi's size, and added it to their reconstruction. To Lovejoy, the whole pattern suggested that Ardi must have had a humanlike S-shaped lordotic spine—the sine qua non of walking upright.

The Ardi team reached a conclusion that defied the principle of parsimony: the other great apes each independently evolved their stiff spines *after* splitting from our common ancestors. To their surprise, the investigators found different patterns of how vertebrae were distributed in the spinal regions in our cousin species, suggesting they evolved stiff backs by different pathways. The number of vertebrae in apes and humans varies slightly between ape species and *within* each species (for example, the average human has

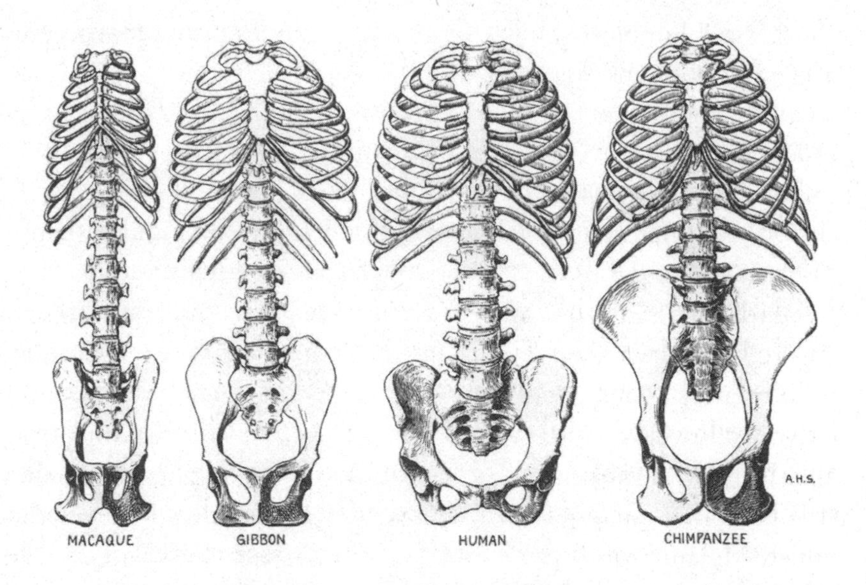

The comparative anatomy of the spine, ribs, and pelvis. (*From left to right:*) macaque monkey, gibbon, human, and chimpanzee. In the chimp, the tall pelvis blades trap the lower lumbar vertebrae and stiffen the back.

thirty-three individual segments, but some individuals have as few as thirty-one or as many as thirty-five).

Another body of evidence came from fossils. The record of ostensible human ancestors vanished beyond 6 million or 7 million years ago, but not so with Miocene apes older than 15 million years. Most Miocene apes walked on all fours with long, flexible horizontal spines. Two African fossil species, *Proconsul* and *Nacholapithecus,* had six lumbar vertebrae. Modern monkeys native to Africa and Asia have seven. Lovejoy and his colleagues became convinced that the long, flexible lumbar spine was the true primitive condition—not the stiff backs of modern apes—and humans never went through a stiff-backed stage. The Ardi scientists thus moved toward a conclusion that seemed so counterintuitive that they might scarcely have believed it a few years earlier: the human flexible spine and mobile lumbar region were closer to the ancestral state. Such findings echoed those about the foot and many other parts of the skeleton.

Could standing erect actually have required less drastic spinal remodeling than everybody had assumed?

Cue the dancing monkeys.

Japan has a thousand-year tradition of street monkey shows known as *Sarumawashi* starring macaques, a species native to the Japanese archipelago. From a young age, macaques are trained to stand on their hind legs and sometimes walk three kilometers per day. Researchers reported that trained macaques developed a "human-like gait." Unlike chimps or gorillas, they did not rock from side to side or plod with bent knees and bent hips. Instead, they strode with good balance, fully extended legs, and even developed lordotic spines (albeit by changing the soft tissues between the vertebrae, not the actual bones). Unlike modern apes, monkeys had lumbar spines unrestricted by the pelvic blades. Some new-world monkeys from the Americas such as spider monkeys also could walk upright with lordotic spines. Admittedly, monkeys are far more distantly related to humans than African apes, making them quite imperfect models for early humanity. Nonetheless, Lovejoy suspected they illustrated the spinal requirements for upright locomotion: a broad sacrum, short hip bones, and flexible spines. The Ardi reconstruction shared all these features, and they theorized the common ancestor of humans and African apes did, too.

Before Ardi, Lovejoy envisioned an ancestor like an African ape that underwent a major makeover to refashion a stiff back into a mobile lordotic spine. Even more complicated, his old theory obliged him to imagine that our lineage had a flexible back during the Miocene, lost it during a stiff-backed stage, then "re-evolved" a flexible spine. Ardi told him: *wrong.* He also had assumed that the lumbar spine must have been one of the earliest targets of natural selection to enable bipedality in human ancestors. Again, everybody had it backward: the lumbar region was more a target of natural selection to enable *climbing* in apes. The stiff back of a chimp or gorilla couldn't have given rise to our spine. To the contrary, he judged that a rigid back likely would have prevented human ancestors from ever standing up in the first place. Inducing chimps to shuffle along with bent knees and bent hips wasn't an accurate reconstruction of the evolution of human locomotion. Generations of researchers had been barking up the wrong tree.

ARDI CARRIED LOVEJOY BACK TO THE MYSTERY THAT BEGAN HIS CAREER. Why would a quadrupedal Miocene ape stand on two legs? He had a one-word answer—sex.

Two major adaptations allied Ardi with the human family—upright locomotion and canine reduction. Why would one lineage of primate sacrifice its ability to take shelter in the trees and impair its ability to fight? Why surrender features essential for survival?

Most theories sought to explain bipedality as some kind of direct advantage such as seeing over tall grass, energy efficiency, minimizing sun exposure, freeing the hands for tools, and so on. Having been trained by artificial joint engineers, Lovejoy saw bipedality as an invitation for disaster—blown knees and hips, thrown backs, torn hamstrings, and more. If speed and efficiency had been the targets of natural selection, our ancestors would have remained four-limbed walkers like every other ape and monkey. Upright walking was a lousy form of getting around—but a convenient way to *carry* things.

The real driver, he believed, was a mating strategy: human ancestors took to two feet to become monogamous. In biology, monogamy simply means pair bonding, not marriage in our modern sense. There are many examples of monogamy in the animal kingdom (and yes, some cheat on partners or divorce) including birds, wolves, beavers, swans, dart frogs, and gibbons. In Lovejoy's theory, males foraged cooperatively with other males, carried food back to the home base, and provisioned mates and children. Both sexes became child-rearing partners in multifamily troops and thereby increased survival of offspring. With this sexual division of labor, the nuclear family was born in what Lovejoy termed "sex for food exchanges." The reduction in canine weaponry testified that fighting ability was no longer a prime criterion for natural selection. Quite the opposite: females preferred provisioners rather than aggressors. Long before humans domesticated other animals, they domesticated *themselves*. In this social revolution, bipedality was just a means to an end—a compromise in locomotion for a bigger payoff in reproduction. Two-legged walking offered an escape from the demographic trap that was driving other apes to extinction.

In biological theory, there are two contrasting reproduction strategies known as r-selection and K-selection, which basically boils down to quantity versus quality.* The names derived from the abbreviations for reproductive rate (r) and carrying capacity (K). Species that practice r-selection are like mass producers of cheap goods: they beget many offspring with little investment of energy. They breed like rabbits—which indeed are prime examples of r-selection, as are bacteria and weeds. In contrast, K-selection species are like artisan breeders who invest more labor in fewer products. K species mature late, require prolonged child rearing, live long, and maintain populations near the carrying capacity of the local environment. They include larger animals such as elephants, tortoises, and, most important, apes. But apes doomed themselves by venturing too far out on the limb of K strategy. A female chimp does not reach sexual maturity until age ten and, with a five-year interval between births, does not replace herself and her mate until about age twenty-one. Humans are K-species in the extreme—big-brained apes that birth helpless infants who

---

* The theory of r- and K-selection was popularized in the 1960s and 1970s by biologists such as Robert MacArthur and E. O. Wilson, drew criticism as too simplistic, and spurred development of more complex models. Nonetheless, it remains a classic textbook concept. Lovejoy reframed his theory in terms of "adaptive suites," or packages of traits with a common reproductive strategy.

must be carried for the first year of life. Yet somehow human ancestors spread like weeds. "Orangutans are going extinct, gorillas are going extinct, and chimps are going extinct," Lovejoy noted. "But hominids are stinking everywhere. Explain that."

Natural selection boils down to one question: who produces more surviving offspring? The biology of any creature reflects its mode of reproducing itself. In primates, mate competition manifests itself in body size, canine fangs, female ovulation, and male sperm production. Our cousin species all have devised different strategies to be fruitful and multiply. Chimps live in promiscuous societies of swingers: males and females engage in brief hookups with multiple partners. Females signal estrus with sexual swelling around the genitals and attract multiple suitors. (Jane Goodall observed one female chimp who was followed by up to fourteen males and copulated fifty times in a single day.) Male competition produces bayonet-like fangs and moderately dimorphic bodies (males are about 30 percent larger than females). Males compete not only for females but *within* females. Chimps and bonobos fire more than a billion seed per ejaculation—up to an order of magnitude more than humans. Chimp testes are about three times larger than humans' and four times larger than gorillas'. Typical of polygamous species, chimp sperm swim faster and stronger than those of their two closest relatives. Chimps also have evolved the nifty trick of a sperm plug—the semen congeals to block the seed of any rival who comes along afterward.

To be sure, there are differences between the two species of the genus *Pan*—the common chimp (*Pan troglodytes*) and the bonobo (*Pan paniscus*). The two cousins separated about 2 million years ago (perhaps due to the formation of the Congo River). Among common chimps, alpha males try to monopolize females, but the sexual politics is complicated by secret dalliances, long-distance affairs, general promiscuity, and the male habit of raiding neighboring troops. Among bonobos, males are mellower and females have higher social status. Some primatologists have personalized chimps as "demonic males" and bonobos as "hippy apes."

Primatologist Frans de Waal summed up the differences: "The chimpanzee resolves sexual issues with power; the bonobo resolves power issues with sex."

Gorillas are very different. The dominant "silverback" male lords over a harem of females and young. Male competition occurs before mating—hence the paltry testes production (chimps produce about two hundred times more sperm). Natural selection apparently favored hulking males with canine daggers who could vanquish other males. Males weigh up to four hundred pounds, more than twice as much as females. Among mountain gorillas, dominant males sire 85 percent of the infants. (Bachelors are exiled to all-male bands, but occasionally one will take out the alpha male, kill his children, and become the new king of the harem.)

Humans are distinct from all our ape cousins. Men are only about 15 percent larger than women, and our puny canine teeth show almost no sex size differences—all of which suggests that violent mate competition was not a very powerful selective force. Human sperm production is much lower than chimps (but higher than gorillas). To Lovejoy, all these traits suggested pair bonding. In 1981, Lovejoy introduced his monogamy theory in an essay titled "The Origin of Man" in *Science* and proposed that nuclear families, canine reduction, and bipedality formed a single adaptive package at the root of the human family. He suggested that monogamy evolved in groups of nuclear families, a social structure unknown in other apes (gibbons are monogamous in single-pair families, not multifamily groups).

The theory ignited great controversy. Feminists dismissed it as a male fantasy. Skeptics objected that provisioners would be cuckolded by sneaky rivals who lurked back at the nest. Wiseacres joked about a food-for-sex scenario proffered by a guy named *Lovejoy*. Primatologists pointed out errors in some of Lovejoy's claims about ape reproduction. (He had assumed that chimpanzee-like sexual swelling was primitive, but concealed ovulation turned out to be the ancestral condition.)

Lovejoy's theory had been inspired by his study of fossils.

Ironically, one immediate objection focused on Lucy herself. *Australopithecus afarensis* females like Lucy were small, while other members of her species were much larger, presumably males—which many interpreted as gorilla-like differences. Many saw the dimorphism of Lucy's species as gorilla-like polygyny, not monogamy. On the other hand, *afarensis* was quite unlike a gorilla because its small canines exhibited few differences between sexes. Thus Lucy's species presented a paradox: the bodies seemed to tell one story and the teeth told another. Lovejoy had an explanation—everybody had been wrong about Lucy.

Scientific consensus had long accepted the idea that Lucy's species was highly dimorphic—a theory championed by Tim White and Don Johanson in the 1970s to explain why *Au. afarensis* represented one species instead of two. Or was it an illusion of a sparse fossil record? As the fossil sample expanded, the tiny Lucy proved to be an outlier at the bottom of the size distribution. Most fossils are individual pieces with no obvious sex attribution, and Lovejoy contended that sexual dimorphism became a self-fulfilling prophecy: small fossils were *assumed* to be female, and large ones *assumed* to be male. Lovejoy, his student Philip Reno, and other colleagues took an expanded *afarensis* sample, employed new statistical methods, and reached a controversial conclusion: Lucy's species was moderately dimorphic like modern humans—and thus consistent with pair bonding. (Opponents remained unconvinced; Bill Jungers of Stony Brook University dismissed the alternative theory as "Owen and a couple of his acolytes.")

Three decades after he first introduced his monogamy theory, Lovejoy proffered it again as an explanation for *Ardipithecus*. Based on the limited sample of fossils, the scientists estimated that sexual dimorphism of *Ardipithecus* was between that of humans and chimps. Lovejoy theorized that Ardi provided a glimpse of the social revolution that founded our lineage. This adaptive package, he reasoned, also would have included elements not visible in the fossil record, such as our poor smell sensitivity. Why would our ancestors evolve a duller sense of smell? So partners couldn't detect fertility cycles and kept up the sex-for-food arrangement year-round.

No aspect of Ardi would be more speculative or divisive. All the other papers listed multiple authors, but the monogamy theory bore only Lovejoy's name. He would venture onto that limb by himself.

Chapter 26

NEITHER CHIMP NOR HUMAN

Human evolution is largely a story of bones and genomes. Sometimes these two sources of information play together nicely, and sometimes they seem to tell conflicting stories. For the last half century, molecular science has dictated the shape of the family tree and where fossil hunters hung their fossils on the branches. Halfway through the analysis of the Ardi skeleton, science passed another historic milestone: In 2001, the first draft of the human genome was published—the genetic code of our species, transcribed into three billion pairs of nucleotide bases. The genome is the entire set of DNA with the genetic instructions to build the animal from fertilized egg through every stage of life. Four years later, the Chimpanzee Sequencing and Analysis Consortium published the genome of our closest relative. Comparative biology advanced into a new level of granular detail of genes and even single bases of nucleotides.

Optimists envisioned the decoded genome as a master blueprint of life and dreamed of "cracking the code" to reveal the basis of human distinctions from big brains to language. As in cryptography, however, intercepting an enciphered message only saddles us with the harder tasks of breaking the code and understanding what it says. Transcription remains a far cry from decipherment. Sequencing the genome did not signal arrival at a promised land of evolutionary biology. Instead, reaching that summit only revealed more mountains to climb and greater levels of complexity. But this much became clear: many old notions were too simplistic.

Even the fundamental concept of the gene had to change. The word *gene* was proposed by Wilhelm Johannsen in 1909 (drawing on concepts articulated by Gregor Mendel in the nineteenth century) and became the basic unit of heredity. For nearly half a century, however, it remained a little more than a placeholder for unknown mechanisms. In 1953, James Watson and Francis Crick published the double helix structure of DNA, and the gene soon came to be defined as a sequence of DNA that encoded a protein. For decades, biologists imagined genetic code as a linear entity and took a reductionist approach of looking for specific genes directly responsible for specific traits. In the new millennium, genomics (as the systematic study of genes came to be known) blew apart such notions. Contrary to past belief, genes were not self-contained sections of the genome surrounded by wastelands of junk DNA. Instead, genes operated in distributed systems of coding and regulatory elements (enhancers, repressors, silencers, insulators), cis-regulatory elements (promoters, transcription factor binding sites), and overlapping domains. With tweaks in regulation, a single line of code might produce varied effects in different parts of the body. Genes were complex bureaucracies, not autocrats.

Comparative genomics redefined what it meant to be nearly 99 percent identical to a chimpanzee. Studies affirmed the close genetic relationship: a 1.2 percent difference between *Homo* and *Pan*. (In other words, humans and chimps differ on about one of one hundred positions on the genome while any two humans differ on about one in one thousand.) But the much-touted 1 percent difference turned out to be misleading. The original figure referred to only "orthologous" sites where matching parts of the chimp and human genomes could be lined up directly, yet in many places the genomes do *not* line up because large sections of code have been inserted, deleted, and moved in both lineages. These "indels" or "structural variants" elevate the true difference to about 5 percent. The rearrangements are so extensive, and the regulation of these genes so variable, that such percentage comparisons have little meaning.

Within a decade, more genomes were sequenced for all the great apes plus other primate species, and the human-chimp similarity ceased to seem so striking. The macaque monkey turned out to be 93.5 percent identical to a human (or 90.8 percent when indels are included). Even the mouse genome turned out to be 85 percent identical to a human in coding sections. "For many, many years, the 1 percent difference served us well because it was under-appreciated how similar we were," Pascal Gagneux, a zoologist at the University of California at San Diego, told *Science* in 2007. "Now it's totally clear that it's more a hindrance for understanding than a help." In the early days of the molecular era, scientists were stunned to discover the small genetic difference between humans and African apes. In the era of genomics, they encountered yet another surprise: those small differences gave rise to enormous differences in biology.

Comparative genomics also challenged the old idea that African apes are more evolutionarily conserved—more primitive, in other words—than humans. In genetic trees, "branch lengths" indicate the number of nucleotide changes on the genome of each lineage since their split from a common ancestor (the longer the branch, the more genetic change has occurred). In coding regions, human and chimp branch lengths are nearly identical, suggesting both lineages have undergone similar amounts of genetic changes. In noncoding regions, however, humans showed 3.5 percent *less* change than chimps. "If anything, the human genome is a little more boring when it comes to structural changes," said Evan Eichler, a professor of genome sciences at the University of Washington. "Gorillas and chimps have had some really massive changes that have restructured their genomes, especially at the ends of their chromosomes. Ours has restructured as well, but at first blush the human genome is closer structurally to what we thought the ancestor would have been."

MOLECULAR BIOLOGY—WHICH ONCE CLARIFIED OUR PRIMATE FAMILY TREE— began to complicate it again. In traditional depictions of trees, species branch off at precise points in time. Our real history now appears not so simple. In 2006, researchers from Harvard and MIT reported that the lineages of humans and chimps may have continued to interbreed for more than 1 million years—hardly the precise "split time" imagined in prior days. Newer findings suggest an even longer breakup: 4 million years between the initial divergence and final split. A number of studies now even suggest the gorilla lineage was *still* interbreeding with ancestors of humans and chimps when the latter two began to diverge. This means that speciation was a protracted saga of multiple populations isolating and remixing—a very messy and drawn-out divorce without a single moment when any sole last common ancestor of humans and chimps begat two lineages that immediately went in separate ways. As a consequence, different parts of our genome have different histories, a phenomenon known as "incomplete lineage sorting." In practical language, this means that one gene may have a different "family tree" than another gene in the same animal. In most of the genome, humans are most closely related to chimpanzees, but more than one-third of our genome does not match this overall species tree. (In 18 percent of the genome, humans are actually closer to gorillas, and in another 18 percent we are equally related to *both* African apes.) The take home: we are unlikely to ever find a precise split time nor any *single* last common ancestor population of humans and African apes. The Holy Grail metaphor for the Last Common Ancestor was apt in unintended ways—a figment of imagination.

At some point, the human and chimp lineages ceased to exchange genes and the split became final. One fertility barrier is all other apes have twenty-four pairs of chromosomes, but human ancestors merged two chromosomes and wound up with twenty-three pairs. Nightmare tales of ape-human hybrids are—so far as we know—mostly fantasy. Not that humans haven't tried: In the 1920s, Soviet zoology professor Il'ya Ivanovich Ivanov failed to create human-ape hybrids by artificial insemination. (Eventually, the Soviets arrested Ivanov, exiled him, and decades later launched some monkeys from his primate nursery into space on Sputnik missions.)

Genomics can tell us only about creatures whose DNA can be collected. It cannot reconstruct extinct creatures nor reveal what ancestors looked like. Nobody could predict a T. rex or an *Ardipithecus*. You could only find them.

In 2005, the first fossils of a chimp ancestor finally appeared. Until that point, thousands of fossils of human ancestors had been discovered in Africa but none of chimps, which might be a consequence of the rarity of apes, poor bone preservation in forests, the eagerness of human fossil hunters to pursue our own lineage, or all those factors. Ironically, the theory of a chimp-like human ancestor took hold not with the strength of the fossil record but with the lack of it. The first record of our ape cousins was four thinly enameled teeth (two molars and two incisors) from the Tugen Hills in Kenya. The teeth were about 545,000 years old—so recent that the incisors were virtually identical to modern chimp teeth and offered little information about how the species evolved. Chimp evolution remained nearly as mysterious as before, and the gorilla entirely unknown.

Then Gen Suwa looked at some teeth and shook things up once again.

In Africa, the fossil record remained almost empty between 7 and 12 million years ago—the crucial window when the ancestors of humans, chimps, and gorillas went through their drawn-out and still-mysterious splits. In the Afar Depression, the oldest sediments lay in the Chorora Formation, a series of exposures on the southwestern rift escarpment. Seeking to probe these dark ages, Gen Suwa, Berhane Asfaw, and Yonas Beyene led a new expedition to Chorora and used space images to locate target sediments. In the final day of the 2006 field season, a young Ethiopian surveyor named Kampiro Kairante spotted a molar tooth on the ground. Suwa judged it to be a possible ancestor of the gorilla. Over the next year, the Chorora team recovered nine teeth from a Miocene ape they named *Chororapithecus abyssinicus*. Suwa and his colleagues concluded the size and proportions of the teeth were "collectively indistinguishable" from a modern gorilla.

But there was a problem. Geologists first thought the *Chororapithecus* deposits were 10 million years old (later revised to 8 million), a couple of million years before there was even supposed to be any such gorilla under then-current split time estimates. Moreover, fossils such as *Sahelanthropus*—which showed traits of the human family by 6 million years ago—provided gnawing evidence that the human-ape divergence must be older than most researchers had thought. The Chorora team concluded the gorilla lineage must have split some 12 million years ago, which also implied that humans and chimps went their separate ways at least 9 million years ago. Later, molecular scientists at the Max Planck Institute for Evolutionary Anthropology in Germany also pushed split times deeper into the past: 7–13 million years for the human-chimp split and 8–19 million years for gorillas. By such new estimates, the separation of humans and chimps might be as much as *twice* as old as previously believed. (Current estimates of the human-chimp breakup range between 6.5 and 10 million years; split times still remain controversial, but most agree they should be revised to be older.)

For years, many paleoanthropologists accepted the earlier molecular studies as oracle because the science carried an aura of objectivity, precision, and near infallibility. In reality, however, such estimates depended on a number of assumptions about the length of each generation, the genetic diversity of ancient populations, and the speed of molecular change—which sometimes resulted in snowballing errors. The endeavor was compromised by a degree of circular reasoning: split-time estimates relied on fossils for calibration, and those estimates were used to make pronouncements about fossils. As doubts grew, one pair of critics denounced some molecular chronologies as "reading the entrails of chickens."

In the new millennium, technological advances enabled scientists to directly measure mutation rates between parents and children—and the rate of molecular evolution turned out to be much slower than previously assumed. The molecular "clock" also did not tick as steadily as people had imagined (mutation rates vary between species, over time, and even within a single genome). Given the slower-than-expected rate of change, more time must have elapsed to account for the genetic differences between humans and apes—yet another finding that pushed the estimated human-chimp split farther back into the Miocene epoch. In this light, it became not altogether surprising that Ardi so differed from our ape cousins: our common ancestors had lived millions of years earlier.

When the Ardi saga began, the fossil hunters imagined they were hot on the trail of the last common ancestor of humans and African apes. Over time, that target became more blurry and receded deeper in the past—all the way back through the darkness of the Miocene gap to the other side when there actually was a deeply ancient fossil record of apes. And they didn't look like modern ones.

ONE DAY, OWEN LOVEJOY GOT IN HIS CAR AND WENT TO VISIT A 15-MILLION-year-old ape. He suspected he had walked past a clue without realizing it.

He had first glimpsed his target years earlier while traveling on the Getty jet. Back in 1996, the Ardi team trooped into the National Museum of Kenya and saw an old friend bent over his newly discovered Miocene fossil skeleton. It was Steve Ward, a fellow veteran of the Lucy team. Ward specialized not in *hominids* but *hominoids,* a taxonomic category that included all living and extinct apes. That small difference in suffix belied a larger difference in professional culture. Scholars of Miocene hominoids formed a cozy coterie that engaged in polite debates about obscure creatures. There were fewer outsize personalities, ego clashes, or bitter arguments about which ones might be ancestral to humans because, frankly, *who the hell knew?* The hominid guys were different beasts—academic gunslingers. "To Owen and Tim," observed Ward, "it's blood sport." On that day, Ward sat in the museum savoring the greatest thrill of his career: a Miocene ape called *Equatorius.* White sidled up and made a wisecrack about Ward's "rift monkey." Excuse me, it was a Miocene *ape,* not a monkey—as his old friend White knew perfectly well. Many anthropologists regarded those Miocene apes as too remote and insufficiently "apelike" (read: like *modern* African apes) to take seriously as informants about human ancestry.

The Ardi team had second thoughts. They had come to conclude that those weird-looking Miocene apes—so unlike anything alive today—offered a better starting point for reconstructing human evolution than modern apes. Lovejoy got back in touch with Ward. Could he please take another look at that rift monkey? Luckily, he didn't have to fly all the way to Africa this time. Ward had casts of *Equatorius* at his lab at the Northeast Ohio College of Medicine (now called Northeast Ohio Medical University), only fifteen minutes down the road. And there were no secrets: he was happy to show them.

Lovejoy climbed into his car and cruised through the Ohio countryside. He loved driving and sometimes kept a radar detector on his dashboard so he didn't get in trouble for liking it *too* much. In his younger days, he owned classic British MGs and Triumphs.

He got married and bought a Volvo. He got divorced and splurged on an Alpha Romeo. He went through a midlife "action car" phase of Porsches and Lamborghinis. Eventually, he decided he was too old for Italian cars and settled down to Mercedes, Audis, and the fine precision of German engineering. (Thanks to his academic stardom, Lovejoy was one of the highest-paid employees at Kent State behind only the president, a few top administrators, and the football and basketball coaches.) Like automobiles, animals exhibited design themes. The old anatomists described it best with a German word—the *bauplan,* the body plan, or construction design. Decipher the *bauplan,* and you could understand the creature's ecological and behavioral adaptations, so even a small part could reveal much about the whole. And that quest led him back to *Equatorius.*

Equatorius shared the common Miocene ape body plan of a creature that clambered on its palms, and Lovejoy wanted to examine its hand. Ward's skeleton preserved a fifth metacarpal, a bone from the palm that joined with the wrist. When Lovejoy examined the cast, he saw it matched the same bone in Ardi—even though the two species lived 10 million years apart. Lovejoy recalled thinking, *I'll be damned. Identical!* The visit reinforced what he and his colleagues had come to believe from multiple lines of evidence—Ardi and humans descended from a palm-walking Miocene ape, not a knuckle walker. And that implied an ancestral body plan with arms and legs close to equal length, bendable wrists, flexible spines with mobile lumbar sections, lever-like feet with the os peroneum sesamoid—and many such features preserved in Ardi.

Lovejoy kept his observations to himself. Like many scientists, Ward had been waiting years for news about *Ardipithecus*—especially since he considered Lovejoy one of the smartest minds in the profession. "Oh my god, nuclear secrets!" Ward recalled with a laugh. "There was a lot of impatience among many of my colleagues—what's taking these guys so goddamn long to produce this fossil?" He was still waiting when Lovejoy said good-bye.

• • •

THE BIG SECRET WAS THAT LOVEJOY AND HIS COLLEAGUES WERE ABOUT TO become heretics. Ever since the era of the molecular revolution and Lucy, many anthropologists had expected the oldest fossils in the human family to converge on an ancestor that resembled a modern chimp or gorilla. Ardi represented a large step closer to that ancestor and—in the judgment of the investigators who had pored over her skeleton for years—she suggested that everybody had it all wrong. Human ancestors *never* went through a stage resembling any ape alive today. Modern African apes were no "time machines." Rather, they were like false leads in a detective story—the suspects who drew attention away from the real culprits. Oddly, some traces of the real Miocene ancestor remained *more* preserved in humans than in living African apes. Moreover, the different species of extant apes weren't nearly so alike as people assumed when they were lumped into a blurry composite of "primitive." Instead, living apes differ greatly in social structures, limb proportions, and even styles of knuckle walking—all of which suggested they became highly specialized after splitting from our common ancestors. These views were a complete abandonment of the paradigm that had dominated thinking about human origins for more than three decades.

But these were not entirely new ideas: some anatomists had made similar predictions a half century earlier. In the late 1940s, Oxford professor Wilfrid E. Le Gros Clark, one of the prominent anatomists of his era, analyzed several species of the Miocene ape *Proconsul* that had been found by Louis and Mary Leakey's team in Kenya. They seemed so unlike today's apes that Le Gros Clark likened them to monkeys (another example of an anatomist invoking familiar analogies to describe things never seen before), yet he recognized they were really early apes. Le Gros Clark judged that such a "generalized" quadruped (in other words, a locomotion jack-of-all-trades, not a specialist) probably was the common ancestor of humans and living apes. He thought that humans "could hardly have been derived" from specialized forms similar to modern knuckle-walking chimps and gorillas or suspensory orangutans. Similarly, Adolph Schultz of Johns Hopkins University (the dissection-obsessed anatomist of the mid-twentieth century whom we met at the Asiatic Primate Expedition and Cold Spring Harbor) also believed modern African apes were very specialized and humans remained more primitive in many aspects of anatomy (with obvious exceptions such as big brains and erect posture). His Johns Hopkins colleague William Straus argued that human ancestors started walking upright early in their evolutionary history and never passed through a stage like modern apes—especially not long-armed suspension. Straus laid out his views in the classic 1949 paper "The Riddle of Man's Ancestry." Half a century later, another anatomist found the article so prescient that he acquired an original signed by the old master himself. "One of my most prized possessions," said Owen Lovejoy.

In the molecular era, such views fell out of favor as evolutionists increasingly looked to modern apes for clues about the precursors of humanity. With the popular approach of cladistics, living species were used to make logical inferences about their common ancestors. In many quarters, the Miocene apes became perceived as an assortment of extinct curios that Harvard scholar David Pilbeam said "throw little if any direct light on hominid origins." Yet in one anatomical detail after another, the Ardi team found it plausible to interpret their skeleton as a descendent of a palm-walking Miocene ape. Similarly, they concluded that chimps and gorillas also likely descended from such a generalized Miocene ancestor—just as the old anatomists suggested a half century earlier.

A few scholars quietly had continued to argue on behalf of such a Miocene ancestor, but the Ardi team would do so more forcefully, backed up by an unprecedented haul of fossils from the early human family. Ardi was not the ancestor everybody expected. She lacked the jutting snout and daggerlike canines of chimpanzees and gorillas. She showed no signs of knuckle-walking ancestry—nor did any known Miocene fossil ape. Her hands, although large and clearly arboreal, were not quite the grappling hook appendages of suspensory apes, but ones better suited for manipulation. Ardi would have been more adept in the trees than any known human ancestors, yet not so acrobatic as chimpanzees. She walked upright but unlike any biped ever known before. She represented something entirely new to science—a creature that her interpreters described as "neither chimpanzee nor human."

BY 2009, THE ARDI TEAM FINALLY WAS READY TO REVEAL ITS HAND—PLUS all the other parts of the skeleton. The team that stubbornly resisted pressure to publish finally was ready to publish. Seventeen years had passed since the discovery of the first tooth of the species and a decade and a half since the excavation of the skeleton. "The different lines of evidence converged into one coherent picture," recalled Suwa. "You're able to understand the paleobiology of *ramidus*."

Not all who began the journey reached its end. Desmond Clark did not live to see the big publication of the species he had hailed as "the find of the century." Clark Howell, who pioneered multidisciplinary fieldwork in Africa, died before unveiling the massive mission to the past that he had inspired. Gadi and the other tribesmen would never hear that the *Kada Daam Aatu* had finally gone to the outside world.

In May 2009, the Ardi team submitted a series of papers to the journal *Science*, which agreed to devote a special issue to *Ardipithecus* like the one it did for the Apollo 11 mission that landed on the moon. Once again, peer review posed a delicate matter. Given its testy relations with the profession, the Middle Awash team routinely sought to exclude certain antagonists due to personal animosities or incompatible philosophies. ("They don't like getting

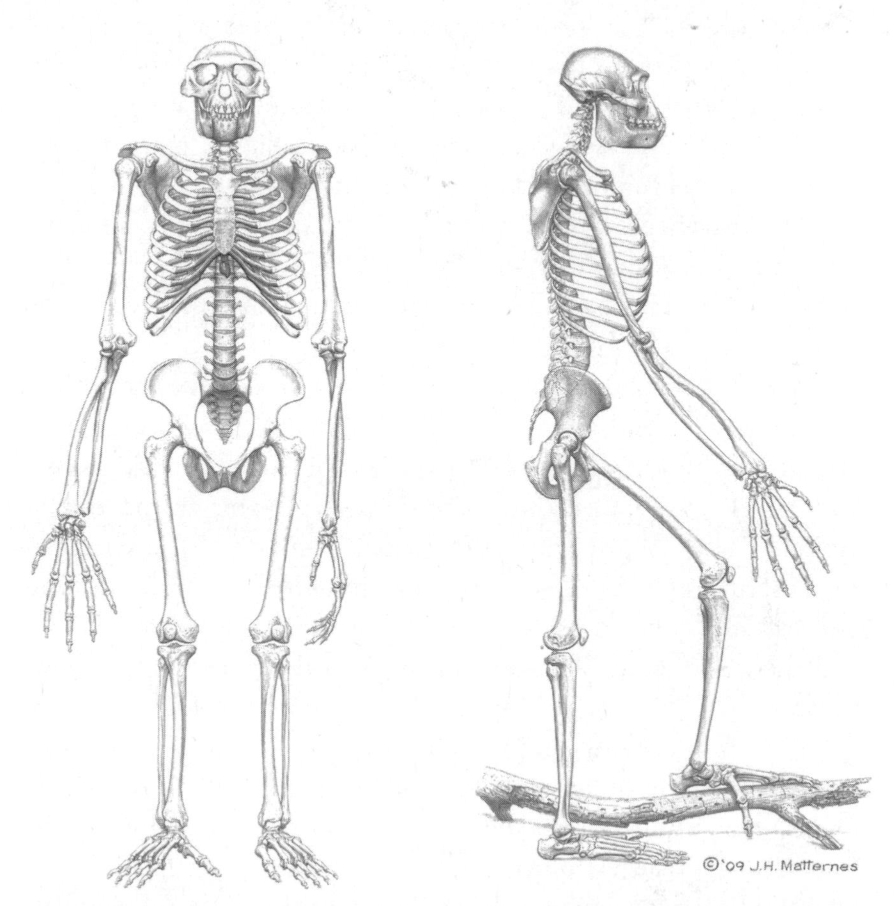

Ardi as imagined by artist Jay Matternes. Broken and missing bones have been drawn to render a complete skeleton.

the long list," chuckled White.) Brooks Hanson, the editor who handled the papers, would not comment on the review process except to state a general policy that each article was scrutinized by at least two editors and two peer reviewers. "Regardless of whether the interpretations were all correct or all persisted, it was still going to be incredibly important material," Hanson reflected afterward. "Putting that out in as much rich detail as possible was a benefit to the broad community." White asserted, "Anybody who works on our project knows the toughest review is our own internal review."

Lovejoy added, "We revised heavily, but I would say we probably self-revised more than anything else. If it goes to the outside world, the feedback will be useless. But within the team, it can be very serious." Others warned that such self-reliance carried risks. "The consequence of that isolationist strategy is you tend to live in an echo chamber," said Bill Kimbel. "You talk to the same people time and time again. You don't hear new and different views. The isolationist strategy has a narrowing effect on your perspective. It's unavoidable."

Shortly before publication, critics again chided the Ardi team for its glacial pace. In September 2009, *Scientific American* ran an editorial demanding limits on the period of exclusive fossil access and called upon journals to force discovery teams to make data available as a condition of publication. The editorial also urged the NSF to "reject without exception those repeat applicants who do not follow the access rules." The editorial singled out White and the Middle Awash team—once again described as "paleoanthropology's Manhattan Project." At the time, the Lucy skeleton toured U.S. museums, and the editorial drew a contrast between the two skeletons—one displayed for the world and the other still under wraps. (One of the staffers who prepared the editorial was Kate Wong, coauthor of Don Johanson's latest book in the Lucy series.) The piece appeared alongside a cartoon of a brute guarding a closed laboratory door from pleading researchers. White dashed off an angry reply, but by the time it was printed his team had answered by other means: the Manhattan Project had dropped its bomb.

Chapter 27
SKELETON FROM THE CLOSET

After fifteen years of secrecy, Ardi finally met the world. In October 2009, the team unveiled its findings in a special issue of *Science* and press conferences in the Ethiopian and American capitals. In Washington, D.C., White, Lovejoy, Yohannes, and Giday spoke before a phalanx of press while the Ethiopian ambassador sat in attendance. "What we celebrate here today," declared White, "are the results of a scientific mission to the very deep past."

With that, the Mission to the Pliocene returned to the present. The investigators described *Ardipithecus* as a representative of the human ancestors who came before Lucy—woodland creatures from 4.4 million years ago with opposable toes, ape-size brains, diamond-shaped canines, and an odd, upright gait—a simian-human combination that nobody had predicted. Said Lovejoy, "What we have from *Ardipithecus* are a host of revelations about the earliest phases of human evolution that Lucy never provided us. It is one of the most revealing hominid fossils I could ever have imagined."

After remaining silent so long, the Ardi team fired full broadside—with eleven papers and 384 pages of dense anatomical descriptions, photos, images from CT scans, and supporting online information. When first announced to the world, *ramidus* had been described as the most chimp-like ancestor ever known. Now it suddenly reappeared to refute the very idea that our ancestor looked much like our closest relatives. Entire schools of thought were dismissed, and the unenlightened were bluntly informed about the errors of their ways. The authors pronounced modern African apes to be "adaptative cul-de-sacs"—evolutionary dead ends, not the route by which we had come. They informed "chimpologists" that they were looking up the wrong trees for human origins, and that their quest to link early humanity to modern apes was *nullified* by Ardi because the last common ancestor looked like *no* modern species. Supposition about knuckle-walking or suspensory ancestors was "now moot." Likewise, the Ardi team asserted that the skeleton "requires a rejection of theories that presume a chimpanzee- or gorilla-like ancestor to explain habitual upright walking." The authors chastised wayward colleagues who trusted computer modeling over comparative anatomy—a "trend accompanied by too many presumptions and too few fossils." All that high-tech wizardry fell short of the near-forgotten anatomists from the previous century who, armed with sharp eyes and scalpels, predicted the common ancestor better than any computer jockey. The savanna paradigm was pronounced dead. Ardi had revealed the new *truth*.

The message went well beyond the skeleton. In the age when splitters proclaimed the human family ever more bushy, these contrarians envisioned a simpler story of a succession of adaptive plateaus running through *Ardipithecus* (if not Ardi's species then her genus), *Australopithecus*, and *Homo*. They suggested all early members of the human family, *Sahelanthropus* from Chad and *Orrorin* from Kenya, probably belonged to the same genus as *Ardipithecus*, and perhaps even a single lineage. The authors presented several scenarios for how *Ardipithecus* might have led to *Australopithecus* and *Homo*, but entertained no hypothesis that put Ardi outside the human family.

The year marked the 150th anniversary of Darwin's *On the Origin of Species*, and the fossil team framed their findings in light of that milestone. In his book, Darwin speculated that someday "light will be thrown on the origin of man and his history." The battery of papers in *Science* proclaimed that a new dawn had broken with Ardi—a human ancestor who "resolves fundamental evolutionary questions persisting since Darwin." White had engineered a masterwork of public relations. Ardi came to life in beautiful, lifelike drawings by anatomical artist Jay Matternes, an old master of natural history artists who also had illustrated the Laetoli footprint walkers and even the Time-Life books on evolution that had captured the imagination of the teenage Tim White. In addition, a Discovery channel documentary (produced under an embargo that lasted much longer than anybody expected) touted "fossil evidence Darwin could only have dreamed of." Alan Walker, a professor from Pennsylvania State University and respected doyen, gave the press an independent evaluation: "This find is far more important than Lucy."

It was academic shock and awe, and scientific peers were stunned. "It's the closest thing we'll ever see to a big science effort in the little field of human evolution," blogged anthropologist John Hawks of the University of Wisconsin, "like Tim White was building a supercollider under everybody's noses." Around the globe in publications large and small, headlines screamed: PREDATING LUCY, FOSSIL SKELETON PUSHES BACK HUMAN ANCESTRY (*New York Times*); FOSSIL UPENDS THEORIES ABOUT EVOLUTION OF HUMAN ANCESTORS

(*Christian Science Monitor*); HELLO MUM!: ARDI OUSTS LUCY AS OUR 1ST ANCESTOR (*Daily Mirror*); MEET ARDI, A MOTHER TO ALL HUMANITY (*Globe and Mail*); A NEW ANCESTOR REDRAWS THE HUMAN FAMILY TREE (*International Herald Tribune*); DISCOVERY MAKES A MONKEY OF OLD BELIEF (*Concord Monitor*). Initially, media coverage remained on message: the origins of upright walking had been found, we didn't evolve from ancestors like chimps, the savanna theory had been mowed down, and so on.*

The fanfare continued for months. The American Association for the Advancement of Science named Ardi "Science Breakthrough of the Year." The Gettys hosted an Ardi party at their San Francisco mansion. The following spring, White was named to the *Time* magazine list of the one hundred most influential people in the world along with Barack Obama, Bill Clinton, Lady Gaga, Oprah Winfrey, Elon Musk, and Steve Jobs. White attended the *Time* gala with his wife, Leslea Hlusko, a fellow Berkeley professor and member of the Ardi team (they had married in 2004). "I joked that I was the mistress and Ardi was the wife," she recalled. "Ardi has been his life as long as I've known him." *National Geographic* put Ardi on the cover and labeled the Middle Awash as "humanity's hometown." Again and again, reporters asked about breakthrough moments and White repeated a mantra: real science was a long-term endeavor and didn't easily reduce to anecdotes, sound bites, or eureka moments.

The team's interpretation of Ardi passed through the media

---

* One news story did stray off message. The Al Jazeera news network reported that western scientists had renounced Darwin's theory of evolution. The leading news network of the Arab world twisted the "not-from-chimps" message into a refutation of evolution itself and quoted an Egyptian geologist expressing relief that "Westerners have begun returning to their senses." The story drew from a Kent State news release with an awkward quote from Lovejoy ("People often think we evolved from apes, but no, apes in many ways evolved from us"). Impatient with the academic fetish for classification, Lovejoy sometimes became reckless with terminology and it got him in trouble—much to the confusion of peers, frustration of teammates, and delight of creationists.

as easily as hominid teeth passed through the digestive tracts of ancient hyenas. For the scientific establishment, however, the skeleton from the closet was causing major indigestion.

A FEW BLOCKS AWAY FROM THE ARDI PRESS CONFERENCE, PROFESSOR BERnard Wood sat in his office at George Washington University and wasn't quite ready to step into the revival tent. "When Tim publishes something," Wood observed, "I think he feels like it's Moses and the tablet—'I've come down from the mountain. Here is the message you've been waiting for. Suck it up.'" As young scientists, Wood and White had worked together in "the Firm"—Richard Leakey's team in Kenya—and the latest news showed just how far they had since diverged. "Just because the tablets were delivered in the form of a press conference, papers, and all the sophisticated public relations manipulations that Tim despises," Wood continued, "there's no reason to assume these ex cathedra announcements are necessarily true. This is one interpretation. There are others."

When the first fragments of the species were found fifteen years earlier, Wood had written an analysis that hailed *ramidus* as a "missing link." At the time, Wood welcomed the discovery as a sign of the "maturity" of the science because the fossils seemed to validate what had been predicted—a chimp-like ancestor. But now the skeleton had waltzed out of the closet to torment everybody with a message that maybe that view of the science wasn't quite so mature after all. In the meantime, Wood's outlook also had evolved. His new world involved metrics, computers, bushy trees, and wariness of evolutionary scenarios based on bones and teeth. He no longer felt so confident making authoritative pronouncements about what had been discovered; instead, he felt chastened by how much more remained unknown. He no longer assumed every discovery represented an ancestor, no longer assumed the human family tree was linear, and no longer trusted morphology—nor the pronouncements of experts. To him, the Ardi group seemed stuck in a time warp.

"I am just not as certain about *anything* as they are about

Ardipithecus," said Wood. "There was a level of certainty that was almost evangelical. They kept using the word *reveals*. I don't think it's coincidence that the Bible uses the world *reveals* a lot. It reminded me of a religious cult."

He thought he might write something about it.

EIGHTEEN DAYS AFTER THE PRESS CONFERENCES, THE ROYAL SOCIETY IN London held a previously scheduled forum on human evolution, which turned into Ardi's first major encounter with the scientific community. The society stands among the most eminent scientific academies in the world, and its fellows have included Charles Darwin, Albert Einstein, and Thomas Henry Huxley. Another fellow was Edward Tyson, who in 1699 made the first scientific description of the chimpanzee, a creature that so boggled his mind that he had to invent new language to describe it. Three centuries later, the society hosted a conference where scientists met another newly revealed creature—*Ardipithecus*. The conference focused on early human evolution and the schedule—drafted before the recent Ardi announcement—began with an awkward first topic: chimpanzees as models for early human ancestors.

The opening speaker was William McGrew, a longtime colleague of the Ardi scientists. A professor of evolutionary primatology at Cambridge University, McGrew had accepted the invitation to kick off the conference based on what he *thought* was a shared belief that our closest relatives could provide insights into human origins. "I have to take a slight detour," he began, "because all of this chimpanzee chasing was deemed to be irrelevant as of two weeks ago with the appearance of the October 2 issue of *Science*."

McGrew was a self-described "chimp chaser" who had spent four decades in the field observing apes. He had joined the research consortium funded by Tim White's last NSF grant—a partnership forged in bygone days when studies of chimp behavior and hominid fossils were seen as complementary. Once Lovejoy and McGrew had been close friends who vacationed together. No longer. The year before, Lovejoy and his student Ken Sayers wrote a provocative article, "The Chimpanzee Has No Clothes," criticizing the "chimpocentrism" and "referential modeling" of primatologists, including McGrew and his wife, Linda Marchant. Then came Ardi—the discovery that Lovejoy insisted "essentially falsifies" models of human origins based on modern apes.

In the past, McGrew had lamented that most paleoanthropologists had never seen a wild chimp. With icy indignation, he issued corrections as he stood at the podium. Chimps do not simply live in rain forests as the Ardi team contended. To the contrary, they dwell in habitats from forests to woodland savannas—like those of *Ardipithecus*. The Ardi team claimed "humans alone" evolved cognition. McGrew said that claim was "certainly a bit of shock to people in Leipzig, Kyoto, St. Andrews, and Emory University in Atlanta who thought they *were* studying chimp cognition." Anything a chimp can do today, McGrew asserted, our last common ancestor could have done, too, millions of years ago. He noted that chimps and bonobos possess brains of similar size as early human ancestors and performed on par with human preschoolers in intelligence tests. Again and again, studies have documented chimps engaged in behaviors previously claimed to be exclusively human—using tools, walking upright, hunting, throwing, spatial navigation, and more. McGrew spent his career documenting chimp culture, or socially learned behavior. Chimps learned by observation, mimicked one another, showed self-consciousness, and even displayed their own version of morality. Chimps also ostracized those who violated social norms—just like humans.

"He won't even talk to me now," Lovejoy said of his former friend. "At the meetings he just ignores me. He thinks we've damaged the cachet of the chimpanzee and therefore we're eroding its chances of survival." (McGrew declined to comment.) At the Royal Society meeting, McGrew closed his presentation with an appeal: "If we have any use for chimpanzees as models, then every one of us is obliged not just to talk about them but to work to try to save them."

The Ardi papers provoked irritation and even fury among some primatologists. During the question session at the Royal Society conference, one primatologist asked why Ardi, even if she herself didn't resemble a chimp, couldn't have descended from an ancestor like one. White and Lovejoy recited several minutes' worth of anatomical detail about canines, molars, the os peroneum sesamoid, the flexor pollicis longus muscle of the thumb, and more. Then White added sarcastically: "So it's just like a chimp except for all those things." Laughter rippled through the audience—but not among those who studied living apes. "You won't take it seriously!" one snapped. Some complained that the Ardi team had erected a straw ape and ripped it apart. A dozen primatologists later wrote a letter to *Science* to dispute the central message of the Ardi barrage that chimps were poor models for our common ancestor and warned that "oversimplistic interpretations of such remarks will devalue exciting progress in synthesizing diverse sources of evidence about our past." Privately, critics were less diplomatic. Some primatologists posted the *Science* back-and-forth correspondence on an academic website with a handwritten note scrawled in the margin beside the Ardi team's argument: "stupid."

More lay at stake than differences of professional opinion. Chimp research faced threats on many fronts. Some researchers had sunk into despair that these remarkable creatures, so hauntingly like us, were vanishing and their closest relatives didn't seem to give a damn. Humans were driving the remaining apes toward extinction with deforestation, development, and poaching. Luminaries such as Jane Goodall, Richard Dawkins, and Jared Diamond formed the "Great Ape Project" to demand rights of "personhood" for captive chimps, bonobos, orangutans, and gorillas—a campaign that depended on *blurring* the line between humans and apes. When science reporter Jon Cohen researched the book *Almost Chimpanzee*, one prominent chimp researcher warned him that conservationists would be angered by any account that emphasized the differences between humans and chimps. By that measure of sensitivity, Ardi was about as welcome as a bush meat hunter. In many branches of science, researchers already were turning away from chimps because of ethical restrictions, practical difficulties, or the realization that the animals turned out to be less informative than expected. (Within a few years invasive biomedical research virtually would cease in the USA and many captive apes would be retired to sanctuaries.) In 2009, the pushback culminated with a book defiantly titled *Not a Chimp*. Ardi added an exclamation point.

The Royal Society conference began to expose the battle lines around the oldest family skeleton. Those guided by genetics and primatology used living apes to make logical inferences about our common ancestors. ("I have no more direct access to the last common ancestor than anyone else in this room," McGrew explained to the audience. "So we have to model. We're forced to do so.") In contrast, those guided by fossils took a ground-up approach—bones revealed what really happened. Those two mind-sets placed faith in different lines of evidence, made different assumptions, and inevitably reached different conclusions. In the first approach, the relationships between humans and apes were irrefutable, but the form of their common ancestors remained conjectural. In the fossil-based method, bones provided rock-hard evidence of past forms, but their places in the family and relationships to humans could only be surmised. Both sides fought to make their work relevant to human origins—and Ardi became the flashpoint.

In short, it was a classic clash of scientific worldviews. Science is not just a quest for new facts; it also is a contest between rival models for interpreting nature. In his 1962 milestone *The Structure of Scientific Revolutions*, historian Thomas Kuhn introduced the idea of the scientific paradigm—the set of theories and methods that unite a research community—and described normal science as "a strenuous and devoted attempt to force nature into the conceptual boxes supplied by professional education." He saw scientific training as a form of indoctrination—"probably more so than any other except perhaps in orthodox theology"—in which rules are hammered into the heads of novitiates. Of course, the crucial difference is that science puts its theories to the test of falsification and adapts to new information. Even so, scientists are not immune from what some psychologists call "motivated reasoning" to affirm their prior beliefs. Adherents stubbornly defend their orthodoxies and wonder how opponents could remain so obtuse. The most dogmatic congratulate themselves for being faithful guardians of science—their version of it—and condemn rivals as unscientific. When disciples see the light, it comes from different sources of illumination. Adversaries talk past each other, and debates become circular and unresolvable. As the quest for knowledge marches on, however, researchers inevitably will make discoveries at odds with current theory and encounter questions that cannot be answered by the old methods. When such anomalies accumulate, the paradigm faces a crisis and must be adjusted or replaced. So it was with Ardi.

At the Royal Society conference, scientists gawked over casts of *Ardipithecus* finally on display after so many years. French paleoanthropologist Michel Brunet had brought along an even older specimen—the replica skull of *Sahelanthropus* from Chad, nicknamed "Toumai," estimated to be about 6 million or 7 million years old. Brunet was an entertaining speaker whose scientific presentations were operatic, punctuated by theatrical lulls and podium-thumping crescendos best QUOTED IN ALL CAPS. Brunet treated his imperfect English as a running joke. With a scarf thrown around his neck, he presented himself as a caricature of haughty Gallic panache. "As you know, I am French," he introduced himself to the audience in London. "Nobody is perfect." He referred to his fossil as *theez guy* and thundered, "*WE NEED MORE TEEM IN ZE FIELD!*"—leaving audiences unsure if he meant more teams, more Tims, or a little of both.

White maintained a playfully antagonistic friendship with *pro-*

fesseur Brunet. Both kept fossils under wraps and refused to reveal anything until they were damn well ready. They sat shoulder to shoulder in labs comparing discoveries (they did show fossils to each other), and shared a certain delight in making mischief. When Brunet announced *Sahelanthropus* in *Nature*, his acknowledgments thanked somebody named Lubaka—Tim White's Afar wildcat.

Brunet thought his skull came close to the *dichotomie*—the split between the human and chimp lineages. White suspected *mon cher collègue* had his dates wrong and *Sahelanthropus* really was about the same age as *kadabba* from the Middle Awash—and all might belong to the same species lineage.

Brunet agreed that *Sahelanthropus*, *Ardipthecus*, and *Orrorin* probably belonged to the same general grade of early human ancestors, but remained less sure about their exact relationships. One lineage? Separate radiations from a common ancestor? The question put him in a quandary because *Ardipithecus* took priority by virtue of being named first. If both species were declared the same, the genus named by Brunet would be absorbed into *Ardipithecus*. Brunet would only answer that more fossils were needed to be sure, and hinted that he might have revelatory specimens, but remained mum for years. An estranged member of his field team published a snapshot that appeared to show parts of a skeleton lying in the desert. "There's this mystery femur that appears to exist but is not published," conference organizer Chris Stringer, an anthropologist at the Natural History Museum in London, later explained. "If you mention it to Brunet, he explodes." Brunet was the rare fossil hunter who irritated access advocates nearly as much as their bête noire—Tim White.

On one essential point the two paleontologists did agree: both their early members of the human family looked surprisingly *unlike* chimpanzees. Like the Ardi team, Brunet felt Miocene apes offered a better starting point for tracing human evolution. "Chimpanzees are very derived—very, very derived," said Brunet. "They are not good for giving us a nice image of the last common ancestor. *No*."

Another voice of support piped up from an unexpected quarter. Bill Kimbel was the director of the Institute of Human Origins—the rival camp in a paleo cold war in Ethiopia—and served as Don Johanson's right-hand man. Kimbel still gritted his teeth when recounting how the feud with the Middle Awash team caused him to lose his permit in Ethiopia for five years. During his estrangement from the Ardi group, Kimbel went through a metamorphosis in his own thinking—a sort of parallel evolution of thought. "From a morphological point of view, I don't think chimpanzees are a particularly good model," Kimbel told the audience. Yes, Lucy's species did share some similarities with chimps, but many of them turned out to be generic features shared by many fossil apes from the Miocene epoch. For too long, scholars had tried to force fossils into a chimp-to-human progression. Debates bogged down because everybody shared a false notion of *primitive*. For years, doubts gathered like a miasma and, finally, they had coalesced into something firm—*Ardipithecus*.

Chapter 28

BACKLASH

Then the real pushback began. Ardi was more than just an unsettling fossil: her description amounted to a wholesale rejection of basic ways of doing business and thinking about human evolution. The scenario posited by the Ardi team obliged everybody to accept a huge amount of parallel evolution in our cousin apes, and many found it too much to swallow. "I frankly don't think Ardi was a hominid, or bipedal," Richard G. Klein, a Stanford University anthropologist, told the *New York Times*. Jack Stern, a professor at Stony Brook University and longtime rival of Lovejoy from the 1980s locomotion wars, guffawed, "As far as I can tell, Owen's interpretation on *Ardipithecus* is likely to be so far wrong as to be laughable." (Stern took some of the criticism personally, especially since he and his colleague Susan Larson had received a $1.9 million NSF grant to study chimps as models of bipedality.) Others insisted the skeleton *was* in fact more like the dreaded chimp than its discovers realized. Primatologist Craig Stanford of the University of Southern California contended that Ardi actually was "very chimpanzee-like" and dismissed the Middle Awash team's arguments as "silly." "To try to rule chimpanzees out as models," Stanford insisted, "requires pretending that early humans were nothing like modern apes, despite all evidence to the contrary."

After being promised the "find of the century" and then waiting until the *next* century, some peers were peeved to be served up theories reminiscent of the 1950s. "This team is about thirty years behind the times," vented anthropologist Martin Pickford, who put Ardi in the chimp family. Some found the Ardi team too eager to narrate a grand saga leading to modern humans. Lovejoy's monogamy theory drew another round of stones.

Despite the deluge of nearly four hundred pages of papers, the complaints about data availability resumed almost immediately. The Ardi team declined to publish some basic measurements of teeth (and wouldn't for another five years). On his anthropology blog, John Hawks compared the omission to government censorship. "IT'S A DATA TABLE WITHOUT ANY DATA," he protested. "What kind of rinky-dink journal is this?" (White said the information would have been provided to anybody who directly contacted the team.)

Eight months after the big publication blitz, *Science* ran a series of critiques by outside scholars, kicking off a new wave of skeptical coverage. In May 2010, *Time* magazine ran a story: "Ardi: The Human Ancestor Who Wasn't?" A few months later, *Scientific American* followed suit with the headline: "Was 'Ardi' Not a Human Ancestor After All?" Blogger John Hawks posted: "Ardipithecus Backlash Begins."

Bernard Wood eventually put his qualms on record. He and Terry Harrison of New York University wrote a lengthy essay in the February 2011 issue of *Nature* that questioned whether Ardi really belonged in the human family. "It's interesting why people want their fossils to be ancestors," Wood told one interviewer. "The reality is that most fossils aren't ancestors. I don't think *Ardipithecus* is ancestral." He advocated the opposite approach: any fossil should be assumed to be something that went *extinct*. A decade earlier, Wood and his student Mark Collard analyzed a large sample of modern apes and monkeys and found that skulls and teeth yielded family trees at odds with ones derived from molecular evidence. The lesson: morphology was misleading. In his eyes, Ardi would have been more interesting if the authors had at least considered that their skeleton might *not* belong to the human family. "Tim just won't go there," said Wood. Instead, he continued, the Ardi team "went into contortions" to plant their discovery near the root of humanity. "Their problem is they made a decision about what *Ardipithecus* was—and then had to squeeze all the evidence into that box," Wood said.

In the *Nature* paper, Wood and Harrison cited two cautionary tales: *Ramapithecus* and *Oreopithecus*, two fossil species formerly touted as human ancestors but later determined to be extinct apes. They questioned the claims that Ardi walked upright for being based on "relatively few" features of the pelvis and foot and "highly speculative" inferences about the spine. "If that's a bipedal foot, then it was designed by a drunken Irishman on Saturday evening," quipped Wood (who, for the record, is British). Speaking at a conference, Wood voiced further doubt: "Most of the inferences about the bipedality are from the pelvis, which is very squashed and looks as if Rush Limbaugh has mistaken it for a toilet seat."

White blew up. The Ardi team had described *entire functional complexes*—not just a few characters. Wood and Harrison cast doubt on the entire suite of papers *without contesting the actual evidence*. "A solicited, anti-*ramidus* six-fucking-page *Nature* paper!" White thundered in his office at Berkeley. "Completely and utterly vacuous. Nothing to it."

The Ardi team was particularly galled because neither Wood nor Harrison had actually examined the fossils, which—to the surprise of peers—*were* finally available for inspection. (No official duplicates or downloadable 3D scans were distributed, so people had to arrange viewings with the discovery team.) At one conference, Berhane Asfaw passed out museum-made casts and invited colleagues to scrutinize them, and he recalled that Wood conspicuously turned away. "The funny thing was," added White, "he already had his manuscript written attacking it for *Nature*." Wood explained: "I can't go look at *Ardipithecus* and have Tim breathing down my neck. . . . It's not a useful conversation."

Nor would it be useful, White decided, to reply directly in *Nature*. Instead, he responded through the media, telling *Science News*, "With no new data, no new ideas, no new methods, no new hypothesis, no new experiments, no new fossils, not even a new classification, this paper will leave everybody wondering what's happened to the peer review process at *Nature*."

In fact, it left people wondering about Ardi. The *Nature* critique branded the beast with a big question mark. If the inconvenient woman did not sit on the human branch of the family tree, then all the paradigm-busting claims would be moot.*

Some skeptics took measurements from the published illustrations and spread rumors that Ardi actually was the long-armed, suspensory, chimp-like creature that the discoverers had disavowed. (In fact, the photo was misleading because the shinbone was missing its far end and the Ardi team's brief anatomical description did not fully describe the damage—and not everyone bothered to check.)

Within a few years, several students had earned Ph.D.s with theses challenging the interpretation of *Ardipithecus* and its place in the human family. But few had the stomach to directly challenge the Middle Awash team. "They just say, 'You get into an argument with Tim White and Owen Lovejoy and you're just going to be battered,'" said Don Johanson. "It's not worth the energy to get involved. There are grad students who come to me who say, 'I'm not going near this!'"

Many took the path of least resistance: they put Ardi in quarantine, and continued business as usual. With no consensus, Ardi could be ignored. Before publication, critics decried the paucity of information about *Ardipithecus*. Afterward, some of the same critics did their best to disregard it and urged others to do the same. "Ardi was more secretive than the government undertakings," continued Johanson. "That developed a lot of animosity to the point where colleagues would say, 'Shit, I'm not even going to read those papers. I'm not going to put up with that. Wait twelve years? For what? An *ape?*'"

But it was selective outrage. Others had kept fossils under wraps just as long with few complaints. In 2014, the hip bone of a Miocene ape from Pakistan was published twenty-four years after its discovery. In 2017, an extraordinary *Australopithecus afarensis* spine from Dikika, Ethiopia, was revealed following thirteen years of painstaking labor by a dogged lab worker to remove it from a sandstone block. A partial foot from the same skeleton was published sixteen years after its discovery. In South Africa, the fossil "Little Foot"—a remarkably well-preserved 3.7-million-year-old skeleton even more complete than Lucy—required more than twenty years of careful extraction from a cave with chisels and airscribes. In such company, the wait for a fragile skeleton like Ardi seemed not so unusual—but the criticism certainly was. Why? Many factors came into play: the anticipation raised by the initial fanfare about the roots of humanity, the expectations placed upon recipients of taxpayer-funded research, the impatience of colleagues lacking sympathy for the difficulties of fieldwork, the combative defiance of the Ardi team, its sheer number of discoveries—and, of course, personal animosities.

The Ardi team found itself on the other side of the locked door. In 2013, one of White's graduate students was denied permission to study the pelvis of the Miocene ape *Oreopithecus*, which had gained renewed interest in the debate about Ardi. The fossil had been declared off-limits because curators at an Italian museum had decided it needed additional restoration and reflagged it as a "new unpublished specimen." It had been discovered in 1958.

Without question, the Ardi team paid the price for its publishing strategy. The moral of the story depended on who you asked. To White, the lesson was that his team should have taken more time to refine and simplify its message because the publication blitz "exceeded bandwidth of just about everybody who had been demanding it." To some peers, however, it was an object lesson of the perils of isolation: the high priests had locked themselves inside the temple too long. By not exposing themselves to a greater diversity of views, they had neglected to sufficiently test their arguments before airing them in public—and suffered the consequences. "If they had adopted a more open strategy early on—with the reasonable qualification that others cannot proceed until the team had the chance to study the stuff—they would not be in the predicament they are in today," observed Bill Kimbel. "People would have been able to see the material and discuss it with them. That's what is supposed to happen."

Ian Tattersall of the American Museum of Natural History added, "Tim is very clever, very industrious, and very hard working. But I think his approach to the distribution of information is ultimately less productive and is going to diminish the proper importance of all the stuff that he has found." Indeed, that reputation sometimes overshadowed how much the team actually had published—literally thousands of pages over the decades. Shortly before the debut of Ardi, the Middle Awash team released a 480-page volume on *Homo erectus* and another 664-page tome on *Ardipithecus kadabba*—both dense compendiums more than a decade in the making (and not exactly clickbait material).

Critics depicted Ardi and her discoverers in the same motif: dead-end Ethiopian isolates. The Middle Awash team swam against one professional current after another. "Everything we're about in the Middle Awash for the last thirty years is about comprehensive, inclusive, multidisciplinary, measured, slow extraction of evidence," White fumed. "Every trend in science, journalism, funding, all of it, goes against that." Increasingly, they even spoke different languages. For decades, the creatures on the human side of the split from chimpanzees had been known as *hominids*. In the new millennium, most of the scientific community adopted new terminology, so the human family became *hominins*. The Middle Awash team remained in the shrinking minority that retained the old term. In more ways than one, they were fossil men.

One fall morning in 2011, Lovejoy sat in his office in Kent State beside a desktop littered with bones. To him, the fallout over Ardi—more recently, the lack of it—was further proof that the quote-*profession*-unquote remained hidebound by old habits of thinking. "The true shocker—we present all this stuff and it's mostly ignored," he said. "Nobody can give up the damn 'we are chimps.'" He returned to a well-worn lament: few anthropologists possessed the anatomical sophistication to recognize the revelations of *Ardipithecus*. The academic mills produced people trained in narrow specialities who were incapable of recognizing anatomical patterns; specialists could tell you all about their particular tree, but don't ask them to find their way through a forest. In generous moments, he took a more philosophical view: his peers wrestled with the same disbelief that afflicted the Ardi team privately for years. "*Ramidus* suddenly pops up in one day in 2009," added Suwa. "The equivalent amount of *Australopithecus* came out over several decades. That's why people have a big problem. *Ramidus* is so paradigm shifting, so idea renovating, because it came out all of a sudden."

Such discord is nothing usual. The great German physicist Max Planck once observed: "New scientific truth does not triumph by convincing its opponents and making them see the light, but rather because its opponents eventually die, and a new generation grows up that is familiar with it." Leaning back beside his desktop of bones, Lovejoy offered a more cynical paraphrase.

"Science," he said, "progresses with the death of each faculty member."

CRITICS WENT ALL THE WAY TO ARDI'S GRAVE. SHORTLY AFTER ARDI WENT public in the fall of 2009, two scientists appeared at the excavation site in Ethiopia. Royhan and Nahid Gani, two young, married geologists, came to render a second opinion on the environment of *Ardipithecus*. At first, they maintained a friendly correspondence with the Middle Awash team and excitedly reported finding teeth and a spinal column thirty feet from the skeleton site. Given that so few vertebrae had been found, an Ardi spine would be a blockbuster discovery. The Ganis e-mailed a photo of a sinuous lump, over which they superimposed a digital image of a human backbone. "A nice match?!?" asked Royhan. White looked at the photo and made a different diagnosis: a fossilized tree root.

Soon they had larger disagreements. In December 2011, the Ganis published findings in *Nature Communications* that suggested the Ardi team had badly misinterpreted the environment of Aramis. The Ganis concluded Ardi lived on the wooded banks of a major river in an otherwise open savanna. They reported finding sandstones from river channels and published a photo of a well-rounded river boulder.

The Middle Awash scientists felt burned. Their multiyear reconstruction had been challenged by two freshly minted Ph.D.s who had made a single site visit. According to the Ardi team, the river "sandstone" described by the Ganis actually was a volcanic ash that fell in shallow water, and the boulder was one of the rocks carried in by the field team to construct a perimeter around the excavation site—not an original in situ feature of the landscape. Once again, White did not bother to respond to the journal. Not until three years later did he respond to the Ganis—in *another* journal in reply to *another* critic in *another* lament about the "failure of the review process." The Ganis defended their work; White dismissed them with a curt three paragraphs and ominously promised a fuller rebuttal at a time of his choosing. Royhan Gani said his career had suffered as a consequence of White's complaints to his university administrators and his funders—and the revelation that the Ganis had collected fossil teeth without authorization. "The man who knows the most about Darwinian evolution," Gani said ruefully, "plays out the Darwinian idea when it comes to competition in the science—survival of the fittest."

But that was just a first skirmish in a bigger battle about savannas, which had so long been touted as the stage of human evolution. A group led by Thure Cerling of the University of Utah analyzed the Aramis isotope, phytolith, and fossil data, and concluded the Ardi did not live in the tree-thick habitat described by the discovery team. Rather, they contended, Ardi dwelled in a "bush savanna" with less than 25 percent tree cover, and thus lived amid more grass than *any* human ancestor older than 4 million years. Rather than falsify the savanna hypothesis, they argued, Ardi might actually *support* it. "The savanna hypothesis is alive and well," declared Cerling, who had long tied human evolution to the spread of grasslands.

Some of the argument was technical: the Utah scientists and the Middle Awash team relied on different models for interpreting isotopes in the soil. But much was semantic: the combatants could not agree on the meaning of *savanna*. The Ardi team equated savannas with open grasslands, while Cerling's definition allowed for up to 80 percent tree cover, which meant that there was precious little in the fossil record that wouldn't support the savanna theory.

Terminology aside, a larger dispute arose: had the Ardi team fully recognized evidence for grass at Aramis? A few coauthors on the Ardi team expressed misgivings. "I agree with Thure Cerling's interpretation," admitted Raymonde Bonnefille, a French pollen expert on the Middle Awash team, referring to the isotopic and phytolith data. "There's no question about that." Downstream in the Awash Valley at Gona, another team reported *Ardipithecus* ranging into more open areas. "We think Ardi lives in a greater diversity of places than Tim," said Scott Simpson, an anthropologist on both expeditions. "Ardi is comfortable in a range of habitats going from closed to more open—not savannas, but wooded grasslands." Simpson took offense because the Ardi team's 2009 report dismissed the Gona findings as a "mixed assemblage" of fossils from different habitats. "I find that to be a professional insult," snapped Simpson. "I have the highest regard for Tim, but I can't even believe that got into press." Cerling added, "Tim can't stand criticism. He certainly can't *possibly* correct himself."

As the debate burned down to its last embers, one side argued that Ardi lived in a "wooded grassland" and the other a "grassy woodland." The grass signal probably was stronger than the "closed, wooded" environment initially described. That said, White argued that the *totality* of the evidence—including the tooth enamel isotopes, dental wear patterns, opposable toe, and all the other animal fossils—overwhelmingly pointed to a woodland creature who stuck close to trees. He taunted opponents as "savanna-philic" and continued hammering a stake into the heart of a theory that just wouldn't die. "It took 2 centuries," White insisted, "but falsification has arrived."

As the debate grew more personal, it became less about education and more about *negation*. Witness this exchange between White and Manuel Domínguez-Rodrigo, a professor at the University of Madrid, in the February 2014 *Current Anthropology:*

> *"Without original research or new data, Domínguez-Rodrigo attempts to resurrect 'the spirit of the old savanna hypothesis' via word games and revisionist history . . . this attempted resurrection of an obsolete mind-set will stand as a monument to futility."*
>
> —Tim White

> *"By denying this evidence, White exemplifies perfectly Kuhn's idea that when a paradigm is assaulted, supporters of the old guard remain intentionally blind to the mounting evidence or selectively utilize data in order to resist change."*
>
> —Manuel Domínguez-Rodrigo

But the Ardi team was outnumbered in the academic war. The graduate school programs that White criticized for overpopulating the professional ranks had—not surprisingly—filled the science with people who did not share his philosophy.

Nor did some of the big names. From the earliest days of the African bone rush, human origins had been dominated by celebrity fossil hunters—and the most famous ones heaped doubt on the skeleton from Ethiopia.

"I don't buy *Ardipithecus* at all," said Richard Leakey.

Leakey sat in his office at the Turkana Basin Institute at Stony Brook University on Long Island. He considered Ardi, shook his head, and wagged his thumb. "The foot," he explained.

Leakey adhered to the old idea that hominids (he, too, had not yet gotten used to the new term *hominins*) were bipeds with *straight* toes. An opposable toe was a mark of *not* being part of the human lineage. Ardi, he figured, was some kind of arboreal creature off the human branch. Leakey remained vague on specifics. "I haven't studied it," he admitted. His critique went beyond the scientific interpretation. "Sitting on these fossils for ten or fifteen years is unacceptable," said Leakey. Interestingly, Leakey's wife, Meave, was perfectly willing to accept Ardi as a member of the human family, albeit one of many in her vision of a bushy tree. And their daughter Louise later would coauthor scholarship recognizing Ardi as *hominin*.

But the family patriarch remained skeptical. Leakey heard White's continuing criticism of all things Leakey with weary amusement. One thing had not changed since they worked together in Kenya four decades earlier: White always thought he was right. "He impugns motives," said Richard Leakey. "He impugns a lack of objectivity and lack of scientific rigor that I think is probably a little unfair."

Don Johanson also sowed doubts about the skeleton celebrated for being more than a million years older than the one he had discovered. He allowed that Ardi might be a member of the human family, but doubted her species was ancestral to his beloved Lucy or modern humans. "Not everything has descendants. Some things die out," he said. "Perhaps *Ardipithecus* was a separate lineage that really went nowhere." By casting doubt on Ardi, Johanson also questioned the judgment of the very people who had served as his brain trust back in the Lucy era. He noticed that the Middle Awash team's illustrations always showed Ardi standing upright to emphasize her stature as a bipedal human ancestor. Johanson commissioned an artist to paint a more apelike Ardi climbing on all fours in the trees and displayed it at public lectures. At one such event at the American Museum of Natural History, Johanson appeared alongside his former adversary Richard Leakey. As elder statesmen, the two old rivals had made amends (Johanson now jokes that Lucy has become so famous that laypeople mistakenly assume she was discovered by the Leakeys). In a private conversation, Johanson asked Leakey's opinion about Ardi. Leakey responded, "I don't study apes."

And Lee Berger—an anthropologist who took showmanship to new heights—contended, "Ardi is the most brilliant ape ever discovered. It's exactly what the field needed—we need an ape to understand what a hominid is."

All these critics shared one thing in common: none had examined the actual Ardi fossils or casts. Those who did came away with a very different impression.

Chapter 29
HELL YES

Unbelievable.

Not far from Leakey's office on the Stony Brook campus sat an anthropologist named Bill Jungers. He was an ursine man with a fleshy, expressive face and a white goatee. He read the Ardi papers and didn't buy much of it. Two decades earlier in "the great *afarensis* wars," he had butted heads with White and Lovejoy. Jungers and his colleagues nicknamed their adversaries "the bone rubbers club"—old-school gestalt morphologists who placed great confidence in gut judgments and eschewed modern, metric-based analysis.

Like everybody else, Jungers had been waiting for details of the skeleton and finally—*this?* Oh, where to begin? Jungers remained unconvinced Ardi was a human ancestor or stood upright. "That ain't the foot of a biped!" he told *National Geographic.* Ardi had one of the most divergent grasping toes he had ever seen. Why would a creature that supposedly clambered with its body horizontal in the trees walk erect on the ground? He deemed the Ardi pelvis reconstruction a "3D Rorschach test"—an ambiguous shape that reflected the bias of the interpreters. He dismissed the speculation about Ardi's spine—they hadn't even *found* most of the damn thing! Finally, he scoffed at Lovejoy's claim that bipedality arose as a mating strategy—"the wildest and woolliest part of all this Ardi stuff."

After Jungers made a number of skeptical comments in the press, he heard from an old acquaintance from graduate school—Tim White. He suggested Jungers might want to examine the evidence before shooting off his mouth. Jungers was taken aback: Tim White was inviting an outsider into the Manhattan Project? In 2011, Jungers brought a team to Berkeley to look at casts and to Ethiopia to study the originals. Then his mind blew up.

To his astonishment, Jungers found himself agreeing with claims that he initially dismissed as absurd. Ardi's feet, hands, and wrists showed features known only in the human lineage. Jungers found the pelvic reconstruction hard to believe until he looked closely at the original fossils. In his judgment, the odd anatomy left only one explanation: Ardi was a climber in the trees and a biped on the ground. Against all expectation, he found himself concluding that the evidence of upright walking was even stronger than what the Ardi team had described. A member of the human family? "Hell, yes!" thundered Jungers. He abandoned the notion that Ardi was a dead-end ape lineage outside the human clade. "That's crazy talk," he said. "That's people who haven't looked at it or paid much attention."

Jungers did have a few differences of opinion. He felt the Ardi team was right about how the animal moved on the ground but wrong about how she climbed through the trees. He found the wholesale dismissal of chimps a little overzealous. Yet Jungers stood in awe of the reconstruction. With so many fragile pieces, it was no wonder the Ardi team had taken so long. On the aggregate, he flew home with a radically different perspective. "I took a little grief from my colleagues when I came back and said, 'You know, they're right,'" Jungers recalled. "Maybe people didn't want to hear that."

Other scholars suffered similarly painful conversion experiences. Matt Cartmill of Boston University, another white-haired sage of the science, initially dismissed the barrage of Ardi papers as too preposterous to believe. "I was indignant," said Cartmill. He corresponded with his old friend Lovejoy and went to Ohio to examine casts of Ardi. "I came to appreciate that a lot of the things he was saying are true," Cartmill admitted in 2013. "I haven't come all the way into the fold, but I've come all the way up to the gate and I'm standing outside bleeding resentfully."

Then there was Bill Kimbel. For years, he had watched the Ardi team come and go in Ethiopia with little idea what they were up to. Meanwhile, Kimbel had kept busy with his own major discoveries—the first skulls of Lucy's species. One focus of research concerned the base of the cranium, a piece of anatomy that distinguished early humans from African apes. Kimbel and his Israeli colleague Yoel Rak showed the humanlike cranial base could be traced back beyond 3 million years to *Au. afarensis*. Did it also go back another million years to Ardi?

It did. When Kimbel examined *Ardipithecus,* he recognized a cranial base that almost could be swapped into Lucy's species. More corroboration came from the humanlike aspects of the Ardi foot and pelvis. "The fact that you have two or three independent sources of data pretty much seals the deal for me," said Kimbel. "It's on our side of the split with chimpanzees."

Many people expected Kimbel would be as critical as his colleague Don Johanson. To Kimbel, however, the science was clear. He rekindled an old habit with his estranged friend Tim White—they collaborated. Exhausted by years of rancor, Kimbel decided he was *done* with the old paleo cold war—and he recognized that his peers upstream along the Awash River had made the most important discovery for early human history since Lucy and *Au. afarensis*. Half-jokingly, he compared the moment to Ronald Reagan's famous Berlin Wall speech near the end of the Cold War: "Tear down this wall!" The two teams began a new era of cooperation.* The Middle Awash team even shared unpublished discoveries and

* White made it quite clear that his team still would have nothing to do with Johanson, the IHO founder. Johanson complained that he requested to inspect Ardi but was rebuffed.

Kimbel and White once again became friendly sparring partners. Kimbel and Rak joined the leaders of the Ardi team on a paper describing the evolution of the human cranial base and bolstering the case that *Ardipithecus ramidus* was indeed a member of the human family, the first major endorsement by scholars outside the Ardi team.

Slowly the profession began to face the hard truth that the most hated men in paleoanthropology actually might be right about a lot of things. The science of the Manhattan Project might seem outdated in the modern era, but an atom bomb was still an atom bomb. "Eventually it will be accepted, I'm quite sure," Peter Andrews, an expert on Miocene apes at the Natural History Museum in London, said of Ardi. "Not easily accepted, I'm sure, until Tim is off the scene."

BUT *ACCEPTANCE* IS A RELATIVE TERM IN A DISCIPLINE IN WHICH OPINIONS are far more abundant than skeletons. For example, Jungers and others still doubted Ardi or the last common ancestor of humans and African apes walked on its palms in the trees. In his judgment, Ardi seemed a versatile climber capable of many forms of locomotion—*except* the flat-palmed branch walking (known as palmigrady) described by the Ardi team. Jungers interpreted Ardi's hand as partially suspensory, albeit not so much as a modern chimp. "In their desire to knock chimps off the ancestral perch, they went a little overboard," Jungers said. Likewise, others believed that Ardi climbed with a mostly upright trunk—not a horizontal posture.

With a technique known as 3D geometric morphometrics (a form of comparative anatomy practiced in the digital realm), Jungers and his team spent several years analyzing the Ardi skeleton with access to fossils and CT scans and affirmed some controversial parts of the original description. For example, they concluded the human hand remained closer to the primitive proportions and chimp hands had evolved much more. (Moreover, chimps and gorillas had very different hand proportions and unique evolutionary histories.) Similarly, their analysis of the femur suggested that bipeds likely evolved from ancestors similar to Miocene fossil apes—not ones like modern African apes. Yet another study affirmed the hybrid nature of Ardi's foot: even while retaining a grasping toe, the lateral foot was being remodeled to walk upright on the ground—the oddity that Jungers and so many others initially deemed so unlikely.

Again and again, science confronts the reality of mosaic evolution—the patterns seen in one body part may not apply elsewhere. For example, some studies of soft tissues do suggest that our cousin apes might indeed be more primitive than we are in some aspects of anatomy. Rui Diogo (a student of Bernard Wood) conducted systematic dissections of primate muscles of the head, neck, and forelimbs, and found that chimps and bonobos had accumulated far *fewer* changes since our shared common ancestor. (The ancestral condition was inferred by comparisons of living species.) Bonobos remained particularly close to the ancestral state. Humans, on the other hand, made twice as many muscle changes compared to our two cousin apes. These soft tissue studies sharply contrasted the bone-based conclusions of the Ardi team. Wood said those results should "make anybody stop, take a breath, and say 'let's not make too many assumptions about what the common ancestor looks like.'"

Ardi sparked a tempest-in-a-teapot debate about what locomotor category best captured her strange hybrid style of locomotion. Some of the 2009 descriptions by the Ardi team had created confusion, and many scholars remained unsure just how Ardi moved through the trees or on the ground. To a layperson, the distinctions might seem trivial, but in academia such esoterica sustains careers like fine print fuels the legal industry. Anthropology loves classification, and one oft-cited scheme lists seventy-four types of locomotion among living primates. But extinct animals do not necessarily fall into the same categories. Ardi was a versatile climber that probably did many things in the trees. Her species clambered atop branches, but not as nimbly as a monkey. It could climb vertically but not so powerfully as a chimp. It suspended, but not so much as the long-armed orangutan. It could walk upright, but not like a human. Reflexively, scholars wanted to pigeonhole her: was Ardi pronograde? Orthograde? Palmigrade? Unwilling to shoehorn their discovery into one of the existing modes of locomotion (nor vindicate critics who wanted to do so) the Ardi team invented yet another: *multigrady*. Ardi required a new language.

The Ardi team eventually would publish a shorter summary of their findings in the *Proceedings of the National Academy of Sciences*. The new piece took a less provocative tone than the previous barrage. Instead of blasting their colleagues in the face with a firehose, they offered Ardi in small sips. This time the team sought outside feedback and benefitted from it. The Ardi team no longer dismissed modern apes as irrelevant and instead deemed chimps "highly informative" because they provided a *contrasting* model of another descendant of our common ancestor. The team shared measurements of teeth and presented a revised description of Ardi's locomotion in the trees with some subtle changes in emphasis. The big revelations, however, remained firm. Ardi showed an evolutionary stage between *Australopithecus* and our common ancestors with African apes—a once-unimaginable creature and a reminder that some mysteries could be solved only with fossils.

ARDI ALSO CAST NEW LIGHT ON LUCY. AGAINST THE BACKDROP OF THE MORE primitive *Ardipithecus ramidus*, Lucy's species suddenly seemed much less apelike than it did when people tried to force it into the role of a transitional link to chimps and gorillas. New discoveries added more pieces to the puzzle of *Australopithecus afarensis*. (It remains one of the best-known paleo species thanks to the richness of Hadar and new discoveries by Ethiopian-led teams at

Dikika and Woranso-Mille.) Despite those curved digits, Lucy's species walked with a forward-facing big toe and arched foot. It had a barrel-shaped rib cage like a human, not the apelike, inverted funnel-shaped rib cage as supposed in old days when scholars rebuilt missing parts of the broken skeleton to resemble a chimp. Yes, *Au. afarensis* might have climbed trees, but the pendulum of opinion swung back toward the old view—Lucy belonged to a species of advanced bipeds. "I'll tell you why it's shifting back," observed Bill Kimbel. "It's because we have *Ardipithecus*."

IN 2013, FOUR YEARS AFTER THE *SCIENCE* PAPERS, A GROUP OF OUTSIDE scholars held a public forum devoted to *Ardipithecus* at Boston University. Kimbel presented a compelling case that Ardi was indeed a primitive member of the human family, while Jungers and Cartmill confessed they had been too hasty in dismissing Ardi—a reaction repeated many times across the academic archipelago. Jungers admitted, "I will eat some crow."

The Ardi team was not present to revel in this vindication. At that very moment, the field team was driving into the Ethiopian rift to search for more fossils. They had more mysteries to solve.

Chapter 30
RETURN TO THE ZOO OF THE UNKNOWN

They returned to the badlands with a machine gun.

In December 2013, the Middle Awash team assembled in Addis Ababa, where a miasma of smog shrouded the sprawling capital. It was the beginning of another field season, a ritual repeated every year. In the walled courtyard outside Berhane Asfaw's house, old friends embraced and exchanged Ethiopian greetings of handshakes followed by bumps of shoulders. The group included Alemayehu Asfaw, now a retired grandfather missing half his teeth, and Alemu Ademassu, the longtime museum casting technician, who now walked with a cane. The three dozen crew members represented a sample of multiethnic Ethiopia and included museum workers, a young professor, and soldiers in plainclothes. Two decades earlier, most of the scientists had been Caucasians. Now it was an all-African crew with one exception.

"Line up those poles by length," White shouted as the men packed the cars. "Military style!"

A young man in a dirty apron hustled back and forth to supervise the loading. His name was Wagano Amerga, and a few years earlier he had been a street boy who served Coca-Cola and food in a grass hut restaurant next to the National Museum. Whenever

Berhane and White came to eat, the boy begged to join them in fossil hunting. He told them he was eighteen and they just laughed. *Yeah right. Maybe thirteen.* The child pleaded for years until Berhane gave him a job. Wagano learned how to identify bones and clean fossils, and he proved to be an excellent restorer. Now he shouldered the logistics of their field operation.

In the basement storeroom of Berhane's house, men pulled equipment off shelves. *Leaf blower, paleo pack, rake, chainsaw, drill, snakebite kit.* Everything was containerized for efficiency. Across the cobbled lane, Berhane stood with pencil and checklist in hand overseeing the packing of more vehicles in a courtyard surrounded by high walls and coils of barbed wire.

The team's continuing devotion to fieldwork made them an anomaly. That year, the annual meeting of the American Association of Physical Anthropologists hosted more than one thousand research presentations, and only three reported new fossil discoveries of human ancestors. A field founded on old bones had broadened its attention to other topics such as technological analysis of previous discoveries, studies of DNA, primate locomotion, or sex and aggression in chimps. To this group, however, fieldwork remained the frontier. Old bones would never grow obsolete.

The men spent half a day packing vehicles with enough supplies for five weeks in the desert. They filled pickup trucks with jerricans of fuel and loaded Land Cruisers with crates of melons, cabbages, avocados, tents, sleeping bags, kitchen gear, mess tables, and tubes filled with rolled-up maps and huge satellite photos. The roof racks were piled high and covered in tarps. The crew squeezed into whatever room remained in the front seats. In the afternoon, a caravan of ten vehicles pulled out and began a two-day journey through a changing Ethiopia. The population had surged to 95 million, roughly tripling since Lucy was discovered four decades earlier. They drove past high-rise buildings under construction, cranes towering above the skyline, and an urban railway being built by the Chinese. Traffic was gridlocked, and it took two hours to escape the city.

The caravan followed the old highway down to the Afar, still a crumbling, two lane road congested by shipping container trucks running to and from the port of Djibouti. Oncoming trucks belched black exhaust while rumbling up the mountains. Minivans, packed with passengers, darted between lanes. Donkey carts and three-wheeled *bajaj* taxis ran along the shoulders. Wrecks littered the roadside. As always, the cool highland air gave way to the dry heat of the rift. The caravan drove until darkness, spent the night in a hotel called Genet—from the Amharic for "paradise"—and got underway at sunrise without breakfast. Approaching the bridge over the Awash River, the road dropped into a deep gorge where baboons scrambled down the cliffs and trotted alongside traffic for handouts of food. The highway descended through dry scrubland and the Ayelu volcano grew larger on the horizon. Afars in traditional dress drove herds of longhorn cattle along the shoulder.

Beside White sat Elema, who still worked as the expedition's guide and ambassador to the local people. He was sixty-two, but looked half his age. He had three wives, the youngest seventeen years old, and had a son named after Berhane. He and White howled with laughter as they remembered their first confrontation when Elema stormed into camp with two guns and ordered the team to leave his land. "I'm eating my fucking breakfast and this asshole interrupted!" boomed White as he drove down the highway.

"After 1992, is finished the problem," added Elema. "Me and the Middle Awash project is good friends."

When the fossil expedition first appeared in his territory, Elema was a bush Afar who knew no English—"zero fucking zero," he explained. Now he spoke a patois of paleontological, geological, and scatological lingo. He seemed to know everybody—village elders, police commanders, foreign scientists, and government officials. His mobile phone rang repeatedly in the car. "An Afar with a cell phone," White muttered, shaking his head at the novelty. What was happening to this place?

This year, the expedition would rely on Elema more than ever.

They were heading to the east side of the Awash River, territory they had barely explored for the last twenty years because the locals had a long-standing habit of shooting at them. The caravan passed near Sibahkaietu, the place where Issa peppered the caravan with bullets in 1991 (a blessing in disguise because it forced them to withdraw to the west side of the river, where they found *Ardipithecus*). In 2012, the year before this expedition, the team had ventured into the east side again and, just as they settled down for lunch on the first day, bullets whizzed over their heads. "We could see the tracers flying through the air," recalled Giday WoldeGabriel. A squad of Ethiopian soldiers cornered the shooters in a canyon and forced them to surrender. They were Afars who had mistaken the visitors for government officials and tried to scare them away by shooting over their heads. "We knew them," said Giday. "They had worked for us." That was the reality of their destination: Afar and Issa running around with guns and the fossil team trying not to get caught in the crossfire.

After two days of traveling, the caravan rolled into Ounda Fao, a dusty roadside village of *tukels* and corrugated metal shacks where the bleating of goats mingled with the roar of shipping container trucks. A group of soldiers in camouflage uniforms lolled under a shade tree near a burning pile of plastic containers. They were the team's security detail, a paramilitary squad of three dozen Ethiopian federal policemen with automatic weapons. When it was time to go, car engines fired up, soldiers snapped to their feet, grabbed AK-47 rifles, and jumped into the back of a white pickup. One muscular man draped in long ammo belts placed the squad's machine gun atop the cab. The Middle Awash Research Project was returning to the field with some serious firepower. White gripped the wheel of his Land Cruiser with fingerless driving gloves and donned an Afar police captain's hat. Through aviator sunglasses, he looked at the machine gun that led his column and grinned with delight. "Look, Issas," he said aloud. "You want to mess with *that*?"

The column drove a few kilometers down the highway and

turned onto a dirt track, a ritual known as "leaving the asphalt." For the last two days, the amenities of civilization had dwindled away. Restaurants, gas stations, running water, and finally pavement were left behind. From here, there would be no cell phone coverage, no roads, and *zero fucking zero* infrastructure. This was tribal territory. Somewhere not far from here, their Afar friend Gadi had been mortally wounded in a gunfight with the Issa. "The Issas are completely in control here," said White. "They've got sentries on every point down by the river. An Afar shows his head and he gets it fucking blown off down on that floodplain. There's no fucking joking around."

The cars thumped over a grassy plateau dotted with basalt boulders. Another twenty kilometers of bushwacking lay ahead before they pitched camp at some still-undetermined place. A herd of long-legged camels loped over the grass. A pair of gazelles lifted their heads from grazing and watched warily.

"When was the last shooting in this area?" asked White.

"Last year," answered Elema. "After the field season."

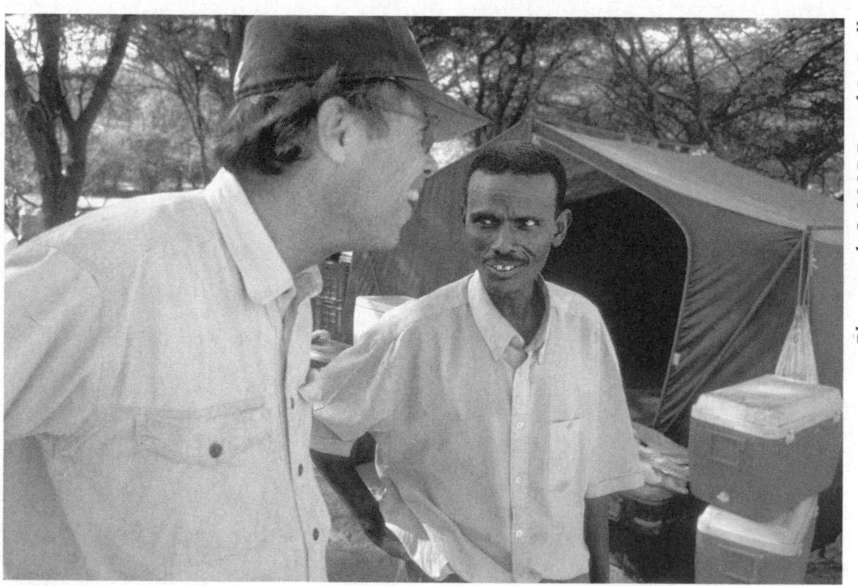

Tim White and Elema.

In the distance, a sere brown hillside crawled with white dots. A herd of goats. Atop the summit stood silhouettes of two thin figures clad in the distinctive white skirts of Issa men. They carried something on their shoulders—shepherd canes or rifles, it was hard to tell from this distance.

White muttered, "They've seen us."

THEY CAMPED AT A SPOT THE AFARS CALLED WILTI DORA—"QUARRELING place." It took hours of searching to find a spot with sufficient shade trees, visibility of the surrounding countryside for security, and no tall grass that attackers could set afire. The first days were devoted to turning a small patch of wilderness into a comfortable home for five weeks. Skip the logistics and the esprit de corps quickly would turn to *merde*—guys would literally be stepping in each other's dung. "If you don't spend the time to put the camp in correctly, it's fine for the first four or five days but then discomfort starts to wipe out the morale," explained White. "Once that happens, our efficiency at finding fossils goes to shit."

Men cleared brush, dug up stumps, and raked ground down to dirt. They dragged branches to the edge of camp and made fences to keep out goats, jackals, hyenas, and baboons. They killed many snakes. They dug an oven in the ground where the cooks would roast goats, cook lasagnas, and bake fresh bread. They set up showers where the crew could rinse away dust and sweat with their daily ration of bathing water—a two-gallon, solar-heated shower bag per person. Soldiers established posts under shady bushes. White marched from place to place, making sure the tent poles stood true vertical and not some cockeyed half-assedness, the gas canisters properly stowed, tools organized, and everything laid according to his master plan. His Ethiopian partners attended to their responsibilities more quietly.

The sun climbed overhead. The landscape seemed to undulate with heat and air buzzed with insects. Three dozen men labored away, and sweat stains darkened dirty shirts. By midday, White began to flag. He pulled off his hat, laid his head on the lunch table, and covered it with a bandana.

He was sick. At the last restaurant stop back in Gewane, he gestured for everybody else to be served first, and he got the last meal—probably bad eggs undercooked by an overworked chef. Just when the camp should be going up, he desperately wanted to lie down. But he would not. The sun dropped lower and his tolerance slipped to zero "We've got six guys working on one fucking tent! We're wasting fucking time. Get a ground cloth. Move it!" He stalked off. "Out of my way!"

As darkness fell, Wagano's crew burrowed an electric generator into a termite mound for sound insulation, buried power lines, and strung up extension cords for lights and outlets. The generator kicked on, and bare bulbs illuminated a canteen table and camp office with banks of radios, GPS units, and computers. Men pitched personal tents by headlamps—the last task of the day because the collective mission came before the self.

White appeared at dinner looking shriveled and exhausted. Once again, he laid his head on the table and shrouded himself with a towel. "He cannot do everything by himself," said Giday. "Not at this age."

ON THE FIFTH DAY, THE CREW FINALLY BEGAN LOOKING FOR FOSSILS.

The team planned to spend the season searching a series of crumbling sandstones that formed along an ancient lake and rivers 2.5 million years ago. This was roughly when our ancestors started resembling us enough to be placed in the genus *Homo*—but just how that happened remained mysterious. Scientists have long argued about what constitutes *Homo* and when it first appeared. When it comes to terminology, we can pinpoint when and where our genus made its debut—1758 in Sweden. That's when Carl Linnaeus classified modern humans as *Homo sapiens*. Beyond that, everything becomes more vague. The criteria to define *Homo* has shifted repeatedly because once-defining traits turned out to be shared with apes or the older genus *Australopithecus*. One thing was clear—the urgent need for more evidence between 3 million and 2.5 million years ago. Several species had been named in that time window, including *Australopithecus garhi* from the Middle Awash, and *Australopithecus sediba* and *Australopithecus africanus* from South Africa. Earlier that year, another team in Ethiopia found a 2.8-million-year-old lower jawbone that became the then-oldest fossil attributed to our genus. With abundant rocks from the Pliocene and Pleistocene epochs, the Middle Awash offered a promising place to further illuminate the mystery.

In the meantime, the team's mission had lost something of its currency. On the surface, the U.S. National Science Foundation celebrated its past investment that yielded so many discoveries in the Middle Awash. In 2010, the agency listed Ardi among its "Sensational 60" scientific advances along with the Google search engine, bar codes, fiber optics, and the discoveries of black holes and the ozone layer. In public reports and congressional testimony, the NSF trotted out Ardi as an example of a taxpayer-supported scientific breakthrough. Behind closed doors, however, NSF peer reviewers were not so complimentary.

In 2011, the Middle Awash team applied for NSF funding for a multiteam investigation of the period between 3 and 2.5 million years ago—the very mission of the expedition this year. The proposal was rejected. One anonymous reviewer complained, "The Middle Awash Project has had a very slow pace of publication of key discoveries and an abysmal record of making casts available, sharing data, and facilitating the access of other researchers to their discoveries. I have seen little evidence that this pattern of action is changing, and it is not something that the scientific community should condone, much less reward." (The same reviewer made comments about White, which the NSF program director deemed too personal and instructed the panel to disregard.) The program had funds for only one award—and it went to another project.

In 2012, the Middle Awash project applied again. This time the consortium did not list White as principal investigator and instead named his teammate Raymond Bernor of Howard University (who previously served as an NSF program director). The reviewers ranked the proposal as "not competitive" and rejected it. Reviewers faulted the application for lacking specifics about hypotheses to be tested and proof they would actually find anything. White complained of an impossible standard: how could applicants state what they would find *before* finding it?

In 2013, the project made a third application, this time to the NSF earth sciences program instead of anthropology. Given the professional animosities, the team added a three-page list of names to exclude as reviewers—essentially a who's who of the profession's most prominent figures and institutions including the Leakeys, Don Johanson, Horst Seidler, Lee Berger, Thure Cerling, Bill McGrew, and many more. The panel gave the proposal strong reviews, but faulted it for lacking a technological component and rejected it once again.

One night in the field camp, White sat at the canteen table and vented.

"They said we lacked a 'modeling aspect' to the project," he said.

"It's an excuse," said Giday.

"Yeah, it's their fucking excuse," snapped White. "We're going to model how fast a fucking tick climbs up my leg to my crotch? What the fuck! *Modeling?*"

Soon White worked himself into indignation and thumped the table.

"The bottom line here," he said, "is they want us to go out and find the fossils and turn them over to people like Bernard Wood who are smart enough to interpret them because we're fucking stupid. And if *we're* not just fucking stupid—maybe that only applies to me—we're also Africans and they don't want that. These people are convinced that we are incompetent."

The story of early human evolution is set in Africa, yet the storytellers remain stubbornly un-African. One survey by the American Association of Physical Anthropologists found that blacks held only 2 percent of faculty positions in biological anthropology (whites held 89 percent). On this expedition, the scientific staff included four African Ph.D.s and the field crew members were all Ethiopians, most trained by Berhane and White. (This year they also were all men.) This emphasis on training Africans, however, was not necessarily an asset in the professional ecosystem. One reviewer of a rejected NSF proposal pointed out that the agency's mandate did not extend to "building infrastructure for Ethiopia and recruiting and training Ethiopian students to compete with Americans for jobs."

"Kenya, Tanzania, South Africa—you don't have real indigenous people collaborating in research like the Middle Awash," added Giday, gesturing down the rows of Africans along both sides of the canteen table. "They don't want to support you unless you give in to those people who don't like to do fieldwork and want all the information to be handed over to them. They block the progress of science, practically."

Across the table, White railed about the inside-the-beltway "cabal" of lab scientists—particularly a recent NSF initiative to invest millions of dollars studying ancient climate and its effect on evolution.

"Little computer jockeys who don't know how to put their butt in the seat of a car, much less tell an Afar from Issa," he fumed. "They sit there and jockey these computers around and they tell us what *might* have happened in the Pliocene if certain conditions applied and you had an El Niño effect in the Hawaiian Islands and what a *ramidus* had coming out of his asshole under the acacia tree. *That's* modeling—arm-waving bullshit. It's inapplicable, and of course we didn't put it in our application."

The Middle Awash team gave up on the NSF and resigned itself to paying its own way. A foundation at Berkeley and private philanthropy kept the project afloat on a reduced budget. (In this respect, the group was not alone: other major American research teams including the IHO and the Turkana Basin Institute also relied on private fund-raising.) White called in favors from longtime collaborators, colleagues paid for travel out of their own pockets, and the Berkeley Geochronology Center discounted the costs of radiometric dating. "The Middle Awash has been, without even a closest second, the most productive paleoanthropological research project ever," said Paul Renne, director of the Geochronology Center. "The fact that this guy can't get funded by the NSF is just a travesty." Renne noted that much of the resentment focuses on White himself: "A lot of people just didn't like him—he disrespected them once, called them a shitty scientist, or whatever."

The philanthropic model offered some advantages. Funding no longer depended on professional peers, and budgets could not be jeopardized due to lack of modeling, dull hypotheses, professional animosities, archaic methods, or the slow pace of publication. "Right now we're prioritizing data acquisition rather than data presentation," said White. "We got criticized when we presented, and

Planning the day's mission.

didn't get any money. Well, OK, we'll acquire and publish when we think we're ready."

Data acquisition started with a lecture at breakfast about avoiding vipers and treating snakebites. Then the crew climbed into cars and headed to a place called Gamedah, a series of rolling hills topped with ledges of sandstone from which tumbled the bones of ancient elephants, hippos, monkeys, and pigs. It had last been searched more than twenty years earlier. This investigation stretched even longer than the one for *Ardipithecus* and was not due to end anytime soon. The team played a long game: if geologists could match rocks from these localities to others elsewhere in the Middle Awash (using techniques such as radiometric dating or geochemical signatures), then all the fauna could be combined into a single sample for comprehensive analysis. That snapshot of deep time would leverage decades of collections on both sides of the river. And there had been lots of collections: the project area had 439 fossil localities spanning 6 million years. All these fit into a master timeline with more than one hundred geologic layers. But all that made very slow science.

When searchers found a fossil worthy of collection (most were not), they planted a yellow flag and called White. Recovering from his illness, White trudged behind his crews. He always carried an ice axe—his all-purpose field tool, probe, pointer, digger, and stool. This day, he used it as a walking cane. Surveyors could sense his approach from behind by the uneven *clink clink clink* over the rocks like a peg-legged captain. "No country for old men," he groaned.

"Flag!"

On the chocolate-colored slopes amid pebbles of crumbling sandstone, a young Ethiopian man named Dawit spotted a chunk of mandible. White congratulated him on finding a hominid. "That was perfect," he said. "Very good job." A two-man crew with a GPS unit read off the exact location and elevation of the specimen. White took a picture of the discoverer with his prize, jotted a few details on paper, and snapped shut his steel notebook case.

"*Bakka!*" he shouted. (Done!) "Another one—let's find another one. There's hominids here!"

So it went day after day. White constantly exhorted his crawlers to go slow and *look*. He kept up a running tutorial familiar to anyone who worked with him: Concentrate on low points where fossils are gathered by water. Always walk upslope, never down. Use your hat to shield the glare of sunlight. When the sun is behind your back, look outside your shadow. When the sun shines in your face, use the glint to highlight shiny fossils. When impatient people got too far ahead, he abruptly turned the crew in another direction and the first became last—a trick he called "strategic misdirection."

"It's like herding goats," White muttered with a mischievous grin. "Except the goats don't know where the grass is."

Savvy veterans walked hunched forward and slowed when they hit ground dense with bone. Neophytes wandered tall and randomly. Invariably, a few ambitious surveyors strayed beyond the prescribed boundaries, and White delivered a tongue lashing that was part rage and part theater. All day long, his voice echoed over the rocky canyons, mixing with the shrill calls of the Issa goat herders on the distant hills.

"Who's out in front? Abebe and Elema? Elema!"

The Bouri chief turned. "Yes sir!"

White stabbed his ice axe in the air.

"I want you to go to back where we found the first fossils this morning, just beyond the police. Get your ass in that gully and find the other collectibles. *Awa-yi! Awa-yi!*"

"Ten-four," answered Elema.

Twenty men slowly inched over the rocky slopes. They collected monkeys, pigs, bovids, hippos, horses, and other fauna. They excavated an exquisitely preserved kudu skull with horns and teeth. All those signaled that they had come to the right place. A 2.5-million-year-old skull or skeleton of a human ancestor might have been just as well preserved—if only somebody could find it.

"OK, Elema, you find nothing?" White shouted.

"Nothing! Zero fucking zero."

"Bizuayehu! Wagano!" White shouted at the two men excavating a kudu skull. "Are you almost done?"

"Almost! Almost!"

"Come to the car fast. Everybody else, *macchina*!"

ONE DAY, AN ETHIOPIAN FOSSIL HUNTER FOUND A STRANGE JAW FRAGMENT with a couple of teeth still attached. White puzzled over the fossil, cracked a faint smile, and summoned the crew. "Now we're going to have an osteology test."

He handed the fossil around.

"I've never seen this before," said Wagano.

"Aardvaark?" offered another.

"Aardvarks eat ants," said White. "These aren't ant-eating teeth."

"I think it is a monkey," said the antiquities officer.

"It's not a monkey," answered White. "Not even a primate. Lucky to be a mammal."

All were stumped.

"What is it, Tim?" begged Yonatan Sahle, a young Ethiopian archeologist. "Just tell us!"

Finally, White ended the suspense: "A giant hyrax. That's like a hundred times more rare than a hominid." A modern hyrax is about the size of a rabbit, but this Pleistocene hyrax was about the size of a German shepherd—another strange creature from the alien country of the past. "Let's hit this valley real hard," White said. "This is the richest we've seen."

BY THE END OF THE EXPEDITION, THE TEAM COLLECTED MORE THAN TWO thousand fossils. Only twenty belonged to the human family, including a broken tooth, part of a leg bone, and a jawbone. That meager haul was typical. Even in one of the most productive fossil territories in the world, human ancestors comprised less than 2 percent of the entire fauna. Take out Aramis, and human ancestors would drop even lower.

Those numbers underscored the rarity of the Ardi discovery. Every morning, a series of hills appeared in the haze across the river. It was the Central Awash Complex, the geological formation where they had found the oldest family skeleton nearly twenty years earlier—and where White longed to return.

"Let me tell you," White said over the breakfast table, voice rising with indignation, "if Louise Leakey had found those fossils at Aramis, it would not have been treated in the same way as our publications. It would have been heralded as the great breakthrough on the direct line to *Homo*!"

He rose from the table and strode away. The day was an hour old, the sun climbing over the rift margin, and already he felt the press of time. There were cars to pack, maps to consult, teams to exhort, and outcrops to search. Odds be damned, maybe this would be the morning when they went into the desert and found another ancient, unknown creature who changed how we think about ourselves.

Chapter 31

NEITHER TREE NOR BUSH

Slowly Ardi, the biped from the Pliocene with the grasping toe, ambled awkwardly into the human family. Scholars began to grapple with all her inconvenient truths. Ardi entered textbooks. Papers on human evolution were obliged to account for her—the oldest skeleton in the human family, more than one million years older than Lucy. Unfortunately, the ancestor everybody expected wasn't the one that turned up. Ardi fit no existing category, boggled imaginations, and required a new vocabulary.

One of Ardi's main lessons is that the simplistic narratives contrived to fill gaps in the fossil record often turn out to be wrong. Consensus can be a poor predictor of who turns out to be right in science. The most enduring work is that which describes things never seen before—human anatomy, fossils of extinct animals, ape behaviors, genetic code, ancient ecosystems, and more—without distorting that novelty to conform to the expectations of the era. But we humans hunger for more than just pure description—we search for meaning and emotionally satisfying endings, and that's when we run astray, because our reach for narrative often exceeds our grasp of facts. In the struggle to comprehend the unfamiliar, we invoke familiar analogies, but nature usually turns out to be more complex than what we imagine in our little brains. The only way to know for sure is to discover.

New scientific truth dawns slowly. The first human ancestor discovered in Africa, *Australopithecus*, sat in scientific purgatory for a quarter century before it was widely recognized as a member of the human family. By that standard, the acceptance of *Ardipithecus* may be happening rapidly. Over time, it ceased to seem so strange. Gradually, the debate shifted from *whether* to fit Ardi into the human story to *how* to do so. Even the people who continued to champion modern apes as "time machines" felt obliged to accept Ardi into the human family (albeit with very different interpretations) and reconcile her with their view of the world.

Yet even as Ardi won more acceptance by the scientific community, she remains little known to the public. She had no hype machine like the one that turned Lucy into a household name. None of her discoverers wrote a popular book. Major museums such as the Museum of Natural History in London and the American Museum of Natural History in New York complained of being unable to obtain replicas for public display. The Ethiopian protections against exploitation of antiquities also wound up curtailing the flow of scientific information. Ironically, the discoverers who fought so long to keep Ardi under wraps ultimately faced the opposite problem—they *couldn't* distribute her because the Ethiopian antiquities administration showed little interest in circulating casts. (The ones the Ardi team showed to other scientists were personal copies, and Ethiopian regulations prohibited the scientists from making duplicates.) The only replicas available were unofficial bootlegs made by a commercial company that were not accurate in fine detail. "People who run these offices don't appreciate the value of distributing these fossils," explained Berhane Asfaw with a sigh. "If I were in charge of these offices, every museum could have gotten this for free."

Modern scientists increasingly saw data as something that came from computers, but there would be no mass distribution of electronic 3D scans for Ardi (some were shared with outside teams on a case-by-case basis). A handful of scholars made pilgrimages to Addis Ababa to examine original fossils or visited the labs of the discovery team to see high-quality casts. The Middle Awash team quietly became more receptive and even allowed outside scholars first crack at publishing discoveries from their collection, such as the oldest spine fossils of *Australopithecus*. As usual, White insisted on verifying every measurement, challenged definitions, uprooted assumptions, and policed his peers to "keep the literature free of bullshit." He issued a blistering rebuke to one young scientist who published a misleading description of Ardi's wrist. "To deal with Ardi, you have to deal with Tim," observed Bill Jungers, "and some people don't want to do that."

Many factors conspired to make *Ardipithecus* the most important fossil that most people have never heard about. A vivid demonstration came in 2015 when President Barack Obama visited Ethiopia. The country decided to show off its treasures. At the National Museum, fossils were pulled from safes and whisked off to the National Palace in a motorcade of matching vehicles (similar to the security measures for Obama himself, noted Berhane, a member of the host delegation). At the state dinner, the hosts steered Obama over the red carpet toward a fossil skeleton. It was a made-for-TV event in which the leader of the free world met the so-called mother of all humanity—all documented by the international press corps. But Ethiopia did not showcase Ardi, its oldest skeleton that had been discovered by a homegrown scholar. Instead, the hosts, led by paleoanthropologist Zeray Alemseged, guided Obama to a younger, more famous fossil discovered by foreigners—Lucy. Later at the state dinner, Obama stood and toasted "Lucy, our oldest ancestor." The television footage captured a sober-faced Berhane Asfaw standing silently in the background at the palace. He had enjoyed a private chat with Obama off camera but would not be the public face of his science. He didn't mind not getting credit: what really mattered was the great publicity for his *country*. Ardi was there too—but unacknowledged.

• • •

ARDI'S DISCOVERERS WENT LOOKING FOR THE ROOTS OF THE HUMAN FAMILY tree. They sought to fill its branches with bones—and certainly succeeded. Others did, too: since Ardi was discovered in 1994, the human family has added about fifteen new species, roughly doubling the number of named taxa. With every announcement, headlines and commentators proclaimed yet more branches in our family tree. In the meantime, the scientific ground beneath that tree jolted in a seismic shift. Just as extinct ancestors and bygone ecosystems may have no analogs in the modern world, the tree of life turns out to look unlike any plant growing on earth today.

For two centuries, the tree has stood as the central metaphor for evolution. In *The Origin of Species*, Darwin spoke of "a great Tree of Life" in which "the limbs divided into great branches, and these into lesser and lesser branches." For the next 150 years, this paradigm gave evolutionary biology its mission: draw the tree, define the branches, and trace them back to common ancestors. Then molecular biology showed that species don't always evolve and diversify as neatly as a tidy branching tree. Sometimes genes pass directly between distantly related species—in other words, hop between branches. (For example, antibiotic resistance genes have jumped among bacteria, and genetically engineered crops have passed genes to weeds.) This phenomenon of "horizontal gene transfer" turned out to be rampant among simple life forms and completely bedeviled attempts to build a universal tree of life (one of all living species on earth) from the ground up. Around the turn of the millennium, some biologists abandoned the tree model in favor of a "web of life."

Higher up the tree among more complex animals like mammals, branches also get tangled. One culprit is incomplete lineage sorting (which we saw in chapter 26). Another is hybridization. Traditionally, evolutionary biologists assumed that animals did not normally mate across species; in other words, branches did not rejoin. In evolutionary lineages, however, they do. Many species—including plants, insects, mammals, and ancient humans—have branches that split and then remix. The closer the relation, the more they can hybridize. As one 2016 paper put it dryly, "We should now perhaps broaden our view of sex across the species boundary."

We can no longer assume that our family, or zoology in general, conforms to any simple model of a single family tree. The tree is not entirely obsolete, but the old, simplistic renderings fail to capture the sheer messiness of lineages that repeatedly split and then rejoin. Even Darwin's finches—the birds of the Galapagos Islands often cited as the canonical example of the branching tree of evolution—show that so-called species can be quite ephemeral, branching off and then merging with others to weave a very tangled web of ancestry. Molecular biology has begun to reveal such an entwined history in our own family. Although genetic information isn't available from old fossils like Ardi and Lucy, scientists have managed to recover DNA from younger bones that have not completely mineralized. The revelations have rewritten the human story yet again.

For the last three decades, anthropology embraced the "Out of Africa" model, which held that modern *Homo sapiens* arose in Africa in the last two hundred thousand years and totally replaced more archaic species such as Neanderthals in Europe and Asia. (Fossils like the Herto skulls—the *sapiens* heads discovered in the flood year—were interpreted within that theory.) In 2010, a consortium of scientists revealed a startling discovery: DNA showed modern humans interbred with Neanderthals. Today, all non-African humans contain about 2 percent Neanderthal DNA (roughly equivalent to the amount we would inherit from an ancestor six generations back). Then a tiny finger-bone fragment from a cave in Siberia provided evidence of another ancient genome, neither human nor Neanderthal, but a third lineage of human ancestor named the Denisovans (surprisingly, this lost lineage was genetically most preserved in descendants far away—modern-day Australian Aboriginals and Papuans). In analyzing these ancient and modern genomes, scientists detected a fourth "ghost lineage" that apparently split more than 1 million years ago but whose physical remains have never been found. More ghost lineages have been detected since then. The bottom line: multiple lineages of the human family coexisted in Africa and Eurasia for hundreds of thousands of years and occasionally interbred. Early humans diversified into creatures so physically different that some are classified as different species, and yet the branches still rejoined. "Many paleontological species were probably capable of hybridization with others," said Evan Eichler, professor of genome sciences at the University of Washington. "Hybridization is the norm, not the exception."

Genomic scientists began to speak of human ancestors not as branches in a tree or bush, but as a "metapopulation"—a web of groups that separated long enough to become biologically distinct but remained close enough to mate when they came back in contact. "Within human populations, there is no simple tree," said David Reich, a geneticist at Harvard University and leading researcher of ancient DNA. "Instead, the truth is it's more like a trellis with mixing, remixing, and separation again and again going back into deep time." This means many old metaphors are dropping like leaves in autumn, and we should not expect to find any archetypal last common ancestor, no single locale of evolutionary Eden, and no lone mother or father of humanity to the exclusion of all others. Our genome is a 3-billion-piece mosaic of ancestors—lots of them. "There was never a single trunk population in the human past," added Reich. "It has been mixtures all the way down."

Our ancestors, even arboreal ones, do not easily fit in trees.

Since the oldest fossils have no surviving DNA, what constitutes an ancient "species" comes down to judgments about morphology—and a matter of taste. It also remains a source of endless controversy. In the field, White kept a pair of bush clippers close at hand. When cars couldn't squeeze between the brush, he and the field crew climbed out of the Land Cruisers and lopped away. He played a similar role back at the academy, ever ready to attack the latest example of species growth, especially in the Pliocene—the time when the genus of Ardi gave way to the genus of Lucy.

Several new species had cropped up alongside *Australopithecus afarensis* between 3 million and 4 million years ago, including *Australopithecus bahrelghazali* in Chad, *Kenyanthropus platyops* in Kenya, and *Australopithecus prometheus* in South Africa. White complained that many of his peers had fallen prey to "taxonomic inflation" and had declared invalid species for the sake of publicity, funding, and career advancement. He lamented that the field had once again lapsed into "typological thinking"—just like their predecessors before the Cold Spring Harbor conference of 1950. He took particular umbrage at *Kenyanthropus platyops*, the flat-faced man from Kenya, named by Meave Leakey and her team. For years, White argued that *K. platyops* might be just a geologically distorted member of *Au. afarensis*, Lucy's species. Following White's critique, the *platyops* team spent years strengthening their case led by paleontologist Fred Spoor, who represented the new breed of data-driven technologist. Spoor analyzed large samples of fossils and primates, employed CT images, made statistical comparisons—and once again concluded that *Kenyanthropus platyops* was a distinct species. At one conference, White aggressively cross-examined Spoor with voice booming through the auditorium; Spoor retorted that his prosecutor worked "data free." "Tim is good at finding fossils—I really have to give him that," Spoor said later. "But he is not the greatest analyzer. The field has become very professional." White dismissed such "professionalism" as techno-pap and derided the Kenyan species as "platyOOPS."

Then species proliferation hit closer to home.

Yohannes Haile-Selassie, the man who found Ardi, was the star protégé of the Middle Awash team. After losing his first field site, Yohannes had been allowed to pick a new location in the Awash

Valley called Woranso-Mille, and it paid off with numerous discoveries. He became the curator of physical anthropology at the Cleveland Museum of Natural History and occupied a basement office with a heavy wall safe that housed Lucy during her stint in exile. In 2011, Yohannes invited Bruce Latimer into his office to see a cast of his latest big find—a partial foot. Being a locomotion expert, Latimer immediately recognized the joint of an opposable toe, just like Ardi. Then Yohannes revealed the punchline: it was 3.4 million years old.

Whoa! That was one million years *younger* than Ardi—and contemporary with Lucy's species. Two members of the human family—one bipedal and one arboreal—coexisted in Ethiopia in the Pliocene. The "Burtele foot" fossil was published in 2012 but not assigned to a species. The senior members of the Ardi team hypothesized that *Ardipithecus* split, with one branch evolving into bipedal *Australopithecus* and another lingering as an arboreal species—with this grasping foot its last known trace.

Then Yohannes made things even more interesting.

In 2015, Yohannes reported a new discovery in *Nature* under the headline: "New Species from Ethiopia Further Expands Middle Pliocene Hominin Diversity." His team named *Australopithecus deyiremeda* (after the Afar word for "close relative") based on teeth and jawbones dated 3.3 million to 3.5 million years old—smack-dab in the heyday of Lucy's species. (The opposable toe was not linked to the new species.) As ever, tree metaphors remained a growth business, and a headline in the *New York Times* proclaimed: THE HUMAN FAMILY TREE BRISTLES WITH NEW BRANCHES.

The announcement caught White by surprise. In the past, Yohannes asked for his feedback when drafting papers, but he heard about this one from a colleague. "He's obviously employing a number of strategies that are not only widespread but currently effective at building a career," White snapped of his former student on the day the article appeared. "When the Middle Awash got no grants after 1999 he has successfully gotten eight, and counting.

He has successfully gotten two *Nature* papers based on diversity claims. Draw your own conclusions."

Back in Cleveland, Yohannes laughed wearily. He anticipated his former mentor would be his harshest critic. "I don't think it's fair to just dismiss everything that's named as a new species as invalid, simply because it doesn't bode well with old ideas from thirty years ago," he said. "We have to abide by whatever the fossil evidence is trying to tell us. That's the whole point."

Yohannes envisaged his project area as a laboratory for studying how ancient species divided and shared their environments. Four years later, he added yet another argument for diversity: a 3.8-million-year-old skull of *Australopithecus anamensis*, the oldest species in that genus. The scholarly consensus previously held that this species was the likely ancestor of the species of Lucy, but Yohannes and his colleagues suggested they were in fact two different lineages (an argument that failed to convince skeptics such as White and Bill Kimbel). Another interesting point was buried deeper in the *Nature* paper: the new skull bore similarities to Ardi and reinforced the membership of *Ardipithecus* in the human family—a quiet affirmation in the same journal where it had been so prominently questioned a few years earlier.

In that weblike metapopulation of Pliocene Africa, there was more than one way to be an *Ardipithecus*. Another trove of bones 4.5 million years old had been found at Gona and kept quiet for fifteen years (critics didn't seem to complain as much this time). The original Ardi walked with her big toe splayed off to the side; the Gona *Ar. ramidus* also had a grasping toe but held it more in line with the other toes. While Ardi pushed off on her second toe, the Gona creature pushed off with its big toe—more like humans. Even so, it probably did not trek long distances because its ankle lacked the impact-absorbing anatomy of later bipeds. But there was no doubt it was one of Ardi's kin. "They would be part of the same species, but it's a curious amount of variation," said Scott Simpson, paleontologist on the Gona team. "That's the crazy thing about fossils—they always seem to have more variation than we see in modern populations."

MORE CRAZY THINGS APPEARED. ANOTHER STRANGE BRANCH OF ANCIENT humanity sparked another epic battle for the soul of paleoanthropology. In 2015, the antithesis of Ardi emerged from a cave in South Africa. The ringmaster of this new fossil extravaganza was Lee Berger, who took inspiration from the Ardi saga—as an example of how *not* to behave.

Berger, an American scientist at the the University of the Witwatersrand in Johannesburg, was determined to transplant the roots of our species back to South Africa (where a fossil-rich region of caves had been branded as "the Cradle of Humankind"). He relished popularity and aspired to be the next Don Johanson, to some dismay of the original one. Berger employed freelance fossil scouts who found a cavern full of more than 1,500 bones—a species later named *Homo naledi*. His first call was to the National Geographic Society, which agreed to fund an expedition and a TV show. Berger invited film crews into the cave, recruited collaborators on Facebook, and published as quickly as possible without geological dating. "Dates are not important for understanding biological relationships and evolutionary relationships," Berger claimed. With apparent seriousness, he continued: "If *naledi* is primitive and fits the biological picture of the ancestor of the entire lineage of *Homo*, then that's what it is—no matter if it's five thousand years old, 5 million, or anything in between." (After the hoopla died down, geologists revealed the remains were between 236,000 and 335,000 years old—more than 2 million years too late to be the progenitor of *Homo*.)

Within three months of finishing the excavation, the *naledi* team submitted twelve papers to *Nature*. When they were not accepted, Berger withdrew the papers and published them in an online journal, *eLife*, only two years after the initial discovery—a remarkably fast turnaround—in conjunction with a TV documentary that proclaimed: "The Rising Star expedition was to be a new kind of paleoanthropology, tailor made for the age of social media and the Internet." The biggest headline arose from the controversial claim that the *naledi* bones were carried into the cavern and represented the earliest evidence of mortuary ritual. Berger distributed casts to colleagues across the world and put 3D files online—another stark contrast with the Ardi team. White retorted they were lousy reproductions lacking the fine detail that any competent morphologist should expect. But no matter—peers were delighted to have them and celebrated the new era of openness. *Naledi* generated an electronic tidal wave of new scholarship. "Length of time spent on something does not equal quality," insisted Berger. "Quality equals quality, regardless of the time it took."

White viewed Berger and *naledi* as the embodiment of everything that had gone wrong in the profession. He wrote an op-ed in the *Guardian* newspaper decrying the collapse of science into entertainment and whipped up a swarm of bees among Berger's social media followers. One posted: "When Tim White dies, I'll intentionally step on his bones." John Hawks, an anthropologist on the *naledi* team, blogged, "Let's face it, the paleoanthropology family has a few cranky uncles who fart at the dinner table just to get a rise out of people." Berger shot back at White: "He's doing everything he warned against. He's an armchair critic and using popular media to promote his ideas, not the scientific literature. He hasn't bothered to look at the evidence. He is exactly everything he criticizes. He is his own Frankenstein monster."

But not only cranky old Uncle Tim was alarmed. Even Don Johanson and Richard Leakey denounced Berger's showmanship. "He's set himself up to basically crash the field," warned Leakey. Bill Kimbel mourned, "The currency of debate has changed. It's less relevant now what someone will say in print than the voice that is the loudest on social media, on the blogs, on Twitter, on Facebook. The entire structure of the field has become deformed." Kimbel put matters in perspective. "I can guarantee you," he said, "that less than fifty years from now *Ardipithecus* will still be with us and *naledi* will be a blip."

• • •

BACK IN OHIO, OWEN LOVEJOY SAT IN HIS STUDY WHERE TWO DECADES EAR-lier he had first glimpsed Ardi, the fossil that upended his world. He was an older man now, with a pair of bifocals resting atop white hair, a gray goatee, and Nike sneakers with soles like marshmallows. Atop his desk sat a prescription pillbox with compartments labeled in large print with days of the week. It contained *Ardipithecus* casts. In this hideaway, he worked in isolation with only his cat for company. Above his desk hung a painting of a lone man rowing against an open sea.

"You keep publishing but you realize you're not going to destroy the myths," he said. "It's like trying to convince Methodists there's no God. They'll take things on faith. Well, the minute you take things on faith, it's all over."

With amusement, Lovejoy pointed to a book on his overcrowded shelf—Jared Diamond's *The Third Chimpanzee*. The title captured the ethos of an era when humanity looked for its origins in living apes. "There is a tendency amongst anthropologists—and I understand it—that you don't want to accentuate what's unusual about humans because we're all anti-creationists," said Lovejoy. "The minute you accept something special about humans, you fall into the grips of creationists. So you say, 'The same principles that created the chimpanzee created the human.' Well, chimpanzees don't drive cars."

IN THE ETHIOPIAN NATIONAL MUSEUM, BERHANE ASFAW WALKED THE HALLS one afternoon with his blazer thrown over his shoulder. He stepped to the window and pointed to the gate where he pleaded with the rebel soldiers who occupied his museum during the civil war a quarter century earlier. That old lab was gone now, replaced by a mammoth complex where he now stood. As a young scientist, he returned home with dreams of building an international center of human origins in his home country. And here it was—a massive museum filled with the history of life. Of course, it wasn't a perfect dream—there were still scarce resources, bureaucratic catfights, administrators who shunned his advice, Internet connections that quit unpredictably, and a casting lab that still wasn't ready to distribute casts into the world—but it was enough to serve the science and nearly thirty international teams working throughout the country. He had no role in government anymore, yet his fingerprints were everywhere. "Without Berhane, it's hard to understand this organization," observed Alemu Ademassu, the longtime museum casting technician. "Many people don't recognize this."

Berhane strode through the museum complex and young workers unlocked doors for him. His passion no longer was a lonely pursuit among Ethiopians. "Now there are so many young guys who are as aggressive as I am," he said. In the basement, a room the size of a basketball court held fossil fauna on shelves provided by the Middle Awash team. Upstairs, their team had helped outfit the fossil preparation rooms and casting lab and White had ensured the conference room table was built by Ethiopian carpenters from indigenous woods. Berhane stepped into a room lined with dozens of heavy safes, gifts from Japan arranged by Gen Suwa. They held many relics of early humanity—Ardi, Lucy, Selam, *kadabba, deyiremeda,* and more yet to be revealed. "In those days, we thought thirty safes was enough," said Berhane. "Now we are almost out of space for hominids."

Soon they would get a bit more crowded. In the morning, he would climb into his Land Cruiser and drive back to the Afar Depression to hunt for more *Ardipithecus,* the creature the world wasn't quite ready to embrace.

Epilogue
SUNSET

For twenty years, something bothered the Ardi team. They suspected the excavation site might contain yet another skeleton—one even better than Ardi herself.

They called it Bigfoot. The first clue had appeared two decades earlier, with a screwup. During the Aramis bone rush of the 1990s, White reviewed one day's collections and discovered an *Ardipithecus* fossil that somebody had failed to recognize and had dropped into the wrong container. He found the junior scientist who he had assigned to supervise the crawlers and delivered one of his dreaded rebukes that his workers called *taking a bullet*. Luckily, he had engineered a system with failsafes against human error and could trace the fossil to a specific spot, where they found more fossils. Among the new pieces was an unusually large metatarsal bone of the foot—hence the nickname. The fossils spilled from a nearby ledge of hard carbonate chock-full of fossils, including an exquisitely preserved monkey skull. Might another Ardi be hidden within, too? A series of excavations turned up more intriguing clues but still no skeleton. In 2016, they decided to return for one more Bigfoot hunt, perhaps the last of their careers.

Meanwhile, the smoldering volcano of Ethiopian politics erupted in the worst political crisis in the twenty-five years since the civil war. The ruling party, the Ethiopian People's Revolutionary

Democratic Front (EPRDF), had followed a Chinese model of an authoritarian state pursuing aggressive economic growth. After the EPRDF proclaimed total victory in the last elections, protests broke out across the country, and more than fifty people died in clashes with police at an Oromo festival in Bishoftu. Activists called for five "days of rage," and mobs attacked foreign-owned farms, factories, and tourist lodges. Protesters stoned a vehicle and killed an American researcher from the University of California at Davis. At that point, White cancelled his plans to bring Berkeley students into the field that year. It was too dangerous.

Ethiopia declared a state of emergency in October 2016, imposed an information blackout, turned off Internet service to mobile phones, blocked social media, and branded opposition groups as terrorists. A government decree prohibited foreign diplomats from traveling outside the capital. More than six hundred and fifty people died in demonstrations, and twenty-five thousand people were thrown in jail.

Despite the crisis, Berhane and White led the Middle Awash team back to the field, just as they always did during the dry season. The caravan drove on a new Chinese-built expressway as fine as any American interstate. Alongside the highway rose kilometer after kilometer of new developments, framed by the distinctive Ethiopian construction scaffolding of eucalyptus trunks. High-rise buildings overlooked traditional villages of *tukel* huts. Pastoralists drove herds of goats along the medians. A newly constructed rail line alongside the highway soon would transport shipping containers between the capital and Djibouti.

Much was changing in the Middle Awash, too. In the old days, the only routes through the wilderness were camel tracks. Nowadays, roads first blazed by the fossil team were traveled by cars of state officials, aid organizations, developers, and even Afars on motorcycles. In the middle of the project territory, the government had constructed a bridge over the Awash River, and journeys that formerly took days were reduced to hours. An oil exploration camp with hundreds of Chinese and Russian contractors threatened to run bulldozers and trucks over deposits at Herto containing early *Homo sapiens* until Elema intervened to detour the machines. In the old days, the scientists worried about security, but increasingly they worried about development. Oryx, cheetahs, wild asses, and ostriches hadn't been seen in years. Furtive African wildcats still occasionally showed themselves.

In Afar villages, new concrete buildings stood beside branch-and-grass huts. For the moment, old tribal enmities seemed to have waned, and Issa sometimes strolled casually through Afar towns. Once again, a detachment of paramilitary police with camouflage uniforms and AK-47s kept guard over the fossil crew. At night, soldiers patrolled the camp perimeter and demanded passwords from anybody approaching in the dark. Afar workers no longer showed up with guns, and the long, curved daggers on their belts had been replaced by cell phone cases.

AT ARAMIS LOCALITY SIX, THE ARDI SKELETON SITE HAD BECOME A PILE OF boulders overgrown with tufts of brown grass. The datum stake—pounded straight into the ground twenty-two years earlier to provide a fixed point of measurement for the excavation—decayed aslant in the eroding hillside.

Another quarrel had arisen over the skeleton. Back in academia, more and more scholars were accepting Ardi as member of the human family, but a counterargument had emerged—that Ardi and humans descended from a chimp-like ancestor. The most detailed summation of this view came from Harvard anthropologists David Pilbeam and Daniel Lieberman. In an exhaustive 120-page analysis, they proposed an "alternative reconstruction" of Ardi from head to toe. The Harvard anthropologists agreed Ardi was a member of the human family, a biped, and not a knuckle walker. Nonetheless, they insisted Ardi could have evolved from a chimp-like, knuckle-walking ancestor who had lived millions of years prior, and believing otherwise required "improbable" amounts of parallel evolution in our ape cousins. Two irreconcilable paradigms lay claim to the oldest family skeleton. Once again, the science was at an impasse and only more fossils would resolve it.

On a slope about two hundred meters away lay the new target—Bigfoot.

A quarter century earlier, White predicted that the work on *Ardipithecus* would never be finished, and it still wasn't. He scribed lines in the dust with his ice axe and sent the crew over the gully. They found more bones—a chunk of hominid skull, a trapezium of the wrist, and a medial cuneiform of the foot. Most Bigfoot fossils were dark from a high concentration of manganese, but somebody found a smaller, lighter-colored fossil. White felt a surge of hope: *maybe more than one carcass.*

The team started another excavation. It would consume a precious field season when the fossil men didn't have many left, but

A crew excavates a hillside in search of the elusive Bigfoot—a skeleton they hoped might be even better preserved than Ardi.

another skeleton jackpot merited the gambling. New bones might reveal what Ardi did not—perhaps a hip, knee or ankle joints, or a heel bone. Judging by the prior discoveries, Bigfoot seemed to be a large male, which would complement the female Ardi.

Once again, Aramis locality six hosted a bustling paleo assembly line. Men dug into the hillside with trowels. Afar men carried buckets of dirt and sieved in clouds of dust. Crews picked through chunks of carbonate and hammered them apart looking for fossils. "It's serious detective work," Berhane Asfaw said as he stood among the excavators. "Most anthropologists don't understand. The moment you find something, they want it."

One morning, a viper slithered out of the rock pile. Afars called the snake *abeesa,* and knew it was deadly (one had killed Elema's brother). The men grabbed a crowbar and goaded the serpent to open ground. Again and again, the viper struck uselessly at the iron bar and exhausted itself under the scorching midday sun. Its slithering became more desultory, its strikes slower, its venom depleted. Then they cut off its head.

Two weeks passed and no skeleton appeared. "Wagano's trowel point could hit a femur at any minute," said White. But that minute passed, then days. Still no Bigfoot.

One day, the excavators heard a loud engine and looked up to see a huge bulldozer rumbling over the badlands. The driver was heading to Aramis to clear a swath of river forest for another agribusiness plantation. Riding on the machine was their old Afar friend Adeni, with an antique wooden rifle, a sack of grain, and bad legs. Decades ago, he had blocked the expedition caravan at gunpoint in a labor negotiation. Now his son worked among the excavators. On this day, the old man guided the bulldozer to his village. Modern Ethiopia had arrived to cultivate the land of the once-wild Aliseras.

When the shadows got long—what the Afars called *ayro korte,* or "short sun"—the team returned to camp, and the cars trailed clouds of dust as they sped across a floodplain toward a blazing sunset. Gazelles sprang across the plains and camels loped out of their path. Afar children—some naked, some in traditional clothing—streamed out of villages and waved.

The twilight of long careers approached, and every day spent chasing Bigfoot was a day not spent doing something else. They had a full docket of unfinished projects, and human scientists, unlike fossils, don't last indefinitely. "We don't do adventurous things anymore," confessed Giday WoldeGabriel. "Certainly we are all much older. Physically, we are not as we were thirty years ago. There are a lot of constraints." For White, one new constraint was family. To the surprise of his peers, the warrior monk of fossils had married, became a father, and started knocking off work early to spend time with his little daughter. He went through the metamorphosis of parenthood at an age when contemporaries like Berhane were having grandchildren. His old Afar wildcat lived long enough to witness the transition and didn't bite when the toddler yanked his tail. But White always spent the holidays apart from his family because the field season remained nonnegotiable. "Maddy is five now," he said of his daughter back home as he stood by the excavation. "This is the last year she's going to think about Santa Claus."

On the day before Christmas, the excavators finally found a fragment of an ancient human—a molar tooth. The first bit of the skeleton? Problem was, the tooth seemed *small*. This didn't look like any Bigfoot. Then came hundreds of fragments, the typical leftovers of bone-crushing hyenas.

Excavations die slowly, without climax, and by the New Year hope had vanished. The explorers came hoping to find the remains of a human ancestor, and they left disappointed that carnivores had found it first.

Another field season was almost gone. They reassembled the carbonate slabs into a mosaic like a huge flagstone patio, but turned them upside down so rains would slowly disintegrate the rock and reveal whatever might be hidden within—another long-term

Photograph by and © Tim D. White 2016

The Ethiopian face of the science: the Middle Awash Field Team, 2016.

strategy for maximizing the harvest of bone. At the bottom of the slope, they erected a stone wall to catch any specimens that tumbled down. Every scrap mattered.

Back at the museum, troves of new discoveries awaited publication. They included a haul of 2.5-million-year-old fossils near the dawn of *Homo*, heaps of 2-million-year-old stone tools and fossils, loads of material between 500,000 and 750,000 years old from the Middle Pleistocene epoch voluminous enough to rival the mission to Aramis, and a 100,000-year-old skeleton (the only one of its period from Africa). "They have got shit in the vault that a lot of people would just dream about—and don't even know about," said Scott Simpson. "It's an embarrassment of riches. The plate is very full." Bernard Wood visited and gawked at the collections (he and White actually enjoyed each other's company when they weren't furious at each other). "There is probably an order of magnitude more evidence that could inform these discussions that has not been published—drawers full of stuff," observed Wood. "They've got crania coming out of the wazoo. The problem is, if you just keep picking up fossils and don't publish them, those fossils really don't exist."

The total: more than thirty-two thousand vertebrate fossils collected over four decades in the Middle Awash plus many more bulk samples, micromammals, birds, fish, and so on. Of those, only 483 belonged to the human family (some single pieces and others skeletons like Ardi) representing at least seven different species. The searchers also recovered truckloads of their handiwork: more than twelve thousand stone tools.

In Ethiopia, the "publish or perish" mandate did not apply. The antiquities administration remained quite willing to allow field teams to publish at their own pace and resist outside pressure to open the safes. "Not everybody wants to do the housework," said Yonas Desta, director general of the ARCCH. "There are 'smart' scientists who just want to make a visit and use technology." Sitting in his office above the National Museum, the director waved his hand dismissively and mimicked the nasal voice of the stereotypical foreign lab scientist making demands for fossils: "It's mostly Tim's project. Who the hell does he think he is? Laboratory access—I want to have that upon my arrival. I have to have vacation in three weeks. I want to visit where I want in Ethiopia. Does your lab have good air conditioning? How's the Internet?" The director reverted to his normal voice. "Those are the kinds of e-mails I receive," he continued. "We have to protect people like Tim who have the *guts*"—he flashed his teeth—"to survive in the game for thirty years, in forty-two degrees Celsius."

Survival also depended on friends in the field. As the crew excavated for Bigfoot, the local Afar administration threw a Christmas party for the scientific team and two hundred people showed up, including sheiks, government dignitaries, dancers, and a TV crew. Over the last quarter century, the scientific team had provided the local people with jobs, medical care, and even saved the life of a boy who had been attacked by a crocodile. Such favors were not forgotten. The desert that foreign explorers once called the "hell-hole of creation" had been rebranded as a world-known laboratory of evolution. The Afar delegation draped the expedition leaders in traditional robes, thanked them for contributing to local development, and presented gifts of Afar knives—ceremonial tokens of a vanishing past.

Despite the disappointment of Bigfoot, the daily surveys at nearby outcrops had turned up about twenty new *Ardipithecus* specimens—maybe enough for another paper, if they took the time to write it up. To the north at Gona, their colleague Sileshi Semaw had the finest *Ardipithecus* palate ever seen. At camp, White rhapsodized over its beauty. "Some fossils have a special character that goes beyond just anatomy and science," he gushed over the breakfast table. "It's the weight, the beauty of the teeth and white bone. They're like jewels, like giant diamonds. Oh, that one is so fucking *perfect*! Every tooth is *perfect*!"

At a conference a few years before, one prominent critic of *Ardipithecus* was offered a preview of this specimen and turned away. Around the camp table, heads shook. How could anybody fail to be awestruck by such a treasure? "It's the anger and the jealousy," said Giday. "He blocks his mind."

"I think he is not a morphologist," added Berhane. "If you are a morphologist, you are curious about anything you see."

The fossil men remained curious. Sooner or later their friend would get around to publishing his palate, and the Middle Awash team would drop new bombshells on their unsuspecting peers. When? Nobody could say exactly—just whenever everything had been nailed down to their satisfaction. It might take years.

Acknowledgments

Nobody in their right mind takes on a project like this. My only excuse for starting is that I was naive. My only explanation for finishing is that I had a lot of help.

My long distraction with this subject can be blamed on Owen Lovejoy. It was Owen who first introduced me to Ardi back when I *thought* I was working on another topic. From our first conversation, he proved a delightful explicator and storyteller. His lifelong involvement with many landmark discoveries in paleoanthropology suggested the threads of a scientific saga.

That led to the field team—the people who gather the primary data for this peculiar science. Tim White proved endlessly generous with time, access, and wisdom. Despite grumbling that he would be "happier with no book at all," he in fact made this one possible. He gave me access to his archive of videos and photos documenting years of fieldwork, peer reviews, correspondence, and much more. There is no more-dedicated worker in the field nor one more demanding. I doubt there is any more hilarious. His patience over the years proved monumental, sometimes exhaustible in the moment but never in the long run. He once told me that his crew went to Aramis expecting to stay for five days and wound up staying five years. Believe me, I get it now.

Giday WoldeGabriel always was a gentle and kindly guide during our many hours together in the car or walking the outcrops. He enabled a journalist to see the Afar Rift in terms of moving parts. He also helped me understand the trials of his generation and grasp the importance of this scientific endeavor for Ethiopians. Berhane Asfaw tolerated my repeated questions even as I spent years assuring him that I was almost done. Other members of the Middle Awash team also deserve special thanks: Scott Simpson, Bruce Latimer, Gen Suwa, Yonas Beyene, Henry Gilbert, Yohannes Haile-Selassie, Leslea Hlusko, Josh Carlson, and Doug Pennington. David Brill enlivened the text with his beautiful photos, the fruit of many years of repeated visits to Ethiopia, and generously shared his memories and even his house when it came time to sort through his voluminous archives.

I made two visits to the Middle Awash camp and I am grateful for the hospitality of the field team. Special thanks to Elema, Alemayehu, Dawit, Dereje, Moussa, Alemu, Kampiro, and Wagano. Abush, the always-reliable mechanic and driver, deserves particular mention for getting me in or out of the field on three separate occasions. Adeni generously invited me to Aramis village and welcomed me with a warmth that rendered the three-way translation unnecessary. In the capital, Yonas Desta, director of the ARCCH, expedited paperwork when other officials seemed determined to slow it down.

The kindness of strangers saved me from disaster more than once. Two secretaries at the Foreign Ministry kindly located my permit application that had been misplaced by another department and went out of their way to make sure it got to the right place. In the Afar Depression, a pair of helpful soldiers stopped all highway traffic until they found a bus willing to transport me to the capital. Later, a policeman did the same in Gewane. Strangers treated a *farenji* with kindness, guided me through the crowded streets of Addis, helped me avoid taxi drivers bent on extortion, and one kind soul even called my hotel to make sure I arrived safely.

Outside the team, I also wish to express gratitude to Don Johanson, Bill Kimbel, Tom Gray, Jon Kalb, Bill Jungers, Bernard Wood, and Richard and Meave Leakey. In this contentious field, I discovered that people who could not tolerate one another often were delightful company by themselves. Many more deserve a few lines of personal praise, but in the interest of brevity (and with some guilt), I must thank them in bulk. Thus we arrive at the point where the author dumps the contents of his address book onto the page.

I am grateful to Alemu Ademassu, Gabe Adugna, Leslie Aiello, Sergio Almecija, Stan Ambrose, Pamela Anderson, Peter Andrews, Alemayehu Asfaw, Zelalem Assefa, Firew Ayele, Bob Beebe, Kaleyesus Bekele, Lee Berger, Raymonde Bonnefille, Steve Brandt, Thure Cerling, Martin Clayton, Marty Cohn, David DeGusta, Alicia DeRosalia, Jeremy DeSilva, Evan Eichler, Dean Falk, Sarah Feakins, Mulugeta Feseha, Alan Fix, John Fleagle, Morris Goodman, Justin Guz, Aliklu Habte, Brooks Hanson, John Harbeson, Bill Hart, King Heiple, Libet Henze, Ralph Holloway, Lyman Jellema, Peter Jones, Kevin Kern, Ethan Key, Kris Krishtalka, Naomi Levin, Dan Lieberman, Reiner Luyken, Tom McAnear, Melanie McCollum, Mary McDonald, Kieran McNulty, Priya Moorjani, Virginia Morell, Jim O'Connell, Marina Ottaway, David Pilbeam, Todd, Preuss, Wes Reeder, David Reich, Paul Renne, Phil Reno, Paul Salopek, Bob Sanders, Vince Sarich, Alwyn Scally, Kathy Schick, Jeffrey Schwartz, David Shinn, William Simmons, Pamela J. Smith, Fred Spoor, Linda Spurlock, Jack Stern, Chris Stringer, Maurice Taieb, Ian Tattersall, Alan Templeton, Nick Toth, Ted Vestal, Emily Vincent, Steve Ward, Mark Weiss, Jacob Wiebel, Martin Williams, Milford Wolpoff, Bernard Wood, Frehiwot Worku, John Yellen, Joan Gosnell, Yuan Tian, and Ajit Varki.

Some spoke to me so long ago they might not even remember. My apologies to anybody I have forgotten. Unfortunately, some passed away before this book was completed and seeing their names on the page stirs a solemn gratitude. That list of names covers only those who interacted with me. Many more educated me with their written work.

I owe profound gratitude to Stephen D. Nash for sharing his beautiful illustrations of primate species, which elevated dry taxonomy to the realm of art. His namesake collection, Nash Collection of Primates in Art and Illustration at the University of Wisconsin, proved to be a gold mine of historic art and illustrations. Similarly, the Wellcome Collection in London also deserves praise for preserving a rich repository of images and scholarship.

My agent, Dan Conaway of Writers House, saw the potential of this book in the early stages and worked to make sure it wound up in the hands of people who shared his excitement. Thanks to the editing team of Henry Ferris, Geoff Shandler, Nick Amphlett, and Molly Gendell. A salute to the rest of the staff at William Morrow, including Dale Rohrbaugh, Mark Steven Long, Liate Stehlik, Ben Steinberg, Kaitlin Harri, Christina Joell, Elina Cohen, Owen Corrigan, and Kyran Cassidy.

A number of friends offered invaluable feedback. Hasok Chang read the draft manuscript and offered keen insight—despite the fact that he was on sabbatical and really should have been laboring to finish two books of his own. Kenny Beckman, James Bandler, and Tim Smith generously offered valuable feedback. Evan Eichler, Sarah Feakins, Tim White, and Paul Renne kindly proofread sections of manuscript. Eamon Dolan provided helpful suggestions and his encouragement lit a beacon during the dark night of middle passage.

More than anything, I would like to thank my family. They supported me throughout an ordeal that lasted far longer than anyone predicted. Thanks to my children, Eli, Alistair, and Siri, the joys of my life who somehow grew up while this book was underway. I owe debts to our extended family: Mom (who always believed I would do this) and Dad (who shared the joy of this book finding a publisher but did not live to see its completion), Marilyn and Tim (who helped in more ways than can be counted here), and Bobbo and Linda. All kept believing and were kind enough not to ask too often why the book wasn't done yet.

This book is dedicated to my wife, Maja. She provided both unconditional support and unsparing critiques. She shouldered the burden of sole parenting when I was consumed by work. She selflessly took it upon herself to master the *Chicago Manual of Style* and the tedium of bibliography when I was swamped on deadline. She summoned me back when I strayed too far into scientific jargon and always advocated for the reader. She also proved to be a keen editor—probably the best of all. For all that, she deserves the last word of thanks.

Notes

ABBREVIATIONS USED IN NOTES

ADST—The Association for Diplomatic Studies and Training

BHL—Bentley Historical Library

MARP—Middle Awash Research Project

NSF—U.S. National Science Foundation

EPIGRAPH

vii ***Ethiopian proverb:*** Cerulli, "Folk Literature," 197.

INTRODUCTION: T. REX

2 ***"so rife with anatomical surprises":*** Lovejoy et al., "Great Divides," 73.

CHAPTER 1: THE ROOTS OF HUMANITY

11 ***the skinny fossil hunter labored:*** Scene details from Powers, "Digging for Old Stones and Bones"; MARP video 1981; Tim White photograph archives.

12 ***not just a chip:*** Milford Wolpoff quoted in Morell, *Ancestral Passions*, 454.

12 ***"collegiality in the department":*** Nelson H. H. Graburn, Department of Anthropology memo, July 27, 1981, White papers.

12 ***"the best in the business today":*** C. Loring Brace, letter to Nelson H. H. Graburn, September 27, 1981. C. Loring Brace papers, BHL, box 8, folder: White, Tim.

13 ***"far beyond the nth degree":*** Steve Ward, interview with author.

13 ***"the beginnings of humankind":*** Johanson and Edey, *Lucy: The Beginnings of Humankind.*

14 ***undercover CIA employees:*** ADST, "Ambassador Owen W. Roberts," 74.

14 ***In the driveway:*** Details from scene drawn from photograph collection of Tim White and MARP video 1981.

14 ***"you have a creature":*** Quoted in Powers, "Digging for Old Stones and Bones."

16 ***"The question is":*** Quoted in Powers, "Digging for Old Stones and Bones."

16 ***"airy fairy model building":*** J. Desmond Clark, letter to Fred Wendorf, November 5, 1979. Wendorf papers, box 18, folder 28.

17 ***"biological and cultural advance":*** Clark, "Africa in Prehistory: Peripheral or Paramount?" 175.

17 ***Without Africa:*** Clark, "Africa in Prehistory: Peripheral or Paramount?" 175.

18 ***"If we find hominids":*** Quoted in Powers, "Digging for Old Stones and Bones."

18 ***a scientific conference:*** Williams, *Nile Waters, Sahara Sands,* 59; Martin Williams, interview with author.

18 ***"quite a lot of it":*** Clark, "An Archaeologist at Work," 217.

18 ***a medieval king:*** Clark, "An Archaeologist at Work," 217–19.

19 ***"bazookas, machine guns, and rifles":*** ADST, "Ambassador Arthur T. Tienken," 31.

20 ***"We'd hear gunshots all night":*** Steve Brandt, interview with author.

20 ***among 360 killed that week:*** Ottaway, "Fighting Up Sharply." (The geologist was Bill Morton, who was killed before his departure to join Clark.)

20 ***reported the box was empty:*** Edward Ullendorff, quoted in Hiltzik, "Does Trail to Ark of Covenant End Behind Aksum Curtain?"

21 ***human behavior had biological origins:*** Sigmon, "Physical Anthropology in Socialist Europe," 130–39.

21 ***his old friend Berhanu Abebe:*** Clark, letter to John Yellen, August 31, 1978, Kalb papers.

21 ***"Ethiopia is intensely nationalistic":*** Yellen, "Report on the Paris and Addis Ababa Conferences," 4.

22 ***"some eastern bloc scientists":*** Clark, letter to John Yellen, August 31, 1978, Kalb papers.

22 ***money for a new museum building:*** Aklilu Habte, letter to J. Desmond Clark, 1977. Wendorf papers, box 18, folder 29. The French faced the same demands and constructed a separate building.

22 ***"considered heretical":*** Quoted in Colburn, "The Tragedy of Ethiopia's Intellectuals," 136.

22 ***a campaign of Red Terror:*** Wiebel, "Revolutionary Terror Campaigns in Addis Ababa, 1976–1978."

23 ***tracked down and liquidated:*** Katz, "Ethiopia After the Revolution."

23 ***opposition group:*** Biographical details from author interviews with Berhane Asfaw and Frehiwot Worku.

24 ***"Jolly good!":*** Berhane Asfaw, "Tributes to J. Desmond Clark."

25 ***"such things in Ethiopia":*** Berhane Asfaw, letter to Higher Education Commissioner, October 11, 1982, Kalb papers.

25 ***"every day is a gift":*** Berhane Asfaw, interview with author.

26 ***"the Ethiopian authorities":*** Clark, letter to John Yellen, December 29, 1981.

26 ***They erected a camp of green tents:*** Camp details from Krishtalka, "Bones from Afar."

26 ***"When friends wrote letters":*** Krishtalka, "Bones from Afar," 18.

26 ***desert floor was covered with bones:*** Krishtalka, "Bones from Afar"; NSF award 8210897 case file.

27 ***"stepping on fossils":*** Quoted in Wilford, "Ethiopia Bones Called Oldest Ancestor of Man."

27 ***"the deepest roots of the Hominidae":*** NSF award 8210897 case file, Summary of Research and Conclusions, 33.

27 ***"the most successful field season":*** Clark, letter to John Yellen, December 29, 1981.

27 ***"study area in the world":*** NSF award 8210897 case file, Summary of Research and Conclusions, 32.

28 ***a big-fanged chimpanzee skull:*** Institute of Human Origins, newsletter winter 1982/83, 1.

29 ***"Fauna here is scrappy":*** White, field notes, November 20, 1981.

CHAPTER 2: BANNED

30 ***a problem to solve:*** A tip of the hat to Erich Fromm, who wrote "Man is the only animal for whom his own existence is a problem which he has to solve," Fromm, *Man for Himself*, 40.

30 ***"most likely to afford remains":*** Darwin, *Descent of Man*, 430.

30 ***explorer Henry Morton Stanley:*** Stanley, *Through the Dark Continent*.

33 ***Y-shaped diagram:*** Johanson and White, "Systematic Assessment."

33 ***"to exist in northern Ethiopia":*** Mayr, "Reflections on Human Paleontology," 233.

33 ***nearly half a million dollars:*** NSF award 8210897 case file.

34 ***"tower of strength":*** J. Desmond Clark, letter to Fred Wendorf, January 1, 1982. Wendorf papers, box 18, folder 23.

34 ***"a superb field and lab man":*** J. Desmond Clark, letter to Fred Wendorf, January 1, 1982. Wendorf papers, box 18, folder 23.

34 ***to the Ministry of Culture:*** Clark and White, letter to Dr. Steven Brush, November 7, 1982. NSF award 8210897 case file.

34 ***"The principal problem":*** Korn, "Archeological Expedition Stalled," Kalb papers.

34 ***from the Ministry of Culture:*** J. Desmond Clark, letter to Dr. Stephen Brush, December 10, 1982, NSF award 8210897 case file.

34 ***a larger role in the research:*** J. Desmond Clark and Tim White, letter to Dr. Stephen Brush, November 7, 1982, NSF award 8210897 case file.

34 ***science and technology:*** Berhane Asfaw, letter to Higher Education Commissioner October 11, 1982, Kalb papers.

34 ***"falsely accused me":*** Berhane Asfaw, letter to Higher Education Commissioner, October 11, 1982, Kalb papers.

35 ***"perceived as 'ripping off'":*** Clark and White, letter to Dr. Steven Brush, November 7, 1982, NSF award 8210897 case file.

35 ***"telephones in every room were bugged":*** Williams, *Nile Waters, Sahara Sands*, 123.

CHAPTER 3: ORIGINS

36 ***"The Ethiopians slept":*** Gibbons, *History of the Decline and Fall of the Roman Empire,* vol. 3, 281.

37 ***sought scientific endorsement:*** Zena, "Archaeology, Politics and Nationalism," 408.

37 ***asked Louis Leakey:*** Morell, *Ancestral Passions,* 275–76. French scholars had made an earlier expedition to Omo Valley in the 1930s and did pioneering archeological work. In 1952, Emperor Haile-Selassie created the Institute of Ethiopian Archaeology in collaboration with French academics. In 1963, a Dutch hydrologist Gerard Dekker found a large collection of stone age tools in Melka Kunture which was subsequently excavated by a French mission under Jean Chavaillon.

37 ***surveying the Afar Depression:*** Details of exploration from Taieb, *Sur la Terre des Premiers Hommes,* and Maurice Taieb, interview with author.

37 ***"an enemy worth killing":*** Munzinger, "Narrative of a Journey," 225.

38 ***"Men must think of blood":*** Nesbitt, *Desert and Forest,* 135.

38 ***"impetuous, excitable":*** Kalb, *Adventures in the Bone Trade,* 21.

39 ***"I met this effervescent guy":*** Don Johanson, interview with author.

39 ***couldn't identify fossils:*** Johanson and Edey, *Lucy,* 124–25.

40 ***"Being strong willed":*** Don Johanson, interview with author.

40 ***he mistook as a hippo rib:*** Account from Johanson and Edey, *Lucy,* 155–59.

40 ***a nearby burial mound:*** Described in Johanson and Edey, *Lucy,* 158–59.

40 ***"They won't believe you":*** Quoted by Johanson and Edey, *Lucy,* 163.

41 ***"popping out of the ground":*** Tom Gray, interview with author.

41 ***"everything back in the box":*** Taieb, *Sur la Terre des Premiers Hommes,* 119.

41 ***the skeleton a foreign name:*** Taieb, *Sur la Terre des Premiers Hommes,* 144.

41 ***"You could not predict":*** Aklilu Habte, interview with author.

42 ***"an unknown anthropology graduate":*** Johanson and Edey, *Lucy,* 185.

43 ***their next of kin:*** U.S. Embassy Ethiopia, "International Afar Expedition," WikiLeaks Cable, October 14, 1976; U.S. Embassy Ethiopia, "International Afar Expedition," WikiLeaks Cable, October 21, 1976.

43 ***he was assassinated:*** Johanson and Edey, *Lucy*, 233.

43 ***"a dimension of competition":*** Johanson, "Anthropologists: The Leakey Family."

44 ***"One species?":*** Conversation quoted in Johanson and Shreeve, *Lucy's Child*, 105.

45 ***"could not proceed without him":*** Johanson and Edey, *Lucy*, 252.

46 ***"somebody else pick on us":*** Johanson and Edey, *Lucy*, 292.

46 ***"Tim kept me in line":*** Johanson and Edey, *Lucy*, 293.

47 ***the new*** afarensis ***species:*** Hinrichsen, "How Old Are Our Ancestors."

47 ***infuriated Mary Leakey:*** Morell, *Ancestral Passions*, 480–81.

49 ***took notes of the discussion:*** Quoted in Reader, *Missing Links*, 384.

50 ***on July 4, 1978:*** Lewin, *Bones of Contention*, 290.

50 ***White killed the snakes:*** Science Friday, "Desktop Diaries: Tim White."

50 ***to catch guinea fowl:*** Peter Jones, interview with author.

51 ***"I never could understand":*** Richard Leakey, interview with author.

51 ***"clear-thinking young man":*** Quoted in Morell, *Ancestral Passions*, 476–77.

51 ***The quest had begun in 1975:*** Details of the footprint discovery from Morell, *Ancestral Passions*, 473–490, and Reader, *Missing Links*, 408–14.

52 ***"At this stage":*** Mary Leakey, *Disclosing the Past*, 177.

52 ***"Mary getting impatient":*** Peter Jones, interview with author.

52 ***"literally run a kilometer":*** Tim White, interview with author.

53 ***"my own camp":*** Mary Leakey, *Disclosing the Past*, 182.

54 ***PLEASE OMIT MY NAME:*** Quoted in Morell, *Ancestral Passions*, 489; Lewin, *Bones of Contention*, 285.

54 ***"clashes in the field":*** Mary Leakey, letter to Tim White, April 26, 1979, quoted in Morell, *Ancestral Passions*, 500.

54 ***a wild goose chase:*** Morell, *Ancestral Passions*, 500.

54 ***published a paper in* Science:** Johanson and White, "Systematic Assessment of Early African Hominids."

54 ***A front page headline:*** Rensberger, "New-Found Species."

55 ***his new species name:*** Johanson and Edey, *Lucy*, 290.

55 ***"Your description of Australopithecus afarensis":*** Ernst Mayr to Dr. D. C. Johanson, letter June 7, 1979, Mayr papers.

55 ***a subspecies of* africanus:** Lewin, *Bones of Contention*, 269, 290.

55 ***"right the first time":*** Rensberger, "Rival Anthropologists Divide."

56 ***"king of anthropology":*** Lewin, *Bones of Contention*, 171.

56 ***"I consider you a scoundrel":*** Richard Leakey, letter to Donald Johanson, May 18, 1983, quoted in Morell, *Ancestral Passions*, 529.

56 ***"The Leakeys took thirty years":*** Quoted in Johanson and Shreeve, *Lucy's Child*, 181.

57 ***"What we find in them":*** Johanson and Edey, *Lucy*, 375.

CHAPTER 4: THE FALSIFIER

58 ***outgrew their childlike curiosity:*** Science Friday, "Desktop Diaries: Tim White"; White, "At Large in the Mountains"; interviews with author.

58 ***"As Southern California filled":*** White, "At Large in the Mountains," 203.

59 ***"He would sit":*** Wes Reeder, interview with author.

60 ***"a real disciplinarian":*** Tim White, interview with author.

60 ***"Snake recognition":*** White, "A View on the Science," 287.

60 ***"pushing the envelope":*** Jim O'Connell, interview with author.

61 ***"In southern Ethiopia":*** Noxon, *The Man Hunters*.

61 ***"God, it's like the textbook!":*** Alan Fix, interview with author.

61 ***"He was supremely confident":*** Alan Fix, interview with author.

61 ***"He was dumbfounded":*** Alan Fix, interview with author.

61 ***"Tim was extremely smart":*** Milford Wolpoff, interview with author.

61 ***"He lived physical anthropology":*** Bill Jungers, interview with author.

62 ***"his work is almost monastic":*** C. Loring Brace, letter to Dr. S. L. Washburn, February 18, 1977, C. Loring Brace papers, BHL, box 8, folder: White, Tim.

62 ***"He is a cautious scholar":*** C. Loring Brace, letter to Dr. S. L. Washburn, February 18, 1977, C. Loring Brace papers, BHL, box 8, folder: White, Tim.

62 ***deprived himself of food:*** Tim White, interview with author.

62 ***"If he has any weakness":*** Recommendation for Tim White, C. Loring Brace, February 13, 1976, C. Loring Brace papers, BHL, box 8, folder: White, Tim.

63 ***"worthy of Machiavelli":*** Morell, *Ancestral Passions*, 298.

63 ***"I was constantly being reminded":*** Morell, *Ancestral Passions*, 288.

63 ***issued standing instructions:*** Leakey, *One Life*, 172.

63 ***"Once the plane touches down":*** Rensberger, "The Face of Evolution," 54.

64 ***White damaged a fossil:*** Morell, *Ancestral Passions*, 477.

64 ***"spoiling a photo-op":*** Tim White, interview with author, and "Ladders, Bushes, Punctuations, and Clades," 136.

64 ***sites must be managed carefully:*** White, "A View on the Science," 290.

65 ***"Tim got very angry":*** Richard Leakey, interview with author.

65 ***"Tim absolutely refused":*** Richard Leakey, interview with author.

65 ***"I lost my temper":*** Tim White, journal, September 8, 1975.

65 ***denied slamming any doors:*** Tim White, e-mail to author, October 31, 2019.

65 ***to stifle dissent:*** Tim White, letter to Roger Lewin, June 6, 1986.

65 ***"the P.R. guy that Don is":*** Tom Gray, interview with author.

66 ***"irreconcilable differences in opinion":*** Phyllis Dolhinow, letter to anthropology department executive committee, February 13, 1981.

66 ***"some 800 students":*** Anton, "Of Burnt Coffee and Pecan Pie," 36.

67 ***Washburn had written a manifesto:*** Washburn, "The New Physical Anthropology."

67 ***"the most fundamental mistake":*** Washburn, "Fifty Years of Studies on Human Evolution," 25.

67 ***"primitive, nineteenth-century science":*** Washburn, "Fifty Years of Studies on Human Evolution," 30.

67 ***"He always felt persecuted":*** William Simmons, interview with author.

68 ***"graduate student instructor revolt":*** Tim White, interview with author.

68 ***one of White's friends:*** The topic of discussion was "The Origin of Man" by Owen Lovejoy.

68 ***"Tim White came to the seminar":*** S. L. Washburn, letter to William S. Simmons (executive committee), February 10, 1981.

68 ***"White's behavior":*** S. L. Washburn, letter to William S. Simmons (executive committee), February 10, 1981.

69 ***A department memo:*** Nelson H. H. Graburn, Berkeley Department of Anthropology memo, July 27, 1981, White papers.

69 ***"the personalities involved":*** Nelson H. H. Graburn, Berkeley Department of Anthropology memo, July 27, 1981, White papers.

70 ***"I wrote down what he told me":*** Tim White, e-mail to author, June 8, 2012.

70 ***asked what he hoped to find:*** White, discussion at 1983 IHO conference.

70 **nothing in particular:** White, discussion at 1983 IHO conference.

70 ***temper their conclusions with odds:*** Washburn, "The Evolution Game," 558.

70 ***"the human foible of false certainty:*** Washburn, "The Evolution Game," 558.

70 ***zero interest in playing games:*** Washburn and White, discussion at 1983 IHO conference.

70 ***good at math or theory:*** Johanson and Edey, *Lucy*, 292.

70 ***"reminiscent of a pygmy chimpanzee":*** Boaz and Cramer, "Fossils of the Libyan Sahara," 39.

71 ***a* National Geographic *cover:*** Tim White, presentation, 1983 IHO conference.

71 ***"enhance your science":*** White, "A View on the Science," 231.

71 ***before finding fossils:*** White and Suwa, "Hominid Footprints at Laetoli," 512.

71 ***"between science and science fiction":*** White, "Ladders, Bushes, Punctuations, and Clades," 141.

CHAPTER 5: THE FARTHEST OUTPOST OF HUMANKIND

72 ***the dictator Comrade Mengistu himself:*** J. Desmond Clark, letter to Dr. Steven Brush, December 10, 1982, NSF SBR award 8210897.

72 ***Ethiopian dissidents:*** Weiner, *Legacy of Ashes*, 456–58.

72 *CAUGHT RED-HANDED HERE: Ethiopian Herald*, February 4, 1984.

73 ***sat in disorganized heaps:*** J. Desmond Clark, letter to John Yellen, October 2, 1978, Wendorf papers, box 18, folder 29; Clark and Howell, "Construction of a Storage/Laboratory Unit in Addis Ababa."

73 ***rolling loose inside cardboard boxes:*** Tim White, interview with author.

73 ***an old paint container:*** Powers, "Digging for Old Stones and Bones"; Tim White, interview with author.

73 ***"the most horrendous condition":*** Tim White, interview with author.

73 ***the U.S. National Science Foundation:*** Clark and Howell, "Construction of a Storage/Laboratory Unit in Addis Ababa"; NSF award 7908342.

73 ***the U.S. State Department:*** Yellen, "John Kalb Reconsideration," Kalb lawsuit exhibit 6.

73 ***a "moral obligation":*** Clark and Howell, "Construction of a Storage/Laboratory Unit in Addis Ababa," 10.

74 ***to produce propaganda:*** Author interviews with Berhane Asfaw, Yonas Beyene, and Tim White.

74 ***Dinknesh joined an exhibit:*** *Ethiopian Herald*, "So that Open Eyes Might See."

74 ***a show to impress the world:*** Ottaway, "Addis Ababa Sees Red on Coup Anniversary."

75 ***The ancient scourge of famine:*** Giogis, "One Man's Love of Power."

75 ***returned home every year:*** Berhane Asfaw, interview with author.

76 ***"didn't want us to come back":*** Frehiwot Worku, interview with author.

76 ***"In his field of study":*** Frehiwot Worku, interview with author.

76 ***"He put in the first core":*** Yonas Beyene, interview with author.

76 ***an important fossil:*** Berhane Asfaw, "A.L. 444 Hominid from Hadar Used in a False Campaign," Walta Information Center, April 7, 2001.

76 ***survey for new antiquities sites:*** Details of inventory project from: Wood, "A Remote Sense for Fossils"; Giday WoldeGabriel et al., "Kesem-Kebena: A Newly Discovered Paleoanthropological Research Area in Ethiopia"; Suwa et al., "Konso-Gardula Research Project"; Berhane Asfaw et al., "Space-Based Imagery in Paleoanthropological Research."

77 ***U.S. National Science Foundation:*** NSF award 8819735, "Paleoanthropological Survey of Ethiopia's Rift System."

78 ***employed around the world:*** For example, see Anemone et al., "GIS and Paleoanthropology."

78 ***Giday WoldeGabriel:*** Biographical details from Giday WoldeGabriel, interviews with author.

81 ***127 days in the field:*** Suwa et al., "The Konso-Gardula Research Project," 3.

81 ***"There were in fact papers":*** From "An Archaeologist at Work," oral history, Tim White, J. Desmond Clark, and Timothy Troy, November 2, 2000, 359.

82 ***A headline in the journal*** Science***:*** Gibbons, "First Hominid Finds from Ethiopia in a Decade."

82 ***The Derg regime collapsed:*** Details from Cohen, "Ethiopia: Ending a Thirty-Year War."

82 ***the chief American negotiator:*** Details from Cohen, "Ethiopia: Ending a Thirty-Year War," 49.

82 ***locked down the museum:*** Berhane Asfaw, interview with author. Additional details from interviews with Alemu Ademassu, Yohannes Haile-Selassie, and Frehiwot Worku.

83 ***"Don't allow anybody":*** Berhane Asfaw, interview with author.

83 ***"How can you go?":*** Frehiwot Worku, interview with author.

84 ***White departed with a small crew:*** Details from interviews with White, Yohannes Haile-Selassie, Zelalem Assefa, and Alemu Ademassu; White, field log, December 15, 1991; White, photograph archives.

86 ***"Many women, children and men":*** White, field log, December 15, 1991.

86 ***"flying bullets":*** Yohannes Haile-Selassie, interview with author.

86 ***"some kind of an Afar army":*** Yohannes Haile-Selassie, interview with author.

86 ***"We were terrified":*** Zelalem Assefa, interview with author.

87 ***"just a one-man house":*** Mulugeta Feseha, interview with author.

87 ***"The problem is that Berhane":*** Close Observer, "What Went Wrong with Prehistory Research (Part V)," April 14, 2001.

87 ***they would be dismissed:*** Berhane Asfaw, interview with author and e-mail to author.

87 ***"We stayed outside":*** Mulugeta Feseha, interview with author.

88 ***"within five feet":*** Don Johanson, interview with author.

89 ***Once he delivered a lecture:*** Kalb, *Adventures in the Bone Trade*, 253.

89 ***a fossil baboon:*** Kalb, *Adventures in the Bone Trade*, 185.

89 ***"If the Government decides":*** Kalb, "Statement Regarding the Rift Valley Research Mission in Ethiopia," 30.

89 ***"being in the CIA":*** Jon Kalb, interview with author; Kalb, *Adventures in the Bone Trade*, 146.

90 ***"did not make distinctions":*** Aklilu Habte, interview with author.

90 ***"It is widely rumored":*** John Yellen, memo, December 27, 1977, "Jon Kalb Reconsideration," 5, Kalb lawsuit exhibit 6.

90 ***"What would he have reported?":*** Wendorf, "Impure Science," 402.

90 ***"damned good man":*** Fred Wendorf, letter to Jon Kalb, May 3, 1977, Kalb lawsuit exhibit 2.

90 ***"connected with Berkeley":*** Kalb, *Adventures in the Bone Trade*, 256.

90 ***"well worth our while":*** Yellen, "Jon Kalb Reconsideration," 1.

90 ***About ten days later:*** J. Desmond Clark, letter to John Yellen, August 31, 1978.

91 ***"formal academic credentials":*** Wendorf, "Impure Science," 403–4.

91 ***"It's very unlikely":*** Quoted in Cherfas, "Grave Accusations Against Lucy Finders," 390.

91 ***different version of events:*** J. Desmond Clark, letter to Fred Wendorf, March 2, 1994, Wendorf papers, box 11, folder 17.

92 ***"Clark's eventual move":*** Charles Redman, memo, October 29, 1982, "Jon Kalb case," 2, Kalb lawsuit exhibit 10.

92 ***"freer rein than normal projects":*** Redman, "Jon Kalb case," 2.

92 ***"soothe Kalb's resentment":*** Redman, "Jon Kalb case," 2.

92 ***"victim of very unfortunate circumstances":*** Charles Redman, letter to Jon Kalb, November 23, 1982, Kalb lawsuit exhibit 13.

92 ***"You know and I know":*** Jon Kalb, letter to Charles Redman, November 27, 1982, Kalb lawsuit exhibit 14.

92 ***the two sides reached a settlement:*** Stipulation of Settlement, December 8, 1987, Kalb lawsuit.

93 ***he sought to "muscle in":*** J. Desmond Clark, letter to Maurice Taieb, February 7, 1979, Wendorf papers, box 18, folder 29.

93 ***expelled geologist:*** Berhanu Abebe, letter to Desmond Clark, September 6, 1978, Wendorf papers, box 18, folder 29.

CHAPTER 6: BADLANDS

94 ***Government officials invited Kalb:*** Kalb, *Adventures in the Bone Trade,* 301–2.

94 ***"They tried to intimidate Tim":*** Berhane Asfaw, interview with author.

94 ***"Three people declined":*** Yonas Beyene, interview with author.

94 ***"The only way to protect yourself":*** Zelalem Assefa, interview with author.

95 ***On the first morning:*** Footage from November 29, 1992, MARP video 1992.

95 ***Afar children took livestock:*** Author interviews with Ahmed Elema, Adeni, and Giday WoldeGabriel.

95 ***clashes with soldiers:*** Yasin, "Political History of the Afar in Ethiopia and Eritrea," 50n13.

95 ***Afar warrior named Elema:*** MARP video archive, 1992; Ahmed Elema, interview with author.

96 ***"Elema was fucking bad ass":*** Ahmed Elema, interview with author.

96 ***He found the strangers:*** Author interviews with Ahmed Elema, Tim White, Giday WoldeGabriel, and Yonas Beyene.

97 ***"always wins":*** Tim White, interview with author.

98 ***shattered a tooth:*** Details from author interviews with Scott Simpson, Bruce Latimer, and Tim White.

98 ***"a Land Cruiser":*** Bruce Latimer, interview with author.

99 *"Making roads, talking to people":* Ahmed Elema, interview with author.

101 *stack of sediments:* Renne et al., "Chronostratigraphy of the Miocene–Pliocene Sagantole Formation."

104 *wondered if Kalb's expedition:* Tim White, field notes, November 20, 1981.

104 *just as frustrated:* Kalb, *Adventures in the Bone Trade*, 198.

104 *"I've got one":* Author interviews with Gen Suwa, Tim White, and Scott Simpson.

104 *working late in the lab:* Biographical details from Gen Suwa, interview with author.

106 *following an Ethiopian surveyor:* Gen Suwa, interview with author.

107 *"it's only a third molar":* Gen Suwa, interview with author.

107 *named Alemayehu Asfaw:* Biographical details from Alemayehu Asfaw, interview with author.

108 *fragment of lower jawbone:* Author interviews with Alemayehu Asfaw, Gen Suwa, and Tim White.

108 *"This one is narrow":* Tim White in *Coincidence in Paradise*.

109 *Scott Simpson was scanning:* Details from Scott Simpson, interview with author.

110 *the same little man:* Author interviews with Tim White and Bruce Latimer.

110 *joint raid against the Issa:* Details from author interviews with Adeni, Ahmed Elema, and Giday WoldeGabriel.

110 *"They forced themselves":* Berhane Asfaw, interview with author.

111 *"their* hata *walking sticks":* Scott Simpson, interview with author.

111 *On the fourth day:* Author interviews with Tim White, Berhane Asfaw, and Scott Simpson; Tim White, field log.

CHAPTER 7: THE ZIPPERMAN'S ASH

112 *cold ash blanketed the ground:* WoldeGabriel et al., "Volcanism, Tectonism, Sedimentation, and the Paleoanthropological Record in the Ethiopian Rift System," 95.

114 *"a sea of imprecisions":* Clark, "Radiocarbon Dating and African Archaeology," 7.

115 ***the preferred method:*** Deino et al., "40Ar/39Ar Dating in Paleoanthropology and Archeology," 63.

115 ***resolved the age of Lucy:*** Walter, "Age of Lucy"; Renne et al., "New Data from Hadar (Ethiopia) Support Orbitally Tuned Time Scale."

115 ***newly calculated results:*** Paul Renne, e-mail to Tim White, November 24, 1994.

117 ***researching blood proteins:*** Summarized in Goodman, "A Personal Account of the Origins of a New Paradigm."

117 ***paleontologist George Gaylord Simpson:*** Simpson did not dispute the molecular findings, but argued that classification should categorize animals not purely by ancestry but also by physical adaptations.

118 ***could reconstruct family trees:*** Zuckerkandl and Pauling, "Evolutionary Divergence and Convergence in Proteins."

118 ***graduate student Vincent Sarich:*** Details recounted by Sarich and Miele in *Race: The Reality of Human Differences,* 110–13.

118 ***published their results:*** Sarich and Wilson, "Immunological Time Scale for Hominid Evolution."

118 ***"they thought we were idiots":*** Vincent Sarich, interview with author.

119 ***"happy as a lark":*** Tim White, interview with author.

119 ***"The atmosphere changed":*** Vincent Sarich, interview with author.

119 ***"general-purpose ape":*** Johanson and Edey, *Lucy,* 344–45.

119 ***99 percent identical to chimps:*** King and Wilson, "Evolution at Two Levels."

120 ***finally was resolved:*** Caccone and Powell, "DNA Divergence Among Hominoids."

120 ***the family tree:*** Ruvolo et al., "Gene Trees and Hominoid Phylogeny."

120 ***"Man the Toolmaker":*** Now monkeys and even New Caledonia crows have been shown to use tools, so the behavior is even less unique than formerly believed.

120 ***"Now we must redefine tool":*** Quoted in Goodall, *Through a Window,* 22.

120 ***Western and Japanese researchers:*** McGrew, "The Cultured Chimpanzee," 44.

122 ***"the ancestral ape":*** Washburn, 1983 IHO locomotion conference.

122 ***"some normal people":*** Diamond, *The Third Chimpanzee,* 2.

124 **To work in our territory:** Berhane Asfaw, interview with author.

CHAPTER 8: UNDER THE VOLCANO

125 ***On the third day:*** Scene details and quotes from MARP video 1993 n. 2.

126 ***scanned his checklist:*** Tim White, "Daily Field Checklist," personal communication.

126 ***"He just keeps going":*** Yohannes Haile-Selassie, interview with author.

126 ***exhaustion in his voice:*** Tim White, e-mail to Owen Lovejoy, January 27, 1995.

127 ***"the best field worker":*** Bruce Latimer, interview with author.

127 ***"This is a remarkable layer":*** Tim White, field notes, December 26, 1993.

128 ***On the fourth day:*** Tim White, field notes, MARP video 1993 n. 2.

128 ***"Tim got attached":*** Giday WoldeGabriel, interview with author.

128 ***"who got respect":*** Berhane Asfaw, interview with author.

130 ***"Doktor Tee, amee":*** Quoted by Tim White, e-mail to author, March 23, 2016.

130 ***"There is a chance":*** Tim White in MARP video 1993 n. 2, December 29, 1993.

130 ***the oldest species:*** One trivial caveat deserves mention. In the 1980s, artist Walter Ferguson named the species *Homo antiquus praegens* and claimed it included some previously unattributed fossils from Kenya that were estimated to be more than 5 million years old. Few scientists took this taxon seriously, and it has been forgotten.

130 ***Alemayehu Asfaw:*** Alemayehu Asfaw and Tim White, interviews with author.

131 ***paper in*** Nature***:*** White et al., "Australopithecus Ramidus, a New Species of Early Hominid."

131 ***"We think this species":*** Tim White on *The MacNeil/Lehrer NewsHour*, September 22, 1994.

132 ***"any less humanlike":*** Gee, *In Search of Deep Time*, 203.

132 ***"hadn't been discovered":*** Gee, *In Search of Deep Time*, 204.

132 ***"The metaphor of a missing link":*** Wood, "Oldest Hominid Yet," 281.

132 ***"It's a woodland or forest dweller":*** Tim White on *The MacNeil/ Lehrer NewsHour*, September 22, 1994.

CHAPTER 9: THE WHOLE THING IS THERE

133 ***Journalists begged:*** White, "Feud over Old Bones."

133 ***from the* Sunday Times:** Fitzgerald, "Rift Valley."

133 ***"I will not participate":*** Quoted in Fitzgerald, "Rift Valley."

133 ***"So paranoid are they":*** Fitzgerald, "Rift Valley."

133 ***"They snubbed me":*** Fitzgerald, "Rift Valley."

133 ***within fifty kilometers:*** White, "Feud over Old Bones."

133 ***"The emotional connection":*** Doug Pennington, interview with author.

134 ***"Afar guys take their camels":*** Berhane Asfaw, interview with author.

134 ***pitching camp:*** Tim White, interview with author; Tim White, field logs, 1994.

135 ***snap a photo:*** Tim White, photograph archives, November 5, 1994.

135 ***"a disciplined crawl":*** Tim White, interview with author; Tim White, photograph archives.

135 ***held a broken bone:*** Author interviews with Yohannes Haile-Selassie and Tim White; White, "Paleontological Case Study."

136 ***gunfire broke out:*** Tim White, field log, November 9, 1994; author interviews with Bruce Latimer, Doug Pennington, and Henry Gilbert.

136 ***"Bullets!":*** author interviews with Bruce Latimer, Doug Pennington, and Henry Gilbert.

136 ***Disgruntled job-seekers:*** Tim White, interview with author.

136 ***tracked down the shooters:*** Tim White, interview with author.

136 ***returned to locality six:*** Discovery details from Tim White, interviews with author; White, "Paleontological Case Study"; Tim White, e-mail to author, October 8, 2014; MARP video 1994 n. 1.

137 ***Then it rained:*** Tim White, field log, November 21, 1994, and December 6, 1994.

137 ***"most of the material":*** Tim White, field log, November 14, 1994.

138 ***days cleaning the tibia:*** Tim White, e-mail to author.

138 ***shortest day of the year:*** Scene details from MARP video 1994 n. 6.

139 ***"placer" mining:*** White, "A Paleontological Case Study," 549.

139 ***"out of a hundred times":*** Tim White, interview with author.

139 ***A painful reminder:*** Details from author interviews with Yonas Beyene, Berhane Asfaw, and Tim White.

140 ***"As we excavate":*** MARP video 1994 n. 6, December 21, 1994.

140 ***"All we can do now":*** MARP video 1994 n. 6, December 21, 1994.

140 **We kill our informants:** Flannery, "The Golden Marshalltown," 275.

141 ***"Look at that canine!":*** MARP video 1994 n. 9, December 31, 1994.

141 ***The skull came out of the ground:*** MARP video 1994 n. 12, January 2, 1995.

142 ***"Need a bandage":*** All dialogue in this scene from MARP video 1994 n. 12, January 2, 1995.

142 ***"Don't touch it":*** MARP video 1994 n.12, January 2, 1995.

143 ***He unslung his gun:*** Details from MARP video 1994 n. 13, December 26, 1994.

143 ***New Year's Eve passed:*** Details and dialogue from MARP video 1994 n. 10, December 31, 1994.

143 ***"You can't take shortcuts":*** MARP video 1994 n. 10, December 31, 1994.

144 ***fresh piles of jackal shit:*** MARP video 1994 n. 12, January 2, 1995.

144 ***"The whole fucking thing":*** MARP video 1994 n. 9, December 31, 1994.

144 ***"Nobody has ever seen":*** MARP video 1994 n. 9, December 31, 1994

144 ***"It's a middle finger!":*** MARP video 1994 n. 8, December 30, 1994.

144 ***"people in the field":*** MARP video 1994 n. 8, December 30, 1994.

145 ***thirty-six other individuals:*** White, "*Ardipithecus ramidus* and the Paleobiology of Early Hominids," 76.

CHAPTER 10: A POISON TREE

146 ***"Johanson steps out":*** Fitzgerald, "Rift Valley."

146 ***"Indiana Jones in Armani":*** McKie, "Bone Idol."

146 ***lunched with journalists:*** McKie, "Bone Idol."

147 ***"Lucy's species was the root":*** Nova, *In Search of Human Origins,* part 1.

147 ***"we have the ancestor of Lucy":*** Tim White on *The MacNeil/Lehrer NewsHour,* September 22, 1994.

147 ***"Johanson is insanely jealous":*** Quoted in Jefferson, "This Anthropologist Has a Style That Is Bone of Contention."

147 ***"very few people in the world":*** Tim White, interview with author.

147 ***"I was introduced to royalty":*** Johanson and Shreeve, *Lucy's Child,* 23.

147 ***"Johansonesque escapades":*** Tim White, interview with author.

148 ***"the substance of Tim":*** Quoted in Kiefer, "The Man Who Loved Lucy."

148 ***"I never liked Johanson":*** Berhane Asfaw, interview with author.

148 ***"quick to remind me":*** Johanson and Wong, *Lucy's Legacy,* 81.

148 ***get out of town:*** Johanson and Wong, *Lucy's Legacy,* 40–44; Don Johanson and Berhane Asfaw, interviews with author.

148 ***"all a smokescreen":*** Don Johanson, interview with author.

149 ***the 1990 visit:*** Tim White, interviews with author.

149 ***Tadesse Terfa rebuked Johanson:*** Johanson and Wong, *Lucy's Legacy,* 81.

149 ***"difficult for me to do that":*** Don Johanson, interview with author.

149 ***"a very black-and-white person":*** Don Johanson, interview with author.

149 ***"always bitterly complaining":*** Steve Brandt, interview with author.

150 ***"like a battlefield":*** Zelalem Assefa, interview with author.

150 ***"considerable egos":*** Swisher et al., *Java Man,* 111.

150 ***"Louis Leakey School of anthropology":*** Johanson and Shreeve, *Lucy's Child,* 29.

150 ***70 percent of its grant money:*** Renne, "Institute of Human Origins Breakup."

150 ***tensions finally blew up:*** Jefferson, "This Anthropologist Has a Style That Is Bone of Contention"; Swisher et al., *Java Man,* 114–17.

150 ***$1 million annual support:*** Details from Swisher et al., *Java Man,* 117–29.

151 *"to minimize embarrassment":* Quoted in Jefferson, "This Anthropologist Has a Style That Is Bone of Contention."

151 *"no longer able or willing":* Berhane Asfaw, "Close Observer Turns to Insult When Confronted with Facts That Exposes His Connection," Walta Information Center, April 5, 2001.

151 *accused the IHO of trespassing:* Gibbons, "Claim Jumpiing Charges"; Petit, "Berkeley Institute in Battle Over Fossils"; Wilford, "Tempers Flare."

151 *"destroy our organization":* Quoted in Petit, "Berkeley Institute in Battle over Fossils."

151 *It would take five years:* Don Johanson and Bill Kimbel, interviews with author.

152 *an IHO event:* Berhane Asfaw, interview with author.

152 *"seems to be the thread":* Don Johanson, interview with author.

152 *"As Tim told me many times":* Don Johanson, interview with author.

153 *"It's very odd":* Don Johanson, interview with author.

153 *Even among the Lucy team:* Maurice Taieb and Bruce Latimer, interviews with author.

153 *Even Maurice Taieb thought so:* Maurice Taieb, interview with author.

153 *spotted the first piece:* Tom Gray, interview with author.

153 *the* Lucy *book claimed:* Johanson and Edey, *Lucy*, 16.

153 *"a goddamn snake":* Tom Gray, interview with author.

153 *"Don knew it was wrong":* Tom Gray, interview with author.

154 *Johanson responded that:* Don Johanson, interview with author.

154 *"incredibly vindictive":* Tom Gray, interview with author.

154 *dropped from the IHO team:* Maurice Taieb, interview with author; Dalton, "The History Man," 269. In recent years, Johanson has acknowledged his debt to Taieb and expressed gratitude.

154 *"Donald Johanson had ego":* Maurice Taieb, interview with author.

154 *on the camp table:* Description and dialogue from MARP video 1994 n. 14, January 5–6, 1995.

CHAPTER 11: THE PLIOCENE RESTORATION

156 *only a single bone:* Tim White, e-mail to Owen Lovejoy, January 22, 1995.

156 ***Acting on a hunch:*** Tim White to Owen Lovejoy, e-mail January 22, 1995, White papers.

157 ***"carnivore hors d'oeuvres":*** Tim White, interview with author.

157 **four *different fossils:*** White and Suwa, "Hominid Footprints at Laetoli: Facts and Interpretations."

157 ***hinted bipedality:*** The far ends of the metatarsals had bulbous, domed joints typical of bipeds that pushed off their toes.

157 ***"a kind of time machine":*** Conversations with History, "On the Trail of Our Human Ancestors: Timothy White."

158 ***"The preparer damage":*** Tim White, interview with author.

159 ***before catching his plane:*** Tim White, e-mails to Owen Lovejoy, January 25–26, 1995.

CHAPTER 12: STANDING UPRIGHT

160 ***"more myths in human evolution":*** Owen Lovejoy, interview with author.

161 ***"anthropologists don't have patients":*** Owen Lovejoy, interview with author.

161 ***The bereaved anatomist:*** Owen Lovejoy, e-mails to Tim White, January 22–February 17, 1995.

161 ***"both equally suspicious":*** Owen Lovejoy, interview with author.

162 ***an opposable big toe:*** Tim White, e-mail to Owen Lovejoy, January 25, 1995.

163 ***"When I was a child":*** Owen Lovejoy, interview with author.

163 ***an unusual pedigree:*** Biographical details from Trina Prufer, e-mail to author, October 21, 2019; Edler, "Life as Artifact"; Prufer, "How to Construct a Model," 107.

164 ***"We dug up 1,500 skeletons":*** Owen Lovejoy, interview with author.

164 ***orthopedic surgeon King Heiple:*** King Heiple, interview with author.

164 ***One skeleton of a young man:*** Pigott, "Bone and Antler Artifacts from the Libben Site," 57.

165 ***imminent extinction of humanity:*** *Daily Kent Stater,* "KSU Anthropologist Predicts Human Extinction."

165 ***"I heard the shots":*** Owen Lovejoy, interview with author.

165 ***Wilfrid E. Le Gros Clark:*** Le Gros Clark, "Penrose Memorial Lecture. The Crucial Evidence for Human Evolution," 169–70.

165 ***"jog trot":*** Napier, "Antiquity of Human Walking," 65.

165 ***"shambling half-run":*** Washburn and Moore, *Ape Into Man*, 112.

165 ***"just bloody apes":*** Quoted in Lewin, *Bones of Contention*, 165.

166 **Australopithecus *walked:*** Lovejoy et al., "The Gait of *Australopithecus*."

167 ***at Lovejoy's house:*** Johanson and Edey, *Lucy*, 162–63.

167 ***"It was gorgeous":*** Owen Lovejoy, interview with author.

167 ***an opposition emerged:*** Stern and Susman, "The Locomotor Anatomy of *Australopithecus afarensis*"; Susman et al., "Arboreality and Bipedality in the Hadar Hominids."

167 ***as a "fairy tale":*** Susman et al., "Arboreality and Bipedality in the Hadar Hominids," 113.

168 ***"capable of walking upright":*** Lovejoy, "Evolution of Human Walking," 125.

168 ***"on top of his desk":*** Bill Kimbel, interview with author.

168 ***an influential article:*** Lovejoy, "Evolution of Human Walking."

169 ***the origins of bipedality:*** Lovejoy, "Origin of Man."

169 ***hanging out with the mechanics:*** Owen Lovejoy and Bruce Latimer, interviews with author.

169 ***A biped couldn't also be arboreal:*** For example, see Lovejoy, "A Biomechanical Review of the Locomotor Diversity of Early Hominids."

170 ***"This guy is a footgrabber":*** Tim White, e-mail to Owen Lovejoy, January 27, 1995.

171 ***"Now ready to admit defeat":*** Owen Lovejoy, e-mail to Tim White, January 29, 1995.

CHAPTER 13: THE WHOLE WORLD WANTS TO KNOW

172 ***dubbed it ML:*** Tim White and Owen Lovejoy, e-mails, January 22–February 17, 1995.

172 ***calling it* the beast:** Tim White and Owen Lovejoy, e-mails, January 22–February 17, 1995.

172 ***"description of this beast":*** Owen Lovejoy, e-mail to Tim White, January 29, 1995.

173 ***a brief announcement:*** White et al., "Corrigendum."

173 ***briefly mentioned the skeleton:*** An announcement in the Ethiopian press and a story in the *Los Angeles Times* already had revealed the discovery of the skeleton.

173 ***"the world right now":*** Quoted in Petit, "One Step at a Time."

173 ***"common ancestor with chimpanzees":*** Quoted in Gore, "First Steps."

173 ***approached and questioned him:*** Details and dialogue from scene from MARP video 1995 n. 2.

173 ***"Middle Awash research team":*** NSF case file award 8210897, Appendix VIII, 81.

175 ***"keep every discovery in secret":*** Negarit Gazeta of the People's Democratic Republic of Ethiopia, "A Proclamation to Provide for the Study and Protection of Antiquities."

175 ***"There was none of this favoritism":*** Tim White, interview with author.

175 ***"it's going to be the* truth*":*** Owen Lovejoy, interview with author.

176 ***"taken to the woodshed":*** Bruce Latimer, interview with author.

176 ***the excavation in miniature:*** Details from MARP video 1995 n. 2.

176 ***"Extremely shattered":*** MARP video 1995 n. 2.

177 ***"no room for error":*** Scott Simpson, interview with author.

177 ***assembled in Addis Ababa:*** Scene details and dialogue from MARP video 1995 n. 2.

177 ***"Are there carnivore holes":*** Dialogue from MARP video 1995 n. 2.

178 ***"the bar in* Star Wars*":*** Quoted in Gore, "First Steps."

178 ***the museum casting technician:*** Scene details and dialogue from MARP video 1995 n. 3.

180 **"afarensis *is the common ancestor":*** Meave Leakey, interview with author.

180 ***announced the new species:*** Leakey et al., "New Four-Million-Year-Old Hominid Species from Kanapoi."

180 ***the scientific consensus:*** Kimbel et al., "Was *Australopithecus anamensis* Ancestral to *A. afarensis*?"

181 ***"may represent Lucy's ancestor":*** Leakey, "The Dawn of Humans," 51.

181 **"ramidus' *claim":*** Quoted in Sawyer, "New Roots for Family Tree."

181 ***"an ancestor to later apes":*** Quoted in Wilford, "New Fossils Reveal the First of Man's Walking Ancestors."

181 ***two pieces of maxilla:*** MARP video 1994 n. 2.

181 ***found in the Turkana Basin:*** Details from White et al., "Asa Issie, Aramis and the Origin of *Australopithecus*."

182 ***"great example of evolution":*** MARP video 1995 n. 3.

182 ***ecological breakout:*** See White et al., "Asa Issie, Aramis and the Origin of *Australopithecus*," 888.

CHAPTER 14: TREES AND BUSHES

184 ***"biological species concept":*** Mayr defined the species as "groups of interbreeding natural populations reproductively isolated from other such groups."

184 ***9 percent of bird species:*** Mallet et al., "How Reticulated Are Species?" 143–44.

184 ***interbreeding for 10 million years:*** Mallet et al., "How Reticulated Are Species?" 143–44.

186 ***about thirty genera:*** Mayr, "Reflections on Human Paleontology," 231.

186 ***"bewildering to other biologists":*** Dobzhansky, "On Species and Races of Living and Fossil Man," 257.

186 ***"never more than one species of man":*** Mayr, "Taxonomic Categories in Fossil Hominids," 112.

186 ***"specialized in despecialization":*** Mayr, "Taxonomic Categories in Fossil Hominids," 116.

187 ***individuals and geographic populations:*** Schultz, "The Specializations of Man and His Place Among the Catarrhine Primates."

187 ***species evolved as* lineages*:*** Simpson, "Some Principles of Historical Biology Bearing on Human Origins."

187 ***main driver of evolution:*** Simpson, *Tempo and Mode in Evolution*, 203.

187 ***"a rather unified group":*** Simpson, "Nature and Origin of Supraspecific Taxa," 270.

189 ***"Looking for ancestors":*** Nelson, "Paleontology and Comparative Biology," 706.

189 ***"extremely imperfect":*** Darwin, *Origin of Species*, 342.

189 ***"punctuated equilibrium":*** Gould and Eldredge, "Punctuated Equilibria: An Alternative to Phyletic Gradualism."

190 ***"a once luxuriant bush":*** Gould, "Ladders, Bushes, and Human Evolution," 31.

190 ***"any leading expert":*** Gould, *Structure of Evolutionary Theory*, 910.

190 ***"the ghosts of Cold Spring Harbor":*** White, "Human Origins and Evolution: Cold Spring Harbor, Déjà Vu," 341.

191 ***"Our family tree":*** White, "Five's a Crowd," R115.

192 ***"riotous profusion of species":*** Tattersall, *The Strange Case of the Rickety Cossack*, 94.

192 ***"dead hand of linear thinking":*** Tattersall, "Once We Were Not Alone," 43.

192 ***"this wonderful sequence":*** Ian Tattersall, interview with author.

192 ***"enthusiastic unilinealist":*** Tattersall, *The Strange Case of the Rickety Cossack*, 161.

192 ***wrote a savage review:*** White, "Monkey Business."

193 ***a "traumatizing influence":*** Tattersall, *Rickety Cossack*, 214.

193 ***"products of human minds":*** Tattersall, *Rickety Cossack*, 184.

193 ***"To take a line of fossils":*** Gee, *In Search of Deep Time*, 116–17.

193 ***"every bit as important":*** Tattersall, *Rickety Cossack*, 180.

194 ***genetically diverse primate species:*** Osada, "Genetic Diversity in Humans and Non-human Primates," 133.

194 ***subpopulation of chimpanzees:*** Osada, "Genetic Diversity in Humans and Non-human Primates," 137.

CHAPTER 15: VOYAGING

195 ***Greek* kynodontes *(dog teeth):*** Wolff, "The Language of Medicine."

195 ***to expand his inquiries:*** Standring, "A Brief History of Topographical Anatomy."

196 ***"this lesser world":*** Quoted in Keele, *Leonardo da Vinci's Elements of the Science of Man*, 197.

196 ***one faithful Galenist:*** Jose, "Anatomy and Leonardo da Vinci," 187.

197 ***"the queen of San Francisco Society":*** West, "Pacific Heights."

197 ***featured in* Architectural Digest*:*** Weaver, "Flying First Class."

198 ***"Opulence beyond all comprehension":*** Owen Lovejoy, interview with author.

198 ***Andreas Cellarius in 1609:*** Weaver, "Flying First Class."

198 ***scores of ape species:*** Andrews, *An Ape's View of Human Evolution;* Begun, *The Real Planet of the Apes.*

200 ***"patently absurd":*** Oliwenstein, "New Foot Steps into Walking Debate."

200 ***The two men went into an office:*** Author interviews with Owen Lovejoy, Bruce Latimer, and Tim White.

200 ***Meanwhile, Lee Berger:*** Details from Berger, *In the Footsteps of Eve.*

200 ***"the East African hegemony":*** Berger, *In the Footsteps of Eve,* 205.

200 ***a "chimpanzee-like" shinbone:*** Berger and Tobias, "A Chimpanzee-like Tibia from Sterkfontein." The paper challenged the theory that East African *afarensis* was ancestral to South African *africanus.*

200 ***"Tim just slaughtered him":*** Meave Leakey, interview with author.

201 ***"say you retract the title":*** Berger, *In the Footsteps of Eve,* 211.

201 ***"At no stop in the tour":*** Tim White, interview with author.

201 ***"There was no model":*** Scott Simpson, interview with author.

201 ***the motor cortex:*** Tubiana et al., *Examination of the Hand and Wrist,* 171.

202 ***the opposite position:*** Aristole, *On the Parts of Animals,* 117–18.

203 ***"The hand supplies":*** Bell, *The Hand,* 38.

203 ***"his present dominant position":*** Darwin, *The Descent of Man,* 141.

203 ***fundamental modes of grasping:*** See Napier, "The Prehensile Movements of the Human Hand," "The Evolution of the Hand," and *Hands.*

204 ***"ultimate refinement in prehensility":*** Napier, "The Evolution of the Hand," 59.

204 ***"intermediate condition":*** Washburn, "Behaviour and the Origin of Man," 23.

204 ***the backs of the fingers:*** Washburn and Moore, *Ape Into Man,* 34–35.

204 ***a nearly complete hand:*** Ardi had bones from both hands. The investigators could reassemble most of the hand skeleton and rely on mirror imaging to duplicate parts missing from the opposite side. A thumb bone came from another individual found nearby.

With this composite, only a few pieces remained missing, such as the pisiform bone of the wrist and some terminal phalanges of the fingertips.

205 ***Ardi bore few signs:*** See Lovejoy et al., "Careful Climbing."

206 ***"Our earliest ancestors":*** Lovejoy et al., "Careful Climbing," 70.

206 ***Simpson and Lovejoy holed up:*** Scott Simpson and Owen Lovejoy, interviews with author.

207 **"nobody *other than Tim*":** Berhane Asfaw, interview with author.

207 ***his friend* Doktor Tee*:*** Details from author interviews with Tim White, Henry Gilbert, Giday WoldeGabriel, and David Brill.

209 ***"The males are solitary":*** Tim White, interview with author.

209 ***"I've never known one":*** Leakey, *Animals of East Africa*, 61.

CHAPTER 16: MISSION TO THE PLIOCENE

210 ***The theory of evolution rests:*** For an overview, see Browne, "History of Biogeography."

211 ***"some race of quadrumanous animals":*** Quoted in Bender et al., "Savannah Hypotheses," 151.

211 ***"more on the ground:"*** Darwin, *Descent of Man*, 140.

211 ***"It is more probable":*** Wallace, *Darwinism*, 459.

211 ***plains were "invigorating":*** Quoted in Bender et al., "Savannah Hypotheses," 163.

211 **Australopithecus africanus *cavorting:*** Dart, "*Australopithecus Africanus*," 199.

212 ***the idea of a Miocene rain forest:*** For a detailed refutation of the classic narrative see Bonnefille, "Cenozoic Vegetation, Climate Changes and Hominid Evolution in Tropical Africa."

212 ***South African scholar C. K. "Bob" Brain:*** See Brain, "Do We Owe Our Intelligence to a Predatory Past?"

212 ***"closed, wooded" habitat:*** WoldeGabriel et al., "Ecological and Temporal Placement," 333.

213 ***long expressed skepticism:*** Both White and Lovejoy expressed doubts about savanna theory in Johanson and Edey, *Lucy*, 339–40. See also Shreeve, "Sunset on the Savanna" and White, "African Omnivores."

213 ***The Central Awash Complex:*** Details from Renne et al., "Chronostratigraphy of the Miocene–Pliocene Sagantole Formation."

213 ***"Here, it's overkill":*** Giday WoldeGabriel, interview with author.

214 ***one hundred thousand years:*** Renne et al., "Chronostratigraphy of the Miocene–Pliocene Sagantole Formation," 869.

215 ***"almost no fossils on the surface":*** David DeGusta, e-mail to author, December 7, 2016.

215 ***six thousand specimens:*** See White et al., "Macrovertebrate Paleontology"; and WoldeGabriel et al., "The Geological, Isotopic, Botanical, Invertebrate, and Lower Vertebrate."

216 ***"People don't give a damn":*** From "An Archaeologist at Work," oral history, Tim White, J. Desmond Clark, and Timothy Troy, November 2, 2000, 369.

216 ***climate-induced pulse:*** For example, see Vrba, "The Pulse That Produced Us."

217 ***the two scientists sat:*** Scene details and dialogue from the documentary *Coincidence in Paradise*.

217 ***Leaf-eating tragelaphines:*** White et al., "Macrovertebrate Paleontology," SOM, 25.

217 ***Vrba and DeGusta developed:*** DeGusta and Vrba, "Method for Inferring Paleohabitats"; David DeGusta, e-mail to author, December 7, 2016.

217 ***forest-adapted Jeeps:*** White et al., "Macrovertebrate Paleontology," 90.

217 ***concentration of* Ardipithecus *fossils:*** The *Ardipithecus* fossils were overwhelmingly concentrated in two places: Locality one (where the first fossils were spotted) and locality six (where the Ardi skeleton was found).

217 ***40 percent of teeth:*** White et al., "Macrovertebrate Paleontology," 88.

217 ***The most common birds:*** Louchart et al., "Taphonomic, Avian, and Small-Vertebrate Indicators," 66.

218 ***areas where* Ardipithecus *dwelled:*** WoldeGabriel et al., "Geological, Isotopic, Botanical, Invertebrate, and Lower Vertebrate Surroundings"; Stan Ambrose, interview with author.

218 ***177 mammal teeth:*** See White et al., "Macrovertebrate Paleontology" and supporting online material.

218 ***"the blind test would be compromised":*** Tim White and Stan Ambrose, interviews with author.

218 ***"I fooled him":*** Tim White, interview with author.

218 ***"not a major force":*** White et al., "Macrovertebrate Paleontology," 67.

219 ***half grasses and sedges:*** Cerling, "Comment on the Paleoenvironment."

219 ***60 percent grass coverage:*** Bonnefille, "Cenozoic Vegetation," 398.

219 ***"not accept our results":*** Raymonde Bonnefille, interview with author.

220 ***Some big events:*** Bonnefille, "Cenozoic Vegetation, Climate Changes and Hominid Evolution," 409.

220 ***defies any simple rendering:*** Kingston, "Shifting Adaptive Landscapes"; Bonnefille, "Cenozoic Vegetation, Climate Changes and Hominid Evolution."

221 ***C_4 grasses displaced C_3 grasses:*** Feakins et al., "Northeast African Vegetation Change."

221 ***"The C_4 expansion happened":*** Sarah Feakins, interview with author.

221 ***"C_3 grasslands in lowland Africa":*** Sarah Feakins, interview with author.

221 ***"should have been dead already":*** Sarah Feakins, interview with author.

221 ***it was a patchwork:*** Bonnefille, "Cenozoic Vegetation"; Kingston, "Shifting Adaptive Landscapes."

221 ***time on the ground:*** Andrews, *An Ape's View of Human Evolution*, xiii–xiv.

CHAPTER 17: HARVEST OF BONE

222 ***the normal rains:*** Wolde-Georgis et al., "The Case of Ethiopia."

223 ***the Indian Ocean Dipole:*** Anderson, "Extremes in the Indian Ocean."

223 ***blazed a new route:*** Tim White, e-mail to author, May 24, 2016.

223 ***On the first day of surveying:*** Tim White, interview with author.

225 ***all modern humans:*** Cann et al., "Mitochondrial DNA and Human Evolution."

226 ***many population lineages:*** Reich, *Who We Are and How We Got Here*, 10; Stringer, "The Origin and Evolution of *Homo Sapiens*."

226 ***"probable immediate ancestors":*** White et al., "Pleistocene *Homo Sapiens* from Middle Awash," 742.

227 ***"It is in the right place":*** Berhane Asfaw et al., "*Australopithecus Garhi*," 634.

227 ***2.6-million-year-old stone tools:*** Semaw et al., "2.5-Million-Year-Old Stone Tools from Gona, Ethiopia."

227 ***"big-toothed chimps":*** AAAS, "New Species of Human Ancestor."

228 ***"the last common ancestor":*** "Project Description," 10, NSF award 9632389.

229 ***called back on the radio:*** Conversation recounted by Yohannes Haile-Selassie and Tim White, interviews with author.

230 ***"close to the common ancestor":*** Yohannes Haile-Selassie, "Late Miocene Hominids," 180.

230 ***"No one thought":*** Quoted in Radford, "Oldest Human Ancestor Discovered."

230 ***a late Miocene menagerie:*** Yohannes Haile-Selassie and Giday WoldeGabriel, *Ardipithecus kadabba*, 17–19.

231 ***called* Orrorin tugenensis*:*** Senut et al., "First Hominid from the Miocene."

231 ***discovery of* Sahelanthropus tchadensis*:*** Brunet et al., "A New Hominid from the Upper Miocene of Chad."

231 ***three evolutionary plateaus:*** Described in White et al., "*Ardipithecus Ramidus* and the Paleobiology of Early Hominids"; and White, "Human Evolution: How Has Darwin Done?"

231 ***three genera of human ancestors:*** Due to the difficulty of differentiating closely related species in a sparse fossil record, some paleontologists such as George Gaylord Simpson have argued that the genus is a more useful taxonomic category than the species.

232 ***"I can take you":*** Berhane Asfaw, "The Origin of Humans," 17.

232 ***"the continent's most important":*** NSF award 931869 case file, "Project Description," 7.

233 ***than they had recognized:*** Begun, "The Earliest Hominins—Is Less More?" 1478.

233 ***"Despite the obstacles":*** NSF award 9632389 case file, "Progress Report, 1997," 1.

233 ***four hundred fossil localities:*** Tim White, e-mail to author, January 30, 2019.

233 ***"The sheer volume of material":*** NSF award 9910344, panel reviews.

234 ***typical carnivore damage:*** White, "Cut Marks in the Bodo Cranium."

234 ***carving up other humans:*** See White, *Prehistoric Cannibalism* and "Once We Were Cannibals."

235 ***"a very cold scene":*** White, *Conversations with History.*

235 ***"rather like ourselves":*** Quoted in Salopek, "Enigmatic War."

CHAPTER 18: BORDER WARS

236 ***turned out to be more* Ardipithecus*:*** Sileshi Semaw et al., "Early Pliocene Hominids from Gona, Ethiopia."

236 ***the Gona team looked up:*** Author interviews with Scott Simpson, Kathy Schick, and Nick Toth.

236 ***"senseless, inscrutable war":*** Salopek, "Enigmatic War."

237 ***he was shot in 2000:*** Markakis, "Anatomy of a Conflict," 451.

237 ***walked into an ambush:*** Markakis, "Anatomy of a Conflict," 445–46.

237 ***wounded in the leg:*** Author interviews with Ahmed Elema, Adeni, and Giday WoldeGabriel.

237 ***"He was trying to get help":*** Berhane Asfaw, interview with author. Additional details from author interviews with Berhane Asfaw, Tim White, Ahmed Elema, and Adeni.

237 ***"We were requested":*** Don Johanson, interview with author

238 ***"at each other's throats":*** Don Johanson, interview with author.

238 ***they exploded volcanically:*** See Yitbarek, "A Problem of Social Capital" and Levine," Ethiopia's Dilemma: Missed Chances."

238 ***"Compromise is not very easy":*** Zelalem Assefa, interview with author.

238 **He who is not vigilantly suspicious:** Yitbarek, "A Problem of Social Capital," 6.

239 ***"with a really big gun":*** Giday WoldeGabriel, interview with author.

239 ***froze to death:*** Seidler et al., "Anthropological Aspects of the Prehistoric Tyrolean Ice Man," 456.

239 ***Ötzi had an arrowhead:*** Holden, "Otzi Death Riddle Solved."

239 ***Yohannes was working:*** Yohannes Haile-Selassie, interview with author.

239 ***"probably a bad move":*** Yohannes Haile-Selassie, interview with author.

239 ***tense confrontation:*** Tim White, Lorenzo Rook, Luca Bondioli, and Roberto Macchiarelli, e-mails, February 2000. White papers.

240 ***used the stage:*** Gibbons, *The First Human*, 180–81.

240 ***"On my word of honor":*** Quoted in Gibbons, *The First Human*, 183.

240 ***"transparent and legal way":*** Quoted in Dalton, "Restrictions Delay Fossil Hunts," 726.

240 ***photographs showing the rival team:*** Haile-Selassie, "Photos May Offer Clues."

240 ***"for two hundred years":*** Jon Kalb, interview with author.

240 ***"There was a big disagreement":*** Yonas Beyene, interview with author.

240 ***"most outrageous claim jumping":*** Berhane Asfaw, "Why the 'Close Observer' Cannot Face the Facts About Paleoanthropology Research in Ethiopia," Walta Information Center, April 28, 2001.

241 ***"He paraded for years":*** Don Johanson, interview with author.

242 ***"Who ever is in charge":*** Zeresenay Alemseged, "Reply to Dr. Berhane Asfaw," Walta Information Center, April 18, 2001.

242 ***"Why is it":*** Berhane Asfaw, "Why the 'Close Observer' Cannot Face the Facts," Walta Information Center, April 28, 2001.

242 ***"No question":*** John Fleagle, interview with author.

242 ***"cities like Vienna":*** Berhane Asfaw, "Why Was the Close Observer's Team of Successful Scientists Suspended," April 12, 2001.

242 ***"history of spreading lies":*** *Addis Tribune*, "AL-444 Safe in Museum."

243 ***a conference in Addis Ababa:*** See Scholl, *Knochenkrieg*. The argument was captured by the video documentary.

243 ***a triumphant return:*** Recounted in Luyken, "Der Krieg der Knochenjäger."

243 ***"well-advised not to reproduce":*** Luyken, "Der Krieg der Knochenjäger."

244 ***sitting with Issa leaders:*** Tillman, *Knochenkrieg*.

244 ***the owner of a local bar:*** Luyken, "Der Krieg der Knochenjäger."

244 ***Was it all a big trick?:*** Quoted in Luyken, "Der Krieg der Knochenjäger."

244 ***"Twenty-five Somalis":*** Jon Kalb, interview with author.

244 ***"I lost everything in my tent":*** Dean Falk, interview with author.

244 ***"They're completely incompetent":*** Tim White, interview with author. In the end, the much-contested territory produced few hominids for the University of Vienna team. By 2008, Seidler's team reported finding six hominid fossils: four molar teeth, one clavicle, and a shattered thighbone. Eventually, the Vienna team abandoned the Ethiopian project and the territory passed to Scott Simpson.

245 ***issued new regulations:*** Ministry of Information and Culture, "Amended Directives."

245 ***"We want more scientific competition":*** Traufetter, "Kabale um die Knochen."

245 ***available to other researchers:*** "Amended Directives," 14. The regulations later were amended so the five-year countdown began only after the completion of fieldwork. In practice, teams could keep their fieldwork "open" for many years and effectively extend the deadline.

245 ***"These corrupt individuals":*** From "An Archaeologist at Work," oral history, Tim White, J. Desmond Clark, and Timothy Troy, November 2, 2000, 361.

245 ***"They want to charge us":*** Exchange from "An Archaeologist at Work," oral history, Tim White, J. Desmond Clark, and Timothy Troy, November 2, 2000, 369–70.

246 ***"If they lose this battle":*** From "An Archaeologist at Work," oral history, Tim White, J. Desmond Clark, and Timothy Troy, November 2, 2000, 371–72.

246 ***"It's a Byzantine process":*** David Shinn, interview with author.

246 ***"revolutionary democracy":*** EPRDF/TPLF, "Our Revolutionary Democratic Goals."

246 ***"pressurized by legal instruments":*** EPRDF/TPLF, "Our Revolutionary Democratic Goals."

246 ***"its belly and its pocket":*** Ethiopian Register, "TPLF/EPRDF Strategies."

246 ***party leaders complained:*** Milkias, "Ethiopia, the TPLF, and the Roots of the 2001 Political Tremor"; Henze, "A Political Success Story," note 16.

247 ***John Yellen:*** Yellen joined Clark's team and conducted an NSF-funded archeology mission at Aduma, a Middle Stone Age site south of Aramis in the Middle Awash.

247 ***the guards told Giday:*** Giday WoldeGabriel, interview with author.

247 ***a familiar figure:*** Author interviews with Giday WoldeGabriel, Tim White, and John Yellen.

247 ***his home province:*** Details from an inside account by a former TPLF member in Berhe, "A Political History of the Tigray People's Liberation Front."

248 ***"see around corners":*** Gettleman, "A Generation Is Protesting."

248 ***"one of the smartest guys":*** David Shinn, interview with author.

248 ***"Whenever we raise concerns":*** Embassy Addis Ababa, "Inferring Prime Minister Meles's Myers-Briggs Type," WikiLeaks cable: 09ADDISABABA729_a, dated March 27, 2009, https://wikileaks.org/plusd/cables/09ADDISABABA729_a.html.

248 ***"You could never corner him":*** Giday WoldeGabriel, interview with author.

248 ***three meals a day:*** Mohammed and de Waal, "Meles Zenawi and Ethiopia's Grand Experiment."

249 ***"To get a permit":*** Giday WoldeGabriel, interview with author.

249 ***"the bodyguards got really prickly":*** Don Johanson, interview with author.

249 ***"simply fiction":*** Tim White, e-mail to author, April 4, 2020.

249 ***"Not one nickel!":*** Quoted in Tim White, interview with author.

249 ***"corruption and bribery":*** John Yellen, interview with author.

249 ***"I'm sure that got back to Addis":*** Don Johanson, interview with author.

249 ***Jara demanded an inventory:*** Quoted in Tim White, e-mails to author, November 2015.

249 ***"you are not willing":*** Quoted in Tim White, e-mails to author, November 2015.

250 ***"It is unbelievable":*** Berhane Asfaw, "Why the 'Close Observer' Cannot Face the Facts about Paleoanthropology Research in Ethiopia," Walta Information Center, April 28, 2001.

250 ***aired his own grievances:*** Jara comments quoted in "Responses to ARCCH accusations," attachment to Tim White, e-mail to Paul Henze, April 23, 2001, Henze papers, box 67, folder 23.

250 ***wrote to Woldemichael Chemu:*** Tim White, letter to Woldemichael Chemu, May 6, 2001, Henze papers, box 67, folder 23.

250 ***The minister urged them to apologize:*** Woldemichael Chemu, letter to Tim White, April 25, 2001, Henze papers, box 67, folder 23.

251 ***it was* revoked:** Jara Haile Mariam, letter to Tim White, April 2001, Henze papers, box 67, folder 23.

251 ***"keep falling below":*** Tim White, e-mail to Paul Henze, April 23, 2001, Henze papers, box 67, folder 23.

251 ***a power struggle was raging:*** See Miklas, "Ethiopia, the TPLF, and the Roots of the 2001 Political Tremor."

252 ***Henze became fascinated:*** See Henze, *Ethiopian Journeys* and *Layers of Time.*

252 ***"I can't even remember":*** Libet Henze, interview with author.

252 ***"always very welcome":*** Meles Zenawi, letter to Paul Henze, December 25, 1991, Henze papers, box 66, folder 4.

252 ***clean up petty corruption:*** See Henze, "Reflections on Development in Ethiopia," 190. In 1995, Meles confessed to Henze, "We underestimated the inertia of the bureaucracy and the deep-rooted tendencies toward protectionism and corruption that existed in it. We should have moved more decisively to shake it up" (see Henze, "A Political Success Story," note 16).

252 ***"Appalling situation!":*** Paul Henze, e-mail to Tim White, May 9, 2001, Henze papers, box 67, folder 23.

252 ***"confrontation needs to be elevated":*** Paul Henze, e-mail to Tim White, May 9, 2001, Henze papers, box 67, folder 23.

253 ***Henze had enough rapport:*** Henze, "Conversation with Ethiopian Ambassador Berhane Gebre Christos June 2001," Henze papers, box 81, folder 8.

253 ***"gave every impression":*** Paul Henze, to Tim White, e-mail May 9, 2001, Henze papers, box 67, folder 23.

253 ***Ethiopia sought to bolster support:*** Paul Henze, e-mail to Tim White, May 9, 2001, Henze papers, box 67, folder 23.

253 ***"Foreign Minister Seyoum":*** Berhane Asfaw, interview with author.

254 ***White urged reporters:*** Tim White, e-mail to Paul Henze, July 15, 2001, Henze papers, box 67, folder 23.

254 ***"The position of the Culture Ministry":*** Paul Henze, e-mail to Brook Hailu, July 15, 2001, Henze papers, box 67, folder 23.

254 ***appear "petty and obstructive":*** Paul Henze, e-mail to Tim White, July 15, 2001, Henze papers, box 67, folder 23.

254 ***White visited Woldemichael Chemu:*** Tim White, e-mail to Paul Henze, July 15, 2001, Henze papers, box 67, folder 23.

254 ***its own bizarre press release:*** Quoted in "*Ardipithecus Ramidus Kadabba*," Walta Information Center, July 20, 2001.

255 ***"He said, 'Don't worry'":*** Giday WoldeGabriel, interview with author.

255 ***Chemu wrote again:*** Woldemichael Chemu, letter to Tim White, July 26, 2001, Henze papers, box 67, folder 23.

255 ***"They made his life hell":*** Don Johanson, interview with author.

255 ***"Many people don't like us":*** Giday WoldeGabriel, interview with author.

CHAPTER 19: BITING THE HAND

256 ***a cover story in*** Nature*:* Richmond and Strait, "Evidence That Humans Evolved from a Knuckle-Walking Ancestor."

256 ***"publishing that thing":*** Tim White, e-mail to Henry Gee, March 23, 2000.

256 ***"flawed evidence and interpretation":*** Tim White, e-mail to Rosalind Cotter, October 5, 2000.

257 ***the secret*** Ardipithecus*:* Richmond and Strait, "Reply to White and DeGusta," 2000.

257 ***"Something is wrong":*** White, "A View on the Science."

258 ***In the field:*** White, "A View on the Science," 291.

258 ***Do not let your ambition:*** White, "A View on the Science," 291.

CHAPTER 20: IN SUSPENSE

260 ***English sculptor Joseph Bonomi:*** Schott, "The Extent of Man," 1519. The measure of arm span includes the width of the chest in addition to the length of both arms.

260 ***classified as "suspensory" primates:*** Fleagle, "Intermembral Index," 661–63.

262 ***as long as her hindlimbs:*** Lovejoy et al., "Great Divides," supplementary information table S3.

262 ***a young Scottish doctor:*** See Keith, *An Autobiography* and "Fifty Years Ago."

264 ***a series of lectures:*** Keith, "Man's Posture: Its Evolution and Disorders."

264 ***longer than their hindlimbs:*** Fleagle, "Intermembral Index," 663.

264 ***long-armed "brachiating" ancestor:*** Keith theorized that the human lineage passed through three stages embodied by modern species: A "Hylobatian" stage like an arm-swinging gibbon, a "Troglodytian" stage like a knuckle-walking chimpanzee or gorilla, and a "Plantigrade" stage like a human biped.

265 ***"our gibbonlike ancestors":*** Coolidge, "Studying the Gibbon for Light on Man's Evolution," 15.

265 ***Professor Adolph Schultz:*** For an overview of his career, see Stewart, "Adolph Hans Schultz."

266 ***"Every day was a seminar":*** Washburn, "Evolution of a Teacher," 4.

266 ***fossil apes from the Miocene:*** White et al., "Neither Chimpanzee nor Human," 4879.

267 ***an NSF review panel:*** NSF award 9318698, Panel Summary.

267 ***"paleoanthropological field work":*** NSF award 9632389, Advisory Panel Summary and peer reviews.

267 ***"until they become accessible":*** Mark Weiss, e-mail to Tim White, February 16, 2000. White papers.

267 ***"uninformed, clearly hostile":*** Tim White, notes on NSF award 9910344 and notes on communications with Weiss, White papers.

267 ***"overwhelmingly enthusiastic":*** Clark Howell, note to Tim White, undated (c. 2000). White papers.

267 ***"the entire research program":*** NSF award 9910344 case file, "Summary of Proposed Research."

267 ***"before the research is ready":*** Tim White notes, NSF award 9910344.

267 ***"He would call up on occasion":*** Mark Weiss, interview with author.

268 ***publicly criticized the agency:*** White, "A View on the Science," 290.

268 ***"A few people go out":*** From "An Archaeologist at Work," oral history, Tim White, J. Desmond Clark, and Timothy Troy, November 2, 2000, 368.

268 ***One morning in February 2002:*** Berhane Asfaw, interviews and e-mail to author, March 28, 2016; author interviews with Alemu Ademassu, Jeffrey Schwartz, and Ian Tattersall.

268 ***"I told him no way":*** Berhane Asfaw, e-mail to author, March 28 2016.

269 ***Schwartz wrote White:*** Jeffrey Schwartz and Tim White, e-mails, March 2–11, 2000.

269 ***"what chutzpah!":*** Jeffrey Schwartz, interview with author.

269 ***director Jara Haile Mariam:*** Don Johanson, e-mail to Jeffrey Schwartz, October 11, 2001.

269 ***"your Laboratory study":*** Jara Haile Mariam, e-mail to Jeffrey Schwartz, January 9, 2002.

269 ***"behind closed doors":*** Jeffrey Schwartz, interview with author.

270 ***"twiddling our thumbs":*** Jeffrey Schwartz, interview with author.

270 ***back at the Hilton:*** Luyken, "Der Krieg der Knochenjäger."

270 ***the lead example:*** Gibbons, "Glasnost for Hominids."

270 ***Ph.D.s per year:*** Figures from Aiello, "The Wenner-Gren Foundation."

270 ***the period of exclusive possession:*** See "Conroy's Manifesto" in "Paleoanthropology Today," 156.

271 ***new era of fossil glasnost:*** Weber, "Virtual Anthropology."

271 ***"eyes fixed to the ground":*** Weber, "Virtual Anthropology," 200.

271 ***More like Napster:*** Tim White, e-mail to Ann Gibbons, July 28, 2002.

271 ***"conducted primary work":*** Tim White, e-mail to Ann Gibbons, July 28, 2002.

272 ***"most of the 'collegial' sources":*** Tim White, e-mail to Ann Gibbons, July 28, 2002.

272 ***"curatorial protectionism":*** Schwartz and Tattersall, *The Human Fossil Record,* Vol. 2, x.

272 ***"trying to hide":*** Schwartz and Tattersall, *The Human Fossil Record,* Vol. 2, x.

272 ***"spirit of inclusivity":*** NSF proposal 0218392, Panel Summary and peer reviews.

272 ***"do not require a grant":*** NSF proposal 0218392, peer reviews.

273 ***"Important hominid fossils":*** NSF proposal 0321893, peer reviews.

273 ***"almost paranoid distrust":*** F. Clark Howell and T. D. White, e-mail to Mark Weiss, May 31, 2003.

273 ***the international press corps:*** Zane, "Ethiopia's Pride in Herto Finds."

274 ***"As a scientist":*** Quoted in Zane, "Ethiopia's Pride in Herto Finds."

274 ***minister of culture, Teshome Toga:*** IRIN News, "Interview with Teshome Toga."

274 ***$10 million structure:*** Pennisi, "Rocking the Cradle of Humanity."

274 ***"on behalf of the country":*** Giday WoldeGabriel, interview with author.

274 ***"hit-and-run projects":*** White, "Ethiopian Paleotourism."

275 ***the emergence of Ethiopian scholars:*** White, "Human Evolution: How Has Darwin Done?" 537.

275 ***"leader in world paleoanthropology":*** White, "Ethiopian Paleotourism."

CHAPTER 21: UNDER THE RADAR

276 ***two men in business suits:*** Details of skeleton export from Yonas Beyene, interview with author; and Paul, *Discovering Ardi*.

276 ***"We were scared":*** Yonas Beyene, interview with author.

277 ***"There was not a single bone":*** Yonas Beyene, interview with author.

277 ***"Idea-renovating discoveries":*** Gen Suwa, interview with author.

277 ***"He has a photographic memory":*** Berhane Asfaw, interview with author.

277 ***"extraordinary bullshit sensor":*** Scott Simpson, interview with author.

278 ***"I don't do anything but work":*** Gen Suwa, interview with author.

278 **the only person I trust:** Tim White, interview with author.

278 ***twenty-one* Ar. ramidus *individuals:*** Teeth details in this section from Suwa et al., "*Ardipithecus Ramidus* Dentition."

281 ***learned to find meals:*** Ungar and Sponheimer, "The Diets of Early Hominins."

281 ***don't share with us is bad teeth:*** Ungar et al., "Evolution of Human Teeth and Jaws: Implications for Dentistry and Orthodontics."

CHAPTER 22: TROUBLE AFOOT

282 ***Lovejoy puzzled over the cast:*** Details from author interviews with Owen Lovejoy.

284 ***even the tiny* ossi petrosi*:*** Clayton and Philo, *Leonardo Da Vinci: The Mechanics of Man*, 34.

284 ***"most distinctly human part":*** Jones, *Structure and Function as Seen in the Foot*, 2.

285 ***the sesamoid on his mind:*** Description from Owen Lovejoy, interview with author.

286 ***bodies from the gallows:*** See Persaud, *History of Anatomy.*

286 ***T. Wingate Todd left an impression:*** Biographical details from Kern, "T. Wingate Todd: Pioneer of Modern American Physical Anthropology."

287 ***"engrossing dissection":*** Kern, "T. Wingate Todd," 9.

287 ***"my greatest thrill":*** Todd, "The Physician as Anthropologist," 588.

288 ***"No collection like this":*** Cobb, "Thomas Wingate Todd," 235.

288 ***a berserk elephant:*** Kern, "T. Wingate Todd," 19.

288 ***the discoverer of the tomb:*** Kern, "T. Wingate Todd," 16.

290 **the village's designated climber:** Susman et al., "Arboreality and Bipedality in the Hadar Hominids," 122.

290 ***a fox hit by another car:*** Bruce Latimer, interview with author.

291 ***"Well, look who's here":*** Bruce Latimer and Bill Kimbel, interviews with author.

292 ***"devoted himself to celibacy":*** Quoted in Montagu, *Edward Tyson*, 420.

292 ***"as if nothing a-kin to them":*** Tyson, *Orang-outang*, 7.

292 ***"more resembling a Man":*** Tyson, *Orang-Outang*, 5.

292 ***"liker a Hand than a Foot":*** Tyson, *Orang-Outang*, 13.

292 **quadrumanus*, or "four-handed":*** Tyson, *Orang-Outang*, 91; Montagu, *Edward Tyson*, 264.

294 ***"It is not pleasing to me":*** Carl Linnaeus, letter to Johann Georg Gmelin, February 25, 1747, quoted in Reid, "Carolus Linnaeus," 27.

294 ***"If man had not been":*** Darwin, *Descent of Man*, 191.

294 ***"nearest allies of men":*** Darwin, *Descent of Man*, 139.

295 ***the "prehuman foot":*** Morton, "Evolution of the Human Foot," 336.

295 ***"There's a procedure":*** Owen Lovejoy, interview with author.

295 ***"We were trying to force":*** Bruce Latimer, interview with author.

296 ***"It was just the way":*** Owen Lovejoy, interview with author.

296 ***a detailed study:*** Manners-Smith, "A Study of the Cuboid and Os Peroneum in the Primate Foot."

296 ***93 percent of humans:*** Edwards, "The Relations of the Peroneal Tendons," 237.

297 ***When an ape perished:*** Bruce Latimer, interview with author.

297 ***the walk-in freezer:*** Owen Lovejoy, interview with author.

298 ***"the custodial people":*** Owen Lovejoy, interview with author.

298 ***Latimer cut down:*** Details of dissection and discussion from Bruce Latimer and Owen Lovejoy, interviews with author.

299 ***"Get a baboon!":*** Owen Lovejoy and Bruce Latimer, interviews with author.

299 ***Latimer had encountered:*** Bruce Latimer, interviews with author.

300 ***"All of a sudden":*** Bruce Latimer, interview with author.

301 ***"You grow up":*** Bruce Latimer, interview with author.

302 ***the arched foot appeared:*** Ward et al., "Complete Fourth Metatarsal."

303 ***"Nothing in biology":*** Dobzhansky, "Nothing in Biology Makes Sense."

CHAPTER 23: TÊTE-À-TÊTE

304 ***conference on human origins:*** Details from Gibbons, "Oldest Femur Wades into Controversy."

305 ***"This is the most fragile":*** Quoted in Gibbons, "Oldest Femur Wades into Controversy," 1885.

306 ***Suwa devoted himself:*** Gen Suwa, interview with author. Details of skull reconstruction from Suwa et al., "*Ardipithecus Ramidus* Skull" and supplementary information.

307 ***kept drowning in* other *work:*** Gen Suwa, interview with author.

307 ***"You can't work like that":*** Gen Suwa, interview with author.

309 ***"tantalizing evidence":*** Suwa et al., "*Ardipithecus Ramidus* Skull," 68e6.

310 ***the Herto man's skull:*** White et al., "Pleistocene *Homo Sapiens* from Middle Awash," 742.

310 ***an* afarensis *skull:*** Kimbel et al., "Cranial Morphology of *Australopithecus afarensis*"; Johanson and Edey, *Lucy*, 349–51.

310 ***"an incredibly protruding face":*** From Smeltzer, *Lucy in Disguise.*

311 ***smashed on the floor:*** Johanson and Edey, *Lucy*, 351.

312 ***"chimpanzee-like morphology":*** White et al., "*Australopithecus Ramidus*, a New Species," 310.

312 ***the "ancestral state":*** White et al., "*Ardipithecus Ramidus* and the Paleobiology of Early Hominids," 78.

312 ***"getting more chimplike":*** Tim White, interview with author.

313 ***a "wild-cat element":*** Quoted in Desmond, *Huxley*, 6.

313 ***"Cutting up monkeys":*** Quoted in Desmond, *Huxley*, xiii.

313 ***Owen's argument was bunk:*** Gross, "Huxley Versus Owen."

313 ***"degree and not of kind":*** Darwin, *Descent of Man*, 105.

314 ***"Human brains are thus":*** Preuss, "The Human Brain," 142.

314 ***"radically anthropomorphized":*** Povinelli, "Behind the Ape's Appearance," 30.

314 ***"After several decades":*** Povinelli, "Behind the Ape's Appearance," 32.

CHAPTER 24: ALL THAT REMAINS

315 ***"a foot and a half!":*** Kovacs, "Making History."

316 ***"Not being practiced":*** C. Owen Lovejoy, letter and report to coroner Elizabeth K. Balraj, MD, December 21, 1999.

316 ***no part suffered:*** Details of pelvis from Lovejoy et al., "The Pelvis and Femur of *Ardipithecus Ramidus*" and supplementary online information; MARP video 1995 n. 2.

317 ***but a* sculpture*:*** Tim White, interview with author.

317 ***Lovejoy rebuilt Lucy's pelvis:*** See Lovejoy, "Evolution of Human Walking."

317 ***"perfected" by the time of Lucy:*** Lovejoy, "Evolution of Human Walking," 118.

317 ***a*** **better** ***biped:*** Lovejoy, "Evolution of Human Walking," 122–23.

318 ***not*** **improvements** ***for locomotion:*** Lovejoy, "The Natural History of Human Gait and Posture, Part 1: Spine and Pelvis."

318 ***She dressed pig carcasses:*** Spurlock, "Burying the Hatchet, the Body, and the Evidence."

318 ***"We'd get it to where":*** Owen Lovejoy, interview with author.

318 ***"He's the most exacting person":*** Owen Lovejoy, interview with author.

319 ***noted by Edward Tyson:*** Tyson, *Orang-Outang,* 74.

319 ***anterior inferior iliac spine:*** MARP video 1995 n. 2; Tim White, e-mail to Owen Lovejoy, February 5, 1995.

320 ***"There was a divide":*** Owen Lovejoy, interview with author.

321 ***"I sat down to write":*** Owen Lovejoy, interview with author.

321 ***the "modern synthesis":*** For an overview, see Carroll et al., *From DNA to Diversity.*

322 ***a "black box":*** Carroll, *Endless Forms,* 7.

323 ***to make*** **any** ***primate:*** Carroll et al., *From DNA to Diversity,* 213

323 ***"Eventually the scales":*** Owen Lovejoy, interview with author.

324 ***anthropology graduate student:*** Marty Cohn, interview with author.

324 ***"the same for both":*** Recounted by Marty Cohn, interview with author.

324 ***"Tabula rasa!":*** Recounted by Marty Cohn, interview with author.

324 ***"For god's sake, stay":*** Recounted by Owen Lovejoy, interview with author.

324 ***"When I saw that":*** Marty Cohn, interview with author.

325 ***"adaptationist" mind-set:*** Gould and Lewontin, "The Spandrels of San Marco."

325 ***"I have to be honest":*** Jack Stern, interview with author.

325 ***"Owen's armchair evo-devo":*** Bill Jungers, interview with author.

326 ***Lovejoy, Cohn, and White published:*** Lovejoy et al., "Morphological Analysis of the Mammalian Postcranium."

326 ***"We tried to bring":*** Marty Cohn, interview with author.

CHAPTER 25: SPINE QUA NON

327 ***low back problems afflict:*** Andersson, "Epidemiological Features of Chronic Low-Back Pain," 581.

327 ***even fossil skeletons are rife:*** Haeusler, "Spinal Pathologies in Fossil Hominins."

328 ***Chimps and gorillas rarely suffer:*** Haeusler, "Spinal Pathologies in Fossil Hominins," 214.

330 ***short, rigid lumbar section:*** Pilbeam, "The Anthropoid Postcranial Axial Skeleton."

330 ***The field team had recovered:*** White et al., "*Ardipithecus Ramidus* and the Paleobiology of Early Hominids," table S2. One of the recovered vertebrae was cervical and the other thoracic. Two other fragments were tentatively listed as vertebrae but were too ambiguous for further identification.

331 ***the true primitive condition:*** Lovejoy et al., "Great Divides"; McCollum et al., "The Vertebral Formula of the Last Common Ancestor"; and Lovejoy and McCollum, "Spinopelvic Pathways to Bipedality."

332 ***a "human-like gait":*** Hirasaki et al., "Do Highly Trained Monkeys Walk Like Humans?" 748.

332 ***a mobile lordotic spine:*** Lovejoy, "Natural History of Human Gait and Posture Part 1," 102.

333 ***to become monogamous:*** Lovejoy, "Origin of Man" and "Reexamining Human Origins."

334 ***"sex for food exchanges":*** Lovejoy, "Reexamining Human Origins," 74e5.

335 ***"Orangutans are going extinct":*** Owen Lovejoy, interview with author.

335 ***Jane Goodall observed:*** Goodall, *Chimpanzees of Gombe,* 446.

335 ***a billion seed per ejaculation:*** Gagneux, "Sperm Count."

335 ***the genus*** **Pan*****:*** For an overview, see Meder, "Great Ape Social Systems."

336 ***"The chimpanzee resolves":*** De Waal, *Bonobo,* 32.

336 ***two hundred times more sperm:*** Fujii-Hanamoto et al., "Comparative Study on Testicular Microstructure," 570.

336 ***85 percent of the infants:*** Bradley et al., "Mountain Gorilla Tug-of-War," 9421.

336 ***Human sperm production:*** Gagneux, "Sperm Count."

336 ***proposed that nuclear families:*** Lovejoy, "Origin of Man."

337 ***moderately dimorphic:*** Reno et al., "Sexual Dimorphism."

337 ***"Owen and a couple of his acolytes":*** Bill Jungers, interview with author.

CHAPTER 26: NEITHER CHIMP NOR HUMAN

339 ***"cracking the code":*** Pollard, "Decoding Human Accelerated Regions."

341 ***The macaque monkey:*** Gibbs et al., "Rhesus Macaque Genome," 223.

341 ***Even the mouse genome:*** National Human Genome Research Institute, "Why Mouse Matters."

341 ***"the 1 percent difference":*** Quoted in Cohen, "Myth of 1%."

341 ***those small differences:*** Some visionaries had predicted as much. In 1975, Mary-Claire King and Allan Wilson of Berkeley suggested the main differences between humans and chimps arose from gene regulation. They posited that small genetic differences could produce major differences in anatomy because evolution moved "at two levels."

341 ***chimp branch lengths:*** Kronenberg et al., "High-Resolution Comparative Analysis of Great Ape Genomes," 2.

341 ***humans showed 3.5 percent:*** Kronenberg et al., "High-Resolution Comparative Analysis of Great Ape Genomes," 2.

341 ***"the human genome is":*** Evan Eichler, interview with author.

342 ***more than 1 million years:*** Patterson et al., "Genetic Evidence for Complex Speciation."

342 ***4 million years between:*** Moorjani et al., "Variation in the Molecular Clock of Primates," 10611.

342 ***the gorilla lineage:*** Prado Martinez et al., "Great Ape Genetic Diversity," 474; Langergraber et al., "Generation Times in Wild Chimpanzees and Gorillas," 15719; Moorjani et al., "Variation in the Molecular Clock of Primates," 10611–12.

342 ***more than one-third:*** Kronenberg et al., "High-Resolution Comparative Analysis of Great Ape Genomes," 2, 4.

343 ***professor Il'ya Ivanovich Ivanov:*** Etkind, "Beyond Eugenics," 207.

343 ***first fossils of a chimp ancestor:*** McBrearty and Jablonski, "First Fossil Chimpanzee."

343 ***incisors were virtually identical:*** McBrearty and Jablonski, "First Fossil Chimpanzee," 106.

344 ***named* Chororapithecus abyssinicus:** Suwa et al., "New Species of Great Ape."

344 ***"collectively indistinguishable":*** Suwa et al., "New Species of Great Ape," 921.

344 ***the human-ape divergence:*** Suwa et al., "New Species of Great Ape," 924.

344 ***pushed split times deeper:*** Langergraber et al., "Generation Times in Wild Chimpanzees and Gorillas Suggest Earlier Divergence Times."

344 ***the human-chimp breakup:*** Moorjani et al., "Variation in the Molecular Clock of Primates."

345 ***"reading the entrails of chickens.":*** Grauer and Martin, "Reading the Entrails of Chickens."

346 ***the Ardi team trooped into:*** Scene details from author interviews with Steve Ward, Owen Lovejoy, and Tim White.

346 ***"To Owen and Tim":*** Steve Ward, interview with author.

347 ***common Miocene ape body plan:*** Ward, "*Equatorius:* A New Hominoid Genus."

347 **I'll be damned. Identical!:** Owen Lovejoy, interview with author.

347 ***"nuclear secrets!":*** Steve Ward, interview with author.

348 ***professor Wilfrid E. Le Gros Clark:*** Walker and Shipman, *The Ape in the Tree*, 37–52. Walker and Shipman called *Proconsul* the best known candidate for common ancestors of modern apes and humans—and outlined arguments similar to those later advanced by the Ardi team.

348 ***such a "generalized" quadruped:*** Le Gros Clark, "New Palaeontological Evidence."

349 ***"could hardly have been derived":*** Le Gros Clark, "New Palaeontological Evidence," 227.

349 ***African apes were very specialized:*** Schultz, "The Specializations of Man," 50–51.

349 ***a stage like modern apes:*** Straus, "The Riddle of Man's Ancestry."

349 *"my most prized possessions":* Owen Lovejoy, e-mail to author, January 13, 2012.

349 *"throw little if any":* Pilbeam, "Genetic and Morphological Records," 155.

350 *"The different lines":* Gen Suwa, interview with author.

350 *"the find of the century":* Clark, "Recent Developments," 168.

350 *"They don't like getting":* Tim White, interview with author.

351 *"Regardless of whether":* Brooks Hanson, interview with author.

351 *"Anybody who works":* Tim White, interview with author.

352 *"We revised heavily":* Owen Lovejoy, interview with author.

352 *"The consequence of that":* Bill Kimbel, interview with author.

352 Scientific American ***ran an editorial:*** *Scientific American*, "Fossils for All."

352 *White dashed off:* White, "Fossils for All" (letter to the editor).

CHAPTER 27: SKELETON FROM THE CLOSET

353 *"What we celebrate":* Remarks from video of press conference.

354 *"adaptative cul-de-sacs":* Lovejoy et al., "Great Divides," 104.

354 ***modern apes was*** **nullified***:* Lovejoy, "Reexamining Human Origins," 74.

354 *was "now moot":* Lovejoy et al., "Careful Climbing," 70e5.

354 *"requires a rejection":* Lovejoy, "Reexamining Human Origins," 74.

354 *a "trend accompanied":* Lovejoy et al., "Great Divides," 104–6.

354 *succession of adaptive plateaus:* White et al., "*Ardipithecus Ramidus* and the Paleobiology of Early Hominids," 64.

355 *"light will be thrown":* Darwin, *Origin of Species*, 488.

355 *"persisting since Darwin":* White et al., "Great Divides," 80.

355 *touted "fossil evidence":* Paul, *Discovering Ardi*.

355 *"This find is far more":* Quoted in Shreeve, "Oldest Skeleton."

355 *"It's the closest thing":* John Hawks weblog, October 1, 2009, "Ardipithecus FAQ."

356 *"Science Breakthrough of the Year":* Alberts, "Breakthroughs of 2009."

356 *The Gettys hosted:* *Haute Living*, "Ardipithecus Research Party."

356 ***White was named:*** *Time*, "The 100 Most Influential People in the World."

356 ***"Ardi was the wife":*** Leslea Hlusko, interview with author.

356 ***"humanity's hometown":*** Table of contents, *National Geographic*, July 2010; Shreeve, "Evolutionary Road."

357 ***"When Tim publishes":*** Bernard Wood, interview with author.

357 ***"Just because the tablets":*** Bernard Wood, interview with author.

357 ***an analysis that hailed:*** Wood, "Oldest Hominid Yet," 281.

357 ***"I am just not as certain":*** Bernard Wood, interview with author.

358 ***"I have to take a slight detour":*** All conference comments from audio recording posted online by Royal Society unless otherwise noted.

359 ***"chimpocentrism":*** Sayers and Lovejoy, "The Chimpanzee Has No Clothes."

359 ***"essentially falsifies" models:*** Lovejoy, "Reexamining Human Origins," 74e1.

359 ***McGrew had lamented:*** McGrew, "New Theaters of Conflict in the Animal Culture Wars," 175.

359 ***"humans alone" evolved cognition:*** Lovejoy et al., "Careful Climbing," 70.

359 ***Anything a chimp can do today:*** See McGrew, "In Search of the Last Common Ancestor."

359 ***"He won't even talk to me now":*** Owen Lovejoy, interview with author.

360 ***"oversimplistic interpretations":*** Whiten et al., "Studying Extant Species to Model Our Past."

361 ***angered by any account:*** Cohen, *Almost Chimpanzee*, 313.

361 ***a book defiantly titled:*** Taylor, *Not a Chimp*.

361 ***"a strenuous and devoted attempt":*** Kuhn, *The Structure of Scientific Revolutions*, 5.

362 ***"probably more so":*** Kuhn, *The Structure of Scientific Revolutions*, 165.

362 ***"motivated reasoning":*** Kunda, "The Case for Motivated Reasoning."

363 ***acknowledgments thanked:*** Brunet, "New Hominid from the Upper Miocene of Chad," 151.

363 ***"mystery femur":*** Chris Stringer, interview with author.

CHAPTER 28: BACKLASH

365 ***"a hominid, or bipedal":*** Quoted in Wilford, "Scientists Challenge 'Breakthrough.'"

365 ***"so far wrong as to be laughable":*** Jack Stern, interview with author.

365 ***"very chimpanzee-like":*** Stanford, "Chimpanzees and the Behavior of *Ardipithecus ramidus*," 142.

365 ***arguments as "silly":*** Stanford, *The New Chimpanzee*, 195.

365 ***"To try to rule chimpanzees":*** Stanford, *The New Chimpanzee*, 195.

366 ***"thirty years behind the times":*** Pickford, "Marketing Palaeoanthropology," 241.

366 ***"IT'S A DATA TABLE":*** John Hawks weblog, October 3, 2009, "Whoa, Who Stole the Data?"

366 **Time** ***magazine ran a story:*** Harrell, "Ardi: The Human Ancestor Who Wasn't?"

366 **Scientific American** ***followed:*** Harmon, "Was 'Ardi' Not a Human Ancestor After All?"

366 ***"Ardipithecus Backlash Begins":*** John Hawks weblog, May 27, 2010.

366 ***belonged in the human family:*** Wood and Harrison, "Evolutionary Context of the First Hominins."

366 **"Ardipithecus** ***is ancestral":*** From "Pikaia Interviews Bernard Wood," YouTube, January 24, 2011.

366 ***"Tim just won't go there":*** Bernard Wood, interview with author.

367 ***"went into contortions":*** Bernard Wood, interview with author.

367 ***"the evidence into that box":*** Bernard Wood, interview with author.

367 ***"relatively few" features:*** Wood and Harrrison, "Evolutionary Context of the First Hominins," 348.

367 ***"on Saturday evening":*** Bernard Wood, interview with author.

367 ***"mistaken it for a toilet seat":*** Wood, "Darwin in the 21st Century."

367 ***"six-fucking-page*** **Nature** ***paper!":*** Tim White, interview with author.

367 ***conspicuously turned away:*** Berhane Asfaw, interview with author.

367 ***"The funny thing was":*** Tim White, interview with author.

368 ***"breathing down my neck":*** Bernard Wood, interview with author.

368 ***"With no new data":*** Quoted in Bower, "Human Ancestors Have Identity Crisis."

368 ***"'going to be battered'":*** Don Johanson, interview with author.

368 ***"the government undertakings":*** Don Johanson, interview with author.

369 ***Miocene ape from Pakistan:*** Morgan et al., "A Partial Hominoid Innominate from the Miocene of Pakistan."

369 **afarensis *spine from Dikika:*** Ward et al., "Thoracic Vertebral Count."

369 ***from the same skeleton:*** DeSilva et al., "A Nearly Complete Foot from Dikika."

369 ***the fossil "Little Foot":*** Clarke, "Excavation, Reconstruction and Taphonomy of the StW 573 *Australopithecus Prometheus* Skeleton."

369 ***the Miocene ape* Oreopithecus:** Tim White and Lorenzo Rook, e-mails, September 2013.

369 ***"new unpublished specimen":*** Lorenzo Rook to Tim White, e-mails, September 2013.

369 ***"exceeded bandwidth":*** Tim White, e-mail to author September 3, 2019.

370 ***"If they had adopted":*** Bill Kimbel, interview with author.

370 ***"Tim is very clever":*** Ian Tattersall, interview with author.

370 ***"Everything we're about":*** Tim White, interview with author.

371 ***"The true shocker":*** Owen Lovejoy, interview with author.

371 ***"one day in 2009":*** Gen Suwa, interview with author.

371 ***"New scientific truth":*** Planck, *Scientific Autobiography*, 33–34.

371 ***"each faculty member":*** Owen Lovejoy, interview with author.

372 ***a spinal column:*** M. Royhan Gani, e-mails to Tim White, January 19, 2010 and February 18, 2010.

372 ***"A nice match?!?":*** M. Royhan Gani, e-mail to Tim White, February 18, 2010.

372 ***the environment of Aramis:*** Gani and Gani, "River Margin Habitat."

372 ***"failure of the review process":*** See Gani and Gani, and White, "On the Environment of Aramis."

372 ***"The man who knows the most":*** Royhan Gani, interview with author.

373 ***25 percent tree cover:*** Cerling et al., "Comment on the Paleoenvironment."

372 ***"The savanna hypothesis":*** Quoted in Gibbons, "How a Fickle Climate," 476.

373 ***"Thure Cerling's interpretation":*** Raymond Bonnefille, interview with author.

373 ***"We think Ardi lives":*** Scott Simpson, interview with author.

373 ***a "mixed assemblage":*** White et al., "Macrovertebrate Paleontology and the Pliocene Habitat of *Ardipithecus Ramidus*," 92.

373 ***"a professional insult":*** Scott Simpson, interview with author.

373 ***"Tim can't stand criticism":*** Thure Cerling, interview with author.

374 ***"closed, wooded":*** White et al., "Macrovertebrate Paleontology," 67.

374 ***"falsification has arrived":*** White, "Reply to Cerling et al.," 472.

374 ***"Without original research":*** White, "Is the 'Savanna Hypothesis' a Dead Concept?" 76.

374 ***"By denying this evidence":*** Domínguez-Rodrigo, "Is the 'Savanna Hypothesis' a Dead Concept?" 78.

375 ***"I don't buy*** **Ardipithecus*****":*** Richard Leakey, interview with author.

375 ***"I haven't studied it":*** Richard Leakey, interview with author.

375 ***"Sitting on these fossils":*** Richard Leakey, interview with author.

375 ***Meave, was perfectly willing:*** Meave Leakey, interview with author.

375 ***as*** **hominin***:* Fernández et al., "Evolution and Function of the Hominin Forefoot."

375 ***"He impugns motives":*** Richard Leakey, interview with author.

375 ***"Not everything has descendants":*** Don Johanson, interview with author.

376 ***"I don't study apes":*** Richard Leakey and Don Johanson, interviews with author.

376 ***"Ardi is the most brilliant ape":*** Lee Berger, interview with author.

CHAPTER 29: HELL YES

377 ***"the bone rubbers club":*** Bill Jungers, interview with author.

377 ***"That ain't the foot of a biped!":*** Quoted in Shreeve, "The Birth of Bipedalism," 66.

377 ***"3D Rorschach test":*** Bill Jungers, interview with author.

377 ***"the wildest and woolliest part":*** Bill Jungers, interview with author.

378 ***"Hell, yes!":*** Bill Jungers, interview with author.

378 ***"That's crazy talk":*** Bill Jungers, interview with author.

378 ***"I took a little grief":*** Bill Jungers, interview with author.

378 ***"I was indignant":*** Cartmill, "Ardipithecus Discussion BU Dialogues."

379 ***"I came to appreciate":*** Cartmill, "Ardipithecus Discussion BU Dialogues."

379 ***"independent sources of data":*** Bill Kimbel, interview with author.

380 ***the human cranial base:*** Kimbel et al., "*Ardipithecus Ramidus* and the Evolution of the Human Cranial Base."

380 ***"I'm quite sure":*** Peter Andrews, interview with author.

380 ***"the ancestral perch":*** Bill Jungers, interview with author.

380 ***a mostly upright trunk:*** For example, see DeSilva et al., "One Small Step."

380 ***closer to the primitive proportions:*** Almécija et al., "The Evolution of Human and Ape Hand Proportions."

381 ***analysis of the femur:*** Almécija et al., "The Femur of *Orrorin Tugenensis.*"

381 ***Yet another study affirmed:*** Fernández et al., "Evolution and Function of the Hominin Forefoot."

381 ***chimps and bonobos had accumulated:*** Diogo et al., "Bonobo Anatomy Reveals Stasis and Mosaicism in Chimpanzee Evolution."

381 ***"stop, take a breath":*** Bernard Wood, interview with author.

382 ***summary of their findings:*** White et al., "Neither Chimpanzee nor Human."

383 ***"I'll tell you why":*** Bill Kimbel, interview with author.

383 **Ardipithecus *at Boston University:*** "BU Dialogues" and "*Ardipithecus* Discussion."

CHAPTER 30: RETURN TO THE ZOO OF THE UNKNOWN

384 ***In December 2013:*** All details and quotations in this chapter from author reporting in Ethiopia unless otherwise noted.

389 ***"quarreling place":*** NSF award 8210897 case file, appendix II, 64.

391 ***2.8-million-year-old lower jawbone:*** Villmoare et al., "Early Homo at 2.8 Ma."

391 ***among its "Sensational 60":*** NSF, "Sensational 60," 10.

391 ***"The Middle Awash Project":*** NSF proposal 1138062, panel summary and reviews.

392 ***ranked the proposal:*** NSF proposal 1241314, panel summary and reviews.

392 ***a three-page list:*** MARP NSF proposal EAR1332712, "Reviewers Not to Include."

392 ***lacking a technological component:*** NSF proposal EAR1332712, reviews and panel summary.

393 ***2 percent of faculty positions:*** Anton et al., "Race and Diversity in U.S. Biological Anthropology," 163.

393 ***"building infrastructure for Ethiopia":*** NSF proposal 1138062, reviews and panel summary.

394 ***"the most productive":*** Paul Renne, interview with author.

394 ***"just didn't like him":*** Paul Renne, interview with author.

395 ***439 fossil localities:*** Tim White, e-mail to author, January 30, 2020.

CHAPTER 31: NEITHER TREE NOR BUSH

400 ***felt obliged to accept Ardi:*** Pilbeam and Lieberman, "Reconstructing the Last Common Ancestor."

400 ***"People who run these offices":*** Berhane Asfaw, interview with author.

401 ***spine fossils of*** *Australopithecus:* Meyer and Williams: "Earliest Axial Fossils from the Genus *Australopithecus*."

401 ***"free of bullshit":*** Tim White, interview with author.

401 ***"To deal with Ardi":*** Bill Jungers, interview with author.

401 ***whisked off to the National Palace:*** Eilperin, "In Ethiopia, Both Obama and Ancient Fossils Get a Motorcade"; Berhane Asfaw interview with author.

402 ***"a great Tree of Life":*** Darwin, *Origin of Species,* 129–30.

402 ***"web of life":*** Lawton, "Uprooting Darwin's Tree"; Doolittle, "Uprooting the Tree of Life."

403 ***branches that split:*** Mallet et al., "How Reticulated Are Species?" 144.

403 ***"broaden our view of sex":*** Mallet et al., "How Reticulated Are Species?" 146.

403 ***any simple model:*** An alternative approach is to replace trees with forests: large data sets of many gene trees can reveal a "consensus" tree. But these represent only statistical central trends, and individual pieces of the genome may have different branching patterns.

403 ***Even Darwin's finches:*** Grants, "Synergism of Natural Selection and Introgression in the Origin of a New Species," 678–79.

403 ***interbred with Neanderthals:*** Green et al., "A Draft Sequence of the Neandertal Genome."

403 ***2 percent Neanderthal DNA:*** Pääbo, "The Contribution of Ancient Hominin Genomes."

403 ***a third lineage:*** Pääbo, "The Contribution of Ancient Hominin Genomes."

404 ***"Many paleontological species":*** Evan Eichler, interview with author.

404 ***as a "metapopulation":*** Pääbo, "The Diverse Origins of the Human Gene Pool," 314.

404 ***"there is no simple tree":*** David Reich, interview with author.

404 ***"a single trunk population":*** Reich, *Who We Are and How We Got Here*, 82.

405 ***At one conference:*** Royal Society, *The First 4 Million Years of Human Evolution*, audio recording for session "Hominin Diversity in the Middle Pliocene of Eastern Africa: The Maxilla of KNM-WT 40,000."

405 ***"Tim is good at finding fossils":*** Fred Spoor, interview with author.

405 ***White dismissed such "professionalism":*** White continually policed his profession. For example, in 2016 scientists from the University of Texas at Austin published a *Nature* paper with the startling claim that Lucy died after falling from a tree, based on CT scans showing fractures in the fossil bones. White diagnosed the ostensible "fall fractures" as typical fossilization cracks, went into the collection at the Ethiopian National Museum and found similar damage in fossils of big animals like an ancient horse and rhino, which are not known to climb trees. "This article," he sniped, "will become another classic example of paleoanthropological story-telling being used as click bait for a commercial journal eager for

media coverage." It wasn't Lucy that had plummeted, he lamented, but scientific standards.

406 ***Yohannes invited Bruce Latimer:*** Bruce Latimer and Yohannes Haile-Selassie, interviews with author.

406 ***The "Burtele foot" fossil was published:*** Yohannes Haile-Selassie et al., "A New Hominin Foot from Ethiopia."

406 ***hypothesized that* Ardipithecus *split:*** White et al., "Neither Chimpanzee Nor Human," 4882.

406 ***a new discovery in* Nature*:*** Yohannes Haile-Selassie et al., "New Species from Ethiopia."

406 ***headline in the* New York Times*:*** Zimmer, "Human Family Tree."

406 ***"a number of strategies":*** Tim White, interview with author.

407 ***"I don't think it's fair":*** Yohannes Haile-Selassie, interview with author.

407 ***3.8-million-year-old skull:*** Yohannes Haile-Selassie et al., "A 3.8-Million-Year-Old Hominin Cranium."

407 ***Another trove of bones:*** Simpson et al., "*Ardipithecus ramidus* Postcrania from the Gona Project Area."

407 ***"a curious amount of variation":*** Scott Simpson, interview with author.

408 ***inspiration from the Ardi saga:*** Lee Berger, interview with author.

408 ***"Dates are not important":*** Lee Berger, interview with author.

408 ***"If* naledi *is primitive":*** Lee Berger, interview with author.

408 ***335,000 years old:*** Dirks et al., "The Age of *Homo naledi*."

409 ***"new kind of paleoanthropology":*** *Nova: Dawn of Humanity.*

409 ***"does not equal quality":*** Lee Berger, interview with author.

409 ***op-ed in the* Guardian*:*** White, "Why Combining Science and Showmanship Risks the Future of Research."

409 ***"When Tim White dies":*** Facebook post, October 25, 2015.

409 ***"a few cranky uncles":*** John Hawks weblog, November 28, 2015.

409 ***"everything he warned against":*** Lee Berger, interview with author.

409 ***denounced Berger's showmanship:*** Don Johanson and Richard Leakey, interviews with author.

409 ***"The currency of debate":*** Bill Kimbel, interview with author.

410 ***"You keep publishing":*** Owen Lovejoy, interview with author.

410 ***"There is a tendency":*** Owen Lovejoy, interview with author.

410 ***walked the halls one afternoon:*** Author interview with Berhane Asfaw at national museum, December 2016.

411 ***"Without Berhane":*** Alemu Ademassu, interview with author.

411 ***"thirty safes was enough":*** Berhane Asfaw, interview with author.

EPILOGUE: SUNSET

413 ***one more Bigfoot hunt:*** Scene details from author reporting trip to Ethiopia, December 2016 unless otherwise indicated. Additional details from interviews with Tim White, Yonas Beyene, Berhane Asfaw, and Giday WoldeGabriel.

415 ***old tribal enmities:*** The tranquility was fleeting. Two years later, the team had to abandon its field camp after Issa gunmen ambushed a nearby military patrol and killed several soldiers. The news media reported a new series of clashes between Afar and Issa.

415 ***"alternative reconstruction":*** Pilbeam and Lieberman, "Reconstructing the Last Common Ancestor."

419 ***"shit in the vault":*** Scott Simpson, interview with author.

419 ***"an order of magnitude":*** Bernard Wood, interview with author.

420 ***thirty-two thousand vertebrate fossils:*** Collection figures from Tim White, e-mail to author, January 30, 2020.

420 ***"to do the housework":*** All comments from Yonas Desta, interview with author.

Bibliography

BOOKS, ARTICLES, AND DISSERTATIONS

Addis Tribune. "AL-444 Safe in Museum: ARCCH Refutes Allegations of a Missing Hominid Fossil." April 6, 2001.

Aiello, Leslie C. "The Wenner-Gren Foundation: Supporting Anthropology for 75 Years." In "The Wenner-Gren Foundation: Supporting Anthropology for 75 Years," edited by Leslie C. Aiello, Laurie Obbink, and Mark Mahoney. Supplement, *Current Anthropology* 57, no. 14 (2016): S211.

Al Jazeera. "Ardi Challenges Darwin's Theory." [in Arabic] October 3, 2009.

Alberts, Bruce. "The Breakthroughs of 2009." *Science* 326, no. 5960 (2009): 1589.

Almécija, Sergio, Jeroen B. Smaers, and William L. Jungers. "The Evolution of Human and Ape Hand Proportions." *Nature Communications* 6 (2015): 7717.

Almécija, Sergio, Melissa Tallman, David M. Alba, Marta Pina, Salvador Moyà-Solà, and William L. Jungers. "The Femur of *Orrorin tugenensis* Exhibits Morphometric Affinities with Both Miocene Apes and Later Hominins." *Nature Communications* 4 (2013): 2888.

American Association for the Advancement of Science (AAAS). "New Species of Human Ancestor." Press release, April 23, 1999.

Anderson, David. "Extremes in the Indian Ocean." *Nature* 401, no. 6751 (1999): 337.

Andersson, Gunnar B. J. "Epidemiological Features of Chronic Low-Back Pain." *Lancet* 354, no. 9178 (1999): 581.

Andrews, Peter. *An Ape's View of Human Evolution*. Cambridge: Cambridge University Press, 2015.

Anemone, R. L., G. C. Conroy, and C. W. Emerson. "GIS and Paleoanthropology: Incorporating New Approaches from the Geospatial Sciences in the Analysis of Primate and Human Evolution." In

"Yearbook of Physical Anthropology," edited by Robert W. Sussman. Supplement, *American Journal of Physical Anthropology* 146, no. S53 (2011): 19.

Antón, Susan C. "Of Burnt Coffee and Pecan Pie: Recollections of F. Clark Howell on His Birthday, November 27, 1925–March 10, 2007." *PaleoAnthropology* (2007): 36–52.

Antón, Susan C., Ripan S. Malhi, and Agustín Fuentes. "Race and Diversity in U.S. Biological Anthropology: A Decade of AAPA Initiatives." In "Yearbook of Physical Anthropology," edited by Trudy R. Turner. Supplement, *American Journal of Physical Anthropology* 165, no. S65 (2018): 158.

"*Ardipithecus ramidus kadabba*." Walta Information Center, July 20, 2001.

Aristotle. *On the Parts of Animals*. Translated by William Ogle. London: Kegan Paul, Trench and Co., 1882.

Asfaw, Berhane. "The Belohdelie Frontal: New Evidence of Early Hominid Cranial Morphology from the Afar of Ethiopia." *Journal of Human Evolution* 16, no. 7–8 (1987): 611.

———. "The Origin of Humans: The Record from the Afar of Ethiopia." In *What Is Our Real Knowledge About the Human Being?* (Pontificia Academia Scientiarum Scripta Varia 109), edited by Marcelo Sánchez Sorondo, 15–20. Vatican City: Pontifical Academy of Sciences, 2007.

Asfaw, Berhane, and Close Observer [pseud.]. "What Went Wrong with Prehistory Research" (a series of exchanges about human origins research in Ethiopia). Walta Information Center, March 17, 2001–April 28, 2001.

Asfaw, Berhane, Cynthia Ebinger, David Harding, Tim White, and Giday WoldeGabriel. "Space-Based Imagery in Paleoanthropological Research: An Ethiopian Example." *National Geographic Research* 6 (1990): 418.

Asfaw, Berhane, Tim White, Owen Lovejoy, Bruce Latimer, Scott Simpson, and Gen Suwa. "*Australopithecus garhi:* A New Species of Early Hominid from Ethiopia." *Science* 284, no. 5414 (1999): 629.

Assefa, Zelalem. "History of Paleoanthropological Research in the Southern Omo, Ethiopia." In *Proceedings of the Eleventh International Conference of Ethiopian Studies*, edited by Bahru Zewde, Richard Pankhurst, and Taddese Beyene, 71–85. Addis Ababa: Institute of Ethiopian Studies, Addis Ababa University, 1994.

Ayala, Francisco J., and John C. Avise. *Essential Readings in Evolutionary Biology*. Baltimore: Johns Hopkins University Press, 2014.

Begun, David R. "The Earliest Hominins—Is Less More?" *Science* 303, no. 5663 (2004): 1478.

Begun, David R. *The Real Planet of the Apes: A New Story of Human Origins*. Princeton, NJ: Princeton University Press, 2016.

Bell, Charles. *The Hand: Its Mechanism and Vital Endowments as Evincing Design*. London: William Pickering, 1833.

Belt, Leonard. *Leonardo the Anatomist*. Lawrence, Kansas: University of Kansas Press, 1955.

Bender, Renato, Phillip V. Tobias, and Nicole Bender. "The Savannah Hypotheses: Origin, Reception and Impact on Paleoanthropology." In "Human Evolution Across Disciplines: Through the Looking Glass of History and Epistemology," edited by Richard G. Delisle. Special issue, *History and Philosophy of the Life Sciences* 34, no. 1–2 (2012): 147.

Berger, Lee R., and Brett Hilton-Barber. *In the Footsteps of Eve: The Mystery of Human Origins*. Washington, D.C.: National Geographic Society, 2000.

Berger, Lee R., and Phillip V. Tobias. "A Chimpanzee-like Tibia from Sterkfontein, South Africa and Its Implications for the Interpretation of Bipedalism in *Australopithecus africanus*." *Journal of Human Evolution* 30, no. 4 (1996): 343.

Berhe, Aregawi. "A Political History of the Tigray People's Liberation Front (1975–1991): Revolt, Ideology and Mobilisation in Ethiopia." Ph.D. diss., Vrije Universiteit Amsterdam, Netherlands, 2008.

Boaz, Noel T., and Douglas L. Cramer. "Fossils of the Libyan Sahara." *Natural History* 91, no. 8 (1982) 34.

Bonnefille, Raymonde. "Cenozoic Vegetation, Climate Changes and Hominid Evolution in Tropical Africa." *Global and Planetary Change* 72, no. 4 (2010): 390.

Bower, Bruce. "Human Ancestors Have Identity Crisis." *Science News*, February 17, 2011.

Bradley, Brenda J., Martha M. Robbins, Elizabeth A. Williamson, H. Dieter Steklis, Netzin Gerald Steklis, Nadin Eckhardt, Christophe Boesch, and Linda Vigilant. "Mountain Gorilla Tug-of-War: Silverbacks Have Limited Control Over Reproduction in Multimale Groups." *Proceedings of the National Academy of Sciences* 102, no. 26 (2005): 9418.

Brain, C. K. "Do We Owe Our Intelligence to a Predatory Past?" Seventieth James Arthur Lecture on the Evolution of the Human Brain. New York: American Museum of Natural History, 2001.

Browne, Janet. "History of Biogeography." In *Encyclopedia of Life Sciences*. Hoboken, NJ: John Wiley and Sons, 2001. https://doi.org/10.1038/npg.els.0003092.

Brunet, Michel, Franck Guy, David Pilbeam, Hassane Taisso Mackaye, Andossa Likius, Djimdoumalbaye Ahounta, Alain Beauvilain et al. "A New Hominid from the Upper Miocene of Chad, Central Africa." *Nature* 418, no. 6894 (2002): 145.

Caccone, Adalgisa, and Jeffrey R. Powell. "DNA Divergence Among Hominoids." *Evolution* 43, no. 5 (1989): 925.

Cann, Rebecca L., Mark Stoneking, and Allan C. Wilson. "Mitochondrial DNA and Human Evolution." *Nature* 325, no. 6099 (1987): 31.

Carpenter, Clarence Ray. "A Field Study in Siam of the Behavior and Social Relations of the Gibbon, (Hylobates lar)." *Comparative Psychology Monographs* 16, no. 5 (1940): 1.

Carroll, Sean B. *Endless Forms Most Beautiful: The New Science of Evo Devo and the Making of the Animal Kingdom*. New York: W.W. Norton and Company, 2005.

Carroll, Sean B., Jennifer K. Grenier, and Scott D. Weatherbee. *From DNA to Diversity: Molecular Genetics and the Evolution of Animal Design*. 2nd ed. Malden, MA: Blackwell Publishing, 2005.

Cerling, Thure E., Naomi E. Levin, Jay Quade, Jonathan G. Wynn, David L. Fox, John D. Kingston, Richard G. Klein, and Francis H. Brown. "Comment on the Paleoenvironment of *Ardipithecus ramidus*." *Science* 328, no. 5982 (2010): 1105.

Cerulli, Enrico. "The Folk-Literature of the Galla of Southern Abyssinia." In *Varia Africana III*, edited by E. A. Hooton and Natica I. Bates, 9–229. Vol. 3 of *Harvard African Studies*. Cambridge, MA: African Department of the Peabody Museum of Harvard University, 1922. http://nrs.harvard.edu/urn-3:FHCL:12674642.

Cherfas, Jeremy. "Grave Accusations Face Lucy Finders." *New Scientist* 97, no. 1344 (February 10, 1983): 390.

Clark, J. Desmond. "Africa in Prehistory: Peripheral or Paramount?" *Man*, New Series, 10, no. 2 (1975): 175.

———. "Digging On: A Personal Record and Appraisal of Archaeological Research in Africa and Elsewhere." *Annual Review of Anthropology* 23 (1994): 1.

———. "Radiocarbon Dating and African Archaeology." In *Radiocarbon Dating: Proceedings of the Ninth International Radiocarbon Conference, Los Angeles and La Jolla, 1976*, edited by Rainer Berger and Hans E. Suess, 7–31. Berkeley: University of California Press, 1979.

———. "Recent Developments in Human Biological and Cultural Evolution." *The South African Archaeological Bulletin* 50, no. 162 (December 1995): 168.

Clarke, Ronald J. "Excavation, Reconstruction and Taphonomy of the StW 573 *Australopithecus prometheus* Skeleton from Sterkfontein Caves, South Africa." *Journal of Human Evolution* 127 (2019): 41.

Clayton, Martin, and Ron Philo. *Leonardo da Vinci: The Mechanics of Man*. London: Royal Collection Trust, 2013.

Cobb, W. Montague. "Thomas Wingate Todd, M.B, Ch. B., F.R.C.S (Eng.), 1885–1938." *Journal of the National Medical Association* 51, no. 3 (1959): 233.

Cohen, Herman J. "Ethiopia: Ending a Thirty-Year War." In *Intervening in*

Africa: Superpower Peacemaking in a Troubled Continent, 17–59. London: Palgrave Macmillan, 2000.

Cohen, Jon. *Almost Chimpanzee: Searching for What Makes Us Human, in Rainforests, Labs, Sanctuaries, and Zoos*. New York: Times Books/Henry Holt and Company, 2010.

———. "Relative Differences: The Myth of 1%." *Science* 316, no. 5833 (2007): 1836.

Colburn, Forrest D. "The Tragedy of Ethiopia's Intellectuals." *The Antioch Review* 47, no. 2 (1989): 133.

Cole, Francis Joseph. *A History of Comparative Anatomy: From Aristotle to the Eighteenth Century*. London: Macmillan and Company, 1944.

Conroy, Glenn Carter. "Paleoanthropology Today." *Evolutionary Anthropology* 6, no. 5 (1998): 155.

Coolidge, Harold Jefferson. "Life of the Gibbon in His Native Home." *New York Times Sunday Magazine*, August 8, 1937.

———. "Studying the Gibbon for Light on Man's Evolution." *New York Times Sunday Magazine*, May 15, 1938.

———. "Trailing the Gibbon to Learn About Man." *New York Times Sunday Magazine*, August 1, 1937.

Daily Kent Stater. "KSU Anthropologist Predicts Human Extinction." February 27, 1970.

Dalton, Rex. "The History Man." *Nature* 443, no. 7109 (2006): 268.

———. "Restrictions Delay Fossil Hunts in Ethiopia." *Nature* 410, no. 6830 (2001): 728.

Dart, Raymond A. "*Australopithecus africanus*: The Man-Ape of South Africa." *Nature* 115, no. 2884 (February 7, 1925): 195.

Darwin, Charles. *On the Origin of Species by Means of Natural Selection: Or the Preservation of Favoured Races in the Struggle for Life*. London: John Murray, 1859.

———. *The Descent of Man, and Selection in Relation to Sex*. London: John Murray, 1871.

De Heinzelin, Jean, J. Desmond Clark, Kathy D. Schick, and W. Henry Gilbert, eds. *The Acheulean and the Plio-Pleistocene Deposits of the Middle Awash Valley, Ethiopia*. Vol. 104 of *Annales Sciences Geologiques*. Tervuren, Belgium: Musee Royal de l'Afrique Centrale, 2000.

De Waal, Frans. *Bonobo: The Forgotten Ape*. Berkeley: University of California Press, 1997.

———. "Was 'Ardi' a Liberal?" *Huffington Post*, March 18, 2010.

DeGusta, David, and Elisabeth Vrba. "A Method for Inferring Paleohabitats from the Functional Morphology of Bovid Astragali." *Journal of Archaeological Science* 30, no. 8 (2003): 1009.

Deino, Alan L., Paul R. Renne, and Carl C. Swisher. "40Ar/39Ar Dating in

Paleoanthropology and Archeology." *Evolutionary Anthropology* 6, no. 2 (1998): 63.

DeSilva, Jeremy M., Corey M. Gill, Thomas C. Prang, Miriam A. Bredella, and Zeresenay Alemseged. "A Nearly Complete Foot from Dikika, Ethiopia and Its Implications for the Ontogeny and Function of *Australopithecus afarensis*." *Science Advances* 4, no. 7 (2018): eaar7723.

DeSilva, Jeremy, Ellison McNutt, Julien Benoit, and Bernhard Zipfel. "One Small Step: A Review of Plio-Pleistocene Hominin Foot Evolution." In "Yearbook of Physical Anthropology," edited by Lyle W. Konigsberg. Supplement, *American Journal of Physical Anthropology* 168, no. S67 (2019): 63.

Desmond, Adrian. *Huxley: From Devil's Disciple to Evolution's High Priest.* Reading, MA: Addison-Wesley, 1997.

Diamond, Jared. *The Third Chimpanzee: The Evolution and Future of the Human Animal.* New York: HarperCollins, 1992.

Diogo, Rui, Julia L. Molnar, and Bernard Wood. "Bonobo Anatomy Reveals Stasis and Mosaicism in Chimpanzee Evolution, and Supports Bonobos as the Most Appropriate Extant Model L for the Common Ancestor of Chimpanzees and Humans." *Scientific Reports* 7 (2017): 608.

Dirks, Paul H. G. M., Eric M. Roberts, Hannah Hilbert-Wolf, Jan D. Kramers, John Hawks, Anthony Dosseto, Mathieu Duval et al. "The Age of *Homo naledi* and Associated Sediments in the Rising Star Cave, South Africa." *eLife* 6 (2017): http://doi.org/10.7554/eLife.2423.

Dobzhansky, Theodosius. "Nothing in Biology Makes Sense Except in the Light of Evolution." *The American Biology Teacher* 35, no. 3 (1973): 125.

———. "On Species and Races of Living and Fossil Man." *American Journal of Physical Anthropology* 2, no. 3 (1944): 251.

Domínguez-Rodrigo, Manuel. "Is the 'Savanna Hypothesis' a Dead Concept for Explaining the Emergence of the Earliest Hominins?" *Current Anthropology* 55, no. 1 (2014): 59.

Doolittle, W. Ford. "Uprooting the Tree of Life." *Scientific American*, February 2000.

Edler, Melissa. "Life as Artifact: Olaf Prufer's Personal Journey Across Continents, Through Time." *Kent State Magazine* 7, issue 1 (Fall 2007): 10.

Edwards, Muriel E. "The Relations of the Peroneal Tendons to the Fibula, Calcaneus, and Cuboideum." *American Journal of Anatomy* 42, no. 1 (1928): 213.

Eilperin, Juliet. "In Ethiopia, Both Obama and Ancient Fossils Get a Motorcade." Weblog post. *Washington Post*, July 27, 2015. https://www.washingtonpost.com/news/post-politics/wp/2015/07/27/in-ethiopia-both-obama-and-ancient-fossils-get-a-motorcade.

Ethiopian Herald. "Counter-Revolutionary Elements Caught Redhanded Here." February 4, 1984.

———. "So That Open Eyes Might See." September 4, 1984.

Ethiopian Register. "TPLF/EPRDF's Strategies to Establish Its Hegemony and Perpetuating Its Rule," abridged English translation of 1993 party document. June 1996.

Etkind, Alexander. "Beyond Eugenics: The Forgotten Scandal of Hybridizing Humans and Apes." *Studies in History and Philosophy of Science Part C: Studies in History and Philosophy of Biological and Biomedical Sciences* 39, no. 2 (2008): 205.

Feakins, Sarah J., Naomi E. Levin, Hannah M. Liddy, Alexa Sieracki, Timothy I. Eglinton, and Raymonde Bonnefille. "Northeast African Vegetation Change Over 12 M.Y." *Geology* 41, no. 3 (2013): 295.

Fernández, Peter J., Carrie S. Mongle, Louise Leakey, Daniel J. Proctor, Caley M. Orr, Biren A. Patel, Sergio Almécija, Matthew W. Tocheri, and William L. Jungers. "Evolution and Function of the Hominin Forefoot." *Proceedings of the National Academy of Sciences* 115, no. 35 (2018): 8746.

Fisher, R. A. *The Genetical Theory of Natural Selection*. Oxford, UK: Clarendon Press, 1930. Reprinted with preface and notes by J. H. Bennett. Oxford: Oxford University Press, 1999.

Fitzgerald, Mary Anne. "Rift Valley: How the Getty Millions Stoked up a Bitter Feud in the African Desert over Man's Origins." *The Sunday Times* (London), March 19, 1995.

Flannery, Kent V. "The Golden Marshalltown: A Parable for the Archeology of the 1980s." *American Anthropologist* 84, no. 2 (1982): 265.

Fleagle, John G. "Intermembral Index." In *The International Encyclopedia of Primatology*, edited by Augustin Fuentes, 661–63. Hoboken, NJ: John Wiley and Sons, 2017.

Fromm, Erich. *Man for Himself: An Inquiry into the Psychology of Ethics*. New York: Rinehart, 1947.

Gagneux, Pascal. "Sperm Count." Center for Academic Research and Training in Anthropogeny (CARTA). Accessed September 25, 2019. https://carta.anthropogeny.org/moca/topics/sperm-count.

Gani, M. Royhan, and Nahid D. Gani. "River-margin Habitat of *Ardipithecus ramidus* at Aramis, Ethiopia 4.4 Million Years Ago." *Nature Communications* 2 (2011): 602.

Gani, Nahid D., and M. Royhan Gani. "On the Environment of Aramis: Concerning Reply of White to Cerling et al. in August 2014." *Current Anthropology* 57, no. 2 (2016): 219.

Gee, Henry. *In Search of Deep Time: Beyond the Fossil Record to a New History of Life*. New York: The Free Press, 1999.

Gettleman, Jeffrey. "'A Generation Is Protesting' in Ethiopia, Long a U.S. Ally." *New York Times*, August 12, 2016.

Gibbon, Edward. *The History of the Decline and Fall of the Roman Empire*, vol. 3. New York: Harper and Brothers, 1841.

Gibbons, Ann. "Claim-Jumping Charges Ignite Controversy at Meeting." *Science* 268, no. 5208 (1995): 196.

———. "First Hominid Finds from Ethiopia in a Decade." *Science* 251, no. 5000 (1991): 1428.

———. "Glasnost for Hominids: Seeking Access to Fossils." *Science* 297, no. 5586 (2002): 1464.

———. "How a Fickle Climate Made Us Human." *Science* 341, no. 6145 (2013): 474.

———. "Oldest Human Femur Wades into Controversy." *Science* 305, no. 5692 (2004): 1885.

———. *The First Human: The Race to Discover Our Earliest Ancestors*. New York: Anchor Books, 2006.

Gibbs, Richard A., Jeffrey Rogers, Michael G. Katze, Roger Bumgarner, George M. Weinstock, Elaine R. Mardis, Karin A. Remington et al. "Evolutionary and Biomedical Insights from the Rhesus Macaque Genome." *Science* 316, no. 5822 (2007): 222.

Gilbert, W. Henry, and Berhane Asfaw, eds. *Homo Erectus: Pleistocene Evidence from the Middle Awash, Ethiopia*. Berkeley: University of California Press, 2008.

Giogis, Dawit Wolde. "One Man's Love of Power Cost One Million Lives." *The Sydney Morning Herald*, July 6, 1987.

Goodall, Jane. *The Chimpanzees of Gombe: Patterns of Behavior.* Cambridge, MA: Belknap Press, 1986.

———. *Through a Window: My Thirty Years with the Chimpanzees of Gombe.* Boston: Mariner Books: 2010.

Goodman, Morris. "Epilogue: A Personal Account of the Origins of a New Paradigm." *Molecular Phylogenetics and Evolution* 5, no. 1 (1996): 269.

Gore, Rick. "The First Steps." *National Geographic*, February 1997.

Gould, Stephen Jay. "Ladders, Bushes, and Human Evolution." *Natural History* 85, no. 4 (1976): 24.

———. *The Richness of Life: The Essential Stephen Jay Gould*. Edited by Steven Rose. New York: W. W. Norton and Company, 2007.

———. *The Structure of Evolutionary Theory*. Cambridge, MA: Belknap Press, 2002.

Gould, Stephen Jay, and Niles Eldredge. "Punctuated Equilibria: An Alternative to Phyletic Gradualism." In *Essential Readings in Evolutionary Biology*, edited by Francisco J. Ayala and John C. Avise, 238–72. Baltimore: Johns Hopkins University Press, 2014.

Gould, S. J., and R. C. Lewontin. "The Spandrels of San Marco and the

Panglossian Paradigm: A Critique of the Adaptationist Programme." *Proceedings of the Royal Society of London* B 205, no. 1161 (1979): 581.

Grant, Peter R., and B. Rosemary Grant. "Synergism of Natural Selection and Introgression in the Origin of a New Species." *The American Naturalist* 183, no. 5 (2014): 671.

Graur, Dan, and William Martin. "Reading the Entrails of Chickens: Molecular Timescales of Evolution and the Illusion of Precision." *Trends in Genetics* 20, no. 2 (2004): 80.

Gray, Henry. *Anatomy, Descriptive and Surgical*. 1901 ed. Reprint. Philadelphia: Running Press, 1974.

Green, Richard E., Johannes Krause, Adrian W. Briggs, Tomislav Maricic, Udo Stenzel, Martin Kircher, Nick Patterson et al. "A Draft Sequence of the Neandertal Genome." *Science* 328, no. 5979 (2010): 710.

Gross, Charles G. "Huxley Versus Owen: The Hippocampus Minor and Evolution." *Trends in Neurosciences* 16, no. 12 (1993): 493.

Haeusler, Martin. "Spinal Pathologies in Fossil Hominins." In *Spinal Evolution*, edited by Ella Been, Asier Gómez-Olivencia, and Patricia Ann Kramer, 213–45. Switzerland: Springer, 2019.

Haile-Selassie, Yohannes. "Late Miocene Hominids from the Middle Awash, Ethiopia." *Nature* 412, no. 6843 (2001): 178.

———. "Photos May Offer Clues over Ethiopian Fossil Site" (letter to editor). *Nature* 412, no. 6843 (2001): 118.

Haile-Selassie, Yohannes, Luis Gibert, Stephanie M. Melillo, Timothy M. Ryan, Mulugeta Alene, Alan Deino, Naomi E. Levin, Gary Scott, and Beverly Z. Saylor. "New Species from Ethiopia Further Expands Middle Pliocene Hominin Diversity." *Nature* 521, no. 7553 (2015): 483.

Haile-Selassie, Yohannes, Stephanie M. Melillo, Antonino Vazzana, Stefano Benazzi, and Timothy M. Ryan. "A 3.8-million-year-old Hominin Cranium from Woranso-Mille, Ethiopia." *Nature* 573, no. 7773 (2019): 214.

Haile-Selassie, Yohannes, Beverly Z. Saylor, Alan Deino, Naomi E. Levin, Mulugeta Alene, and Bruce M. Latimer. "A New Hominin Foot from Ethiopia Shows Multiple Pliocene Bipedal Adaptations." *Nature* 483, no. 7391 (2012): 565.

Haile-Selassie, Yohannes, and Giday WoldeGabriel, eds. *Ardipithecus Kadabba: Late Miocene Evidence from the Middle Awash, Ethiopia*. Berkeley: University of California Press, 2009.

Hansen, Mark. "Believe It or Not." *ABA Journal* 79, no. 6 (June 1993): 64–67.

Harmon, Katherine. "Was 'Ardi' Not a Human Ancestor After All? New Review Raises Doubts." Observations. *Scientific American* (February 16, 2011).

Harrell, Eben. "Ardi: The Human Ancestor Who Wasn't?" *Time* (May 27, 2010).

Haute Living. "*Ardipithecus* Research Party." January 12, 2010. https://hauteliving.com/2010/01/ardipithecus-research-party/19214.

Henze, Paul B. *Ethiopian Journeys: Travels in Ethiopia 1969–72.* London: Ernest Benn, Ltd., 1977.

———. "Is Ethiopia Democratic? A Political Success Story." *Journal of Democracy* 9, no. 4 (1998): 40–54.

———. *Layers of Time: A History of Ethiopia.* New York: Palgrave, 2000.

———. "Reflections on Development in Ethiopia." *Northeast African Studies* 10, no. 2 (2003): 189–201.

Hiltzik, Michael. "Does Trail to Ark of Covenant End Behind Aksum Curtain?" *Los Angeles Times*, June 9, 1992.

Hinrichsen, Don. "How Old Are Our Ancestors." *New Scientist* 78, no. 1105 (1978): 571.

Hirasaki, Eishi, Naomichi Ogihara, Yuzuru Hamada, Hiroo Kumakura, and Masato Nakatsukasa. "Do Highly Trained Monkeys Walk Like Humans? A Kinematic Study of Bipedal Locomotion in Bipedally Trained Japanese Macaques." *Journal of Human Evolution* 46, no. 6 (2004): 739–50.

Hogervorst, Tom, and Evie E. Vereecke. "Evolution of the Human Hip. Part 1: The Osseous Framework." *Journal of Hip Preservation Surgery* 1, no. 2 (2014): 39.

Holden, Catherine. "Ötzi Death Riddle Solved." *Science* 293, no. 5531 (2001): 795.

Institute of Human Origins. Newsletter, vol. 1, no. 1 (Winter 1982/83).

International Human Genome Sequencing Consortium. "Initial Sequencing and Analysis of the Human Genome." *Nature* 409, no. 6822 (2001): 860.

IRIN News. United Nations Office for the Coordination of Humanitarian Affairs. "Interview with Teshome Toga, Youth, Sports and Culture Minister." March 15, 2004. http://www.thenewhumanitarian.org/report/49088/ethiopia-interview-teshome-toga-youth-sports-and-culture-minister.

Jefferson, David J. "This Anthropologist Has a Style That Is Bone of Contention: Dr. Johanson's 'Lucy' Show Sparks Spat, with Scientists and Gettys Filing Lawsuit." *Wall Street Journal*, January 31, 1995.

Johanson, Donald C. "Anthropologists: The Leakey Family." *Time*, March 29, 1999.

Johanson, Donald C., and Tim D. White. "A Systematic Assessment of Early African Hominids." *Science* 203, no. 4378 (1979): 321.

Johanson, Donald C., and Maitland Edey. *Lucy: The Beginnings of Humankind.* New York: Simon and Schuster, 1981.

Johanson, Donald C., and James Shreeve. *Lucy's Child: The Discovery of a Human Ancestor.* New York: William Morrow, 1989.

Johanson, Donald C., and Kate Wong. *Lucy's Legacy: The Quest for Human Origins.* New York: Three Rivers Press, 2010.

Jones, Frederic Wood. *Structure and Function as Seen in the Foot,* 2nd ed. London: Baillière, Tindall and Cox, 1949.

Jose, Antony Merlin. "Anatomy and Leonardo da Vinci." *Yale Journal of Biology and Medicine* 74, no. 3 (2001): 185–95.

Kalb, Jon. *Adventures in the Bone Trade: The Race to Discover Human Ancestors in Ethiopia's Afar Depression.* New York: Copernicus Books, 2001.

Kalb, Jon E., C. J. Jolly, Assefa Mebrate, Sleshi Tebedge, Charles Smart, Elizabeth B. Oswald, Douglas Cramer et al. "Fossil Mammals and Artefacts from the Middle Awash Valley, Ethiopia." *Nature* 298, no. 5869 (1982): 25.

Katz, Donald R. "Ethiopia After the Revolution: Vultures Return to the Land of Sheba." *Rolling Stone,* September 21, 1978.

Keele, Kenneth D. *Leonardo da Vinci's Elements of the Science of Man.* New York: Academic Press, 1983.

Keith, Arthur. *An Autobiography.* London: Watts and Company, 1950.

———. "Hunterian Lectures on Man's Posture: Its Evolution and Disorders." *The British Medical Journal* 1, no. 3246 (March 17, 1923): 451.

———. "Fifty Years Ago." *American Journal of Physical Anthropology* 26, no. 1 (1940): 251.

Kern, Kevin F. "T. Wingate Todd: Pioneer of Modern American Physical Anthropology." *Kirtlandia* 55 (September 2006): 1–42.

Kiefer, Michael. "The Man Who Loved Lucy." *Phoenix New Times,* August 7, 1997.

Kimbel, William H., Charles A. Lockwood, Carol V. Ward, Meave G. Leakey, Yoel Rak, and Donald C. Johanson. "Was *Australopithecus anamensis* Ancestral to *A. afarensis*? A Case of Anagenesis in the Hominin Fossil Record." *Journal of Human Evolution* 51, no. 2 (2006): 134.

Kimbel, William H., Gen Suwa, Berhane Asfaw, Yoel Rak, and Tim D. White. "*Ardipithecus ramidus* and the Evolution of the Human Cranial Base." *Proceedings of the National Academy of Sciences* 111, no. 3 (2014): 948.

Kimbel, William H., Tim D. White, and Donald C. Johanson. "Cranial Morphology of *Australopithecus afarensis*: A Comparative Study Based on a Composite Reconstruction of the Adult Skull." *American Journal of Physical Anthropology* 64, no. 4 (1984): 337.

King, Mary-Claire, and Allan C. Wilson. "Evolution at Two Levels in Humans and Chimpanzees." *Science* 188, no. 4184 (1975): 107.

Kingston, John D. "Shifting Adaptive Landscapes: Progress and Challenges in Reconstructing Early Hominid Environments." In "Yearbook of Physical Anthropology," edited by Sara Stinson. Supplement, *American Journal of Physical Anthropology* 134, no. S45 (2007): 20.

Klenerman, Leslie, and Bernard Wood. *The Human Foot: A Companion to Clinical Studies*. London: Springer-Verlag, 2006.

Kovacs, Jennifer. "Making History: As He Tries to Uncover the Past, Owen Lovejoy is Living a Life to Remember." *The Burr* (Kent State University), Fall 2003.

Krishtalka, Leonard. "Bones from Afar, Oldest Ape Man from Ethiopia." *Terra* (September/October 1983): 18.

Kronenberg, Zev N., Ian T. Fiddes, David Gordon, Shwetha Murali, Stuart Cantsilieris, Olivia S. Meyerson, Jason G. Underwood et al. "High-resolution Comparative Analysis of Great Ape Genomes." *Science* 360, no. 6393 (2018): http://doi.org/10.1126/science.aar6343.

Kuhn, Thomas S. *The Structure of Scientific Revolutions*. 4th ed. Chicago: University of Chicago Press, 2012.

Kunda, Ziva. "The Case for Motivated Reasoning." *Psychological Bulletin* 108, no. 3 (1990): 480.

Langergraber, Kevin E., Kay Prüfer, Carolyn Rowney, Christophe Boesch, Catherine Crockford, Katie Fawcett, Eiji Inoue et al. "Generation Times in Wild Chimpanzees and Gorillas Suggest Earlier Divergence Times in Great Ape and Human Evolution." *Proceedings of the National Academy of Sciences* 109, no. 39 (2012): 15716.

Lawton, Graham. "Uprooting Darwin's Tree." Editorial, *The New Scientist* 201, no. 2692 (2009): 34.

Le Gros Clark, Wilfrid. E. "New Palaeontological Evidence Bearing on the Evolution of the Hominoidea." *Quarterly Journal of the Geological Society of London* 105 (1949): 225.

———. "Penrose Memorial Lecture. The Crucial Evidence for Human Evolution." *Proceedings of the American Philosophical Society* (1959): 159.

Leakey, Louis S. B. *Animals of East Africa*. Washington, D.C.: National Geographic Society, 1969.

Leakey, Mary. *Disclosing the Past: An Autobiography*. Garden City, NY: Doubleday and Company, 1984.

Leakey, Mary Douglas, and John Michael Harris. *Laetoli, A Pliocene Site in Northern Tanzania*. Oxford: Clarendon Press, 1987.

Leakey, Meave. "The Dawn of Humans: The Farthest Horizon." *National Geographic*, September 1995.

Leakey, Meave G., Craig S. Feibel, Ian McDougall, and Alan Walker. "New Four-million-year-old Hominid Species from Kanapoi and Allia Bay, Kenya." *Nature* 376, no. 6541 (1995): 565.

Leakey, Richard. *One Life: An Autobiography.* Salem N.H.: Salem House, 1983.

Levine, Donald. "Ethiopia's Dilemma: Missed Chances from the 1960s to the Present." *International Journal of African Development* 1, no. 1 (2013): 3.

Levine, Donald N. *Wax and Gold: Tradition and Innovation in Ethiopian Culture.* Chicago: University of Chicago Press, 1965.

Lewin, Roger. *Bones of Contention: Controversies in the Search for Human Origins.* New York: Simon and Schuster, 1987.

Lieberman, Daniel E. *Evolution of the Human Head.* Cambridge, MA: Belknap Press, 2011.

Lonsdorf, Elizabeth, Stephen R. Ross, and Tetsuro Matsuzawa. *The Mind of the Chimpanzee: Ecological and Experimental Perspectives.* Chicago: University of Chicago Press, 2010.

Louchart, Antoine, Henry Wesselman, Robert J. Blumenschine, Leslea J. Hlusko, Jackson K. Njau, Michael T. Black, Mesfin Asnake, and Tim D. White. "Taphonomic, Avian, and Small-Vertebrate Indicators of *Ardipithecus ramidus* Habitat." *Science* 326, no. 5949 (2009): 66.

Lovejoy, C. Owen. "A Biomechanical Review of the Locomotor Diversity of Early Hominids." In *Early Hominids of Africa,* edited by C.J. Jolly, 403–29. New York: St. Martin's Press, 1978.

———. "Evolution of Human Walking." *Scientific American,* November 1988.

———. "The Natural History of Human Gait and Posture: Part 1. Spine and Pelvis." *Gait & Posture* 21 (2005): 95.

———. "The Natural History of Human Gait and Posture: Part 2. Hip and thigh." *Gait & Posture* 21 (2005): 113.

———. "The Natural History of Human Gait and Posture: Part 3. The Knee." *Gait & Posture* 25 (2007): 325.

———. "The Origin of Man." *Science* 211, no. 4480 (1981): 341.

———. "Reexamining Human Origins in Light of *Ardipithecus ramidus.*" *Science* 326, no. 74 (2009): 74.

Lovejoy, C. Owen, Martin J. Cohn, and Tim D. White. "Morphological Analysis of the Mammalian Postcranium: A Developmental Perspective." *Proceedings of the National Academy of Sciences* 96, no. 23 (1999): 13247.

Lovejoy, C. Owen, Kingsbury G. Heiple, and Albert H. Burstein. "The Gait of *Australopithecus.*" *American Journal of Physical Anthropology* 38, no. 3 (1973): 757.

Lovejoy, C. Owen, and Melanie A. McCollum. "Spinopelvic Pathways to Bipedality: Why No Hominids Ever Relied on a Bent-Hip–Bent-Knee Gait." *Philosophical Transactions of the Royal Society B: Biological Sciences* 365, no. 1556 (2010): 3289.

Lovejoy, C. Owen, Scott W. Simpson, Tim D. White, Berhane Asfaw, and Gen Suwa. "Careful Climbing in the Miocene: The Forelimbs of *Ardipithecus ramidus* and Humans Are Primitive." *Science* 326, no. 5949 (2009): 70.

Lovejoy, C. Owen, Gen Suwa, Scott W. Simpson, Jay H. Matternes, and Tim D. White. "The Great Divides: *Ardipithecus Ramidus* Reveals the Postcrania of Our Last Common Ancestors with African Apes." *Science* 326, no. 5949 (2009): 73.

Lovejoy, C. Owen, Gen Suwa, Linda Spurlock, Berhane Asfaw, and Tim D. White. "The Pelvis and Femur of *Ardipithecus Ramidus*: The Emergence of Upright Walking." *Science* 326, no. 5949 (2009): 71.

Luyken, Reiner. "Der Krieg der Knochenjäger." (War of the Bone Hunters) *Die Zeit* (Hamburg, Ger.), July 25, 2002.

Mallet, James, Nora Besansky, and Matthew W. Hahn. "How Reticulated Are Species?" *BioEssays* 38, no. 2 (2016): 140.

Manners-Smith, T. "A Study of the Cuboid and Os Peroneum in the Primate Foot." *Journal of Anatomy and Physiology* 42 (1908): 397.

Markakis, John. "Anatomy of a Conflict: Afar and Ise Ethiopia." *Review of African Political Economy* 30, no. 97 (September 2003): 445.

———. *Ethiopia: The Last Two Frontiers*. Suffolk, UK: James Currey, 2011.

Mayr, Ernst. "Reflections on Human Paleontology." In *A History of American Physical Anthropology, 1930–1980*, edited by Frank Spencer, 231–37. New York: Academic Press, 1982.

———. "Taxonomic Categories in Fossil Hominids." In *Cold Spring Harbor Symposia on Quantitative Biology* 15, 109–18. Cold Spring Harbor Laboratory Press, 1950.

McBrearty, Sally, and Nina G. Jablonski. "First Fossil Chimpanzee." *Nature* 437, no. 7055 (2005): 105.

McCollum, Melanie A., Burt A. Rosenman, Gen Suwa, Richard S. Meindl, and C. Owen Lovejoy. "The Vertebral Formula of the Last Common Ancestor of African Apes and Humans." *Journal of Experimental Zoology Part B: Molecular and Developmental Evolution* 314, no. 2 (2010): 123.

McGrew, William C. "The Cultured Chimpanzee: Nonsense or Breakthrough?" *Human Ethology Bulletin* 30, no. 1 (2015): 41.

———. "New Theaters of Conflict in the Animal Culture Wars: Recent Findings from Chimpanzees." In *The Mind of the Chimpanzee*, edited by Elzabeth Lonsdorf, Stephen R. Ross, and Tetsuro Matsuzawa, 168–75. Chicago, IL: University of Chicago Press, 2010.

———. "In Search of the Last Common Ancestor: New Findings on Wild Chimpanzees." *Philosophical Transactions of the Royal Society B: Biological Sciences* 365, no. 1556 (2010): 3267.

McKie, Robin. "Bone Idol: He's Indiana Jones in Armani, He Has Bust-ups on TV, and the Gettys Hate Him. Do You Still Think Fossil-Hunting Is Dull?" *The Observer* (London), March 16, 1997.

Meyer, Marc R., and Scott A. Williams. "Earliest Axial Fossils from the Genus Australopithecus." *Journal of Human Evolution* 132 (2019): 189.

Ministry of Information and Culture of Ethiopia. "Amended Directives for Archeology and Anthropology Study and Research." July 2000. In author's possession.

Mohammed, Abdul, and Alex de Waal. "Meles Zenawi and Ethiopia's Grand Experiment." *New York Times,* August 22, 2012.

Montagu, Ashley. *Edward Tyson, MD, FRS, 1650–1708 and the Rise of Human and Comparative Anatomy in England: A Study in the History of Science*. Philadelphia: American Philosophical Society, 1943.

Moorjani, Priya, Carlos Eduardo G. Amorim, Peter F. Arndt, and Molly Przeworski. "Variation in the Molecular Clock of Primates." *Proceedings of the National Academy of Sciences* 113, no. 38 (2016): 10607.

Morell, Virginia. *Ancestral Passions: The Leakey Family and the Quest for Humankind's Beginnings.* New York: Simon and Schuster, 1995.

Morgan, Michèle E., Kristi L. Lewton, Jay Kelley, Erik Otárola-Castillo, John C. Barry, Lawrence J. Flynn, and David Pilbeam. "A Partial Hominoid Innominate from the Miocene of Pakistan: Description and Preliminary Analyses." *Proceedings of the National Academy of Sciences* 112, no. 1 (2015): 82.

Morton, Dudley J. "Evolution of the Human Foot." *American Journal of Physical Anthropology* 5, no. 4 (1922): 305.

Munzinger, Werner. "Narrative of a Journey Through the Afar Country." *Journal of the Royal Geographical Society of London* 39 (1869): 188.

Napier, John. "The Antiquity of Human Walking." *Scientific American* 216, no. 4 (1967): 56.

———. "The Evolution of the Hand." *Scientific American* 207, no. 6 (1962): 56.

Napier, John R. "The Prehensile Movements of the Human Hand." *The Journal of Bone and Joint Surgery* 38, no. 4 (1956): 902.

National Human Genome Research Institute. "Why Mouse Matters." Accessed September 25, 2019. https://www.genome.gov/10001345/importance-of-mouse-genome.

National Science Foundation. "NSF Sensational 60." Accessed September 27, 2019. https://www.nsf.gov/about/history/sensational60.pdf.

Nelson, Gareth. "Paleontology and Comparative Biology." 1969 presentation reprinted in *Journal of Biogeography* 31 (2004): 702.

Nesbitt, Lewis Mariano. *Desert and Forest: The Exploration of Abysinnian Danakil*. London: Jonathan Cape, 1934. (This book also was published in 1935 under the title *Hell-Hole of Creation*.)

Oakley, Kenneth. *Man the Tool-Maker.* 5th ed. London: British Museum (Natural History), 1952.

Oliwenstein, Lori. "New Foot Steps into Walking Debate." *Science* 269, no. 5223 (July 28, 1995): 476.

Olson, Everett. "George Gaylord Simpson 1902–1984." In *Biographical Memoirs,* 330–53. Washington, D.C.: National Academy of Sciences, 1991.

O'Malley, Charles D., and J. B. de C. M. Saunders. *Leonardo da Vinci on the Human Body.* New York: Henry Schuman Publishers, 1952.

Osada, Naoki. "Genetic Diversity in Humans and Non-Human Primates and Its Evolutionary Consequences." *Genes and Genetic Systems* 90, no. 3 (2015): 133.

Ottaway, David B. "Addis Ababa Sees Red on Coup Anniversary." *Washington Post,* September 13, 1984.

———. "Fighting Up Sharply in Ethiopia." *Washington Post,* March14, 1977.

Pääbo, Svante. "The Contribution of Ancient Hominin Genomes from Siberia to Our Understanding of Human Evolution." *Herald of the Russian Academy of Sciences* 85, no. 5 (2015): 392.

———. "The Diverse Origins of the Human Gene Pool." *Nature Reviews Genetics* 16, no. 6 (2015): 313.

Parrington, John. *The Deeper Genome.* Oxford: Oxford University Press, 2017.

Patterson, Nick, Daniel J. Richter, Sante Gnerre, Eric S. Lander, and David Reich. "Genetic Evidence for Complex Speciation of Humans and Chimpanzees." *Nature* 441, no. 7097 (2006): 1103.

Pennisi, Elizabeth. "Rocking the Cradle of Humanity." *Science* 319, no. 5867 (2008): 1182.

Negarit Gazeta of the People's Democratic Republic of Ethiopia. "Proclamation 36/1989. A Proclamation to Provide for the Study and Protection of Antiquities." October 7, 1989. In author's possession.

Persaud, T. V. N. *A History of Anatomy in the Post-Vesalian Era.* Springfield, Illinois: Charles C. Thomas Publisher, 1997.

Petit, Charles. "Berkeley Institute in Battle Over Fossils." *The San Francisco Chronicle,* April 26, 1995.

———. "One Step at a Time: Paleontologists Are Exulting over New Fossils That Date Back to the Time the Earliest Human Ancestors Stood Upright." *The San Francisco Chronicle,* October 22, 1995.

Pickford, Martin. "Marketing Palaeoanthropology: The Rise of Yellow Science." In *Le Patrimoine Paléontologique Des Trésors du Fond des Temps.* Bucharest: Institut National de Geologiie et Geoecologie Marines (2010): 215.

Pigott, Thomas R. "Bone and Antler Artifacts from the Libben Site, Ottawa Co., Ohio." *Ohio Archaeologist,* 61, no. 4 (Fall 2011): 55.

Pilbeam, David. "The Anthropoid Postcranial Axial Skeleton: Comments on Development, Variation, and Evolution." *Journal of Experimental Zoology* 302B (2004): 241.

———. "Genetic and Morphological Records of the Hominoidea and Hominid Origins: A Synthesis." *Molecular Phylogenetics and Evolution* 5, no. 1 (1996): 155.

Pilbeam, David R., and Daniel E. Lieberman. "Reconstructing the Last Common Ancestor of Chimpanzees and Humans." In *Chimpanzees and Human Evolution*, edited by Martin N. Muller, Richard W. Wrangham, and David R. Pilbeam, 22–141. Cambridge, MA: Belknap Press, 2017.

Planck, Max. *Scientific Autobiography, and Other Papers*. New York: Philosophical Library, 1949.

Pollard, Katherine. S. "Decoding Human Accelerated Regions." *The Scientist*, August 1, 2016.

Porter, Ray, ed., *The Cambridge Illustrated History of Medicine*. Cambridge: Cambridge University Press, 1996.

Povinelli, Daniel John. "Behind the Ape's Appearance: Escaping Anthropocentrism in the Study of Other Minds." *Daedalus* 133, no. 1 (2004): 29.

Powers, Charles T. "Digging for Old Stones and Bones: Team Resumes Quest for Man's Origins in Ethiopia." *Los Angeles Times*, October 14, 1981.

Prang, Thomas Cody. "The African Ape-like Foot of *Ardipithecus ramidus* and Its Implications for the Origin of Bipedalism." *eLife* 8 (2019): e44433.

Preuss, Todd M. "The Human Brain: Evolution and Distinctive Features." In *On Human Nature*, 125–49. London: Elsevier/Academic Press, 2017.

Prufer, Olaf H. "How to Construct a Model: A Personal Memoir." In *Ohio Hopewell Community Organization*, edited by William S. Dancey, Paul J. Pacheco, 105–28. Kent, Ohio: Kent State University Press, 1997.

Quade, Jay, Naomi E. Levin, Scott W. Simpson, Robert Butler, William C. McIntosh, Sileshi Semaw, Lynnette Kleinsasser, Guillaume Dupont-Nivet, Paul Renne, and Nelia Dunbar. "The Geology of Gona, Afar, Ethiopia." In *The Geology of Early Humans in the Horn of Africa*, edited by Jay Quade and Jonathan G. Wynn, 1–31. Boulder: The Geological Society of America, 2008.

Radford, Tim. "Earliest Human Ancestor Discovered." *Guardian*, July 12, 2001.

Reader, John. *Missing Links: In Search of Human Origins*. Oxford: Oxford University Press, 2011.

Reich, David. *Who Ae Are and How We Got Here: Ancient DNA and the New Science of the Human Past*. New York: Pantheon Books, 2018.

Reid, Gordon McGregor. "Carolus Linnaeus (1707–1778): His Life,

Philosophy and Science and Its Relationship to Modern Biology and Medicine." *Taxon* 58, no. 1 (February 2009): 18.

Renne, Paul. "Institute of Human Origins Breakup," letter to editor. *Science* 265, no. 5173 (August 5, 1994): 721.

Renne, Paul, Robert Walter, Kenneth Verosub, Monica Sweitzer, and James Aronson. "New Data from Hadar (Ethiopia) Support Orbitally Tuned Time Scale to 3.3 Ma." *Geophysical Research Letters* 20, no. 11 (1993): 1067.

Renne, Paul R., Giday WoldeGabriel, William K. Hart, Grant Heiken, and Tim D. White. "Chronostratigraphy of the Miocene–Pliocene Sagantole Formation, Middle Awash Valley, Afar Rift, Ethiopia." *Geological Society of America Bulletin* 111, no. 6 (1999): 869.

Reno, Philip L., Richard S. Meindl, Melanie A. McCollum, and C. Owen Lovejoy. "Sexual Dimorphism in *Australopithecus afarensis* Was Similar to That of Modern Humans." *Proceedings of the National Academy of Sciences* 100, no. 16 (2003): 9404.

Rensberger, Boyce. "The Face of Evolution." *New York Times*, March 3, 1974.

———. "New-Found Species Challenges Views on Evolution of Humans." *New York Times*, January 19, 1979.

———. "Rival Anthropologists Divide on Pre-Human Find." *New York Times*, February 18, 1979.

Reznick, David, Michael J. Bryant, and Farrah Bashey. "R- and K-Selection Revisited: The Role of Population Regulation in Life-History Evolution." *Ecology* 83, no. 6 (2002): 1509.

Richmond, Brian G., and David S. Strait. "Evidence That Humans Evolved from a Knuckle-Walking Ancestor." *Nature* 404, no. 6776 (2000): 382.

Ruvolo, Maryellen, Deborah Pan, Sarah Zehr, Tony Goldberg, Todd R. Disotell, and Miranda Von Dornum. "Gene Trees and Hominoid Phylogeny." *Proceedings of the National Academy of Sciences* 91, no. 19 (1994): 8900.

Salopek, Paul. "Enigmatic War Plagues the Cradle of Humanity." *Chicago Tribune*, October 3, 1999.

Salzberg, Steven L. "Horizontal Gene Transfer Is Not a Hallmark of the Human Genome." *Genome Biology* 18, no. 1 (2017): 85.

Sanders, Robert. "160,000-Year-Old Fossilized Skulls Uncovered in Ethiopia Are Oldest Anatomically Modern Humans." UC Berkeley press release, June 11, 2003. https://www.berkeley.edu/news/media/releases/2003/06/11_idaltu.shtml.

Sarich, Vincent, and Frank Miele. *Race: The Reality of Human Differences*. Boulder, CO: Westview Press, 2004.

Sarich, Vincent M., and Allan C. Wilson. "Immunological Time Scale for Hominid Evolution." *Science* 158, no. 3805 (1967): 1200.

Sawyer, Kathy. "New Roots for Family Tree: Oldest Bipedal Human Ancestor Found." *Washington Post,* August 17, 1995.

Sayers, Ken, and C. Owen Lovejoy. "The Chimpanzee Has No Clothes." *Current Anthropology* 49, no. 1 (February 2008): 87.

Scally, Aylwyn, Julien Y. Dutheil, LaDeana W. Hillier, Gregory E. Jordan, Ian Goodhead, Javier Herrero, Asger Hobolth et al. "Insights into Hominid Evolution from the Gorilla Genome Sequence." *Nature* 483 no. 7388 (2012): 169.

Scerri, Eleanor M. L., Mark G. Thomas, Andrea Manica, Philipp Gunz, Jay T. Stock, Chris Stringer, Matt Grove et al. "Did Our Species Evolve in Subdivided Populations Across Africa, and Why Does It Matter?" *Trends in Ecology and Evolution* 33, no. 8 (August 2018): 582.

Schott, G. D. "The Extent of Man from Vitruvius to Marfan." *Lancet* 340 (1992): 1518.

Schultz, Adolph H. "Die Körperproportionen der erwachsenen catarrhinen Primaten, mit spezieller Berücksichtigung der Menschenaffen." *Anthropologischer Anzeiger* 10, 2/3 (1933): 154.

———. "The Physical Distinctions of Man." *Proceedings of the American Philosophical Society* 94, no. 5 (1950): 428.

Schultz, Adolph H. *The Life of Primates.* New York: Universe Books, 1969.

———. "The Specializations of Man and His Place Among the Catarrhine Primates." In *Cold Spring Harbor Symposia on Quantitative Biology* 15, 37–53. Cold Spring Harbor Laboratory Press, 1950.

Schwartz, Jeffrey H., and Ian Tattersall. *The Human Fossil Record, Volume Two: Craniodental Morphology of Genus Homo (Africa and Asia).* Hoboken, NJ: John Wiley and Sons, 2003.

Semaw, Sileshi, P. Renne, J. W. K. Harris, C. S. Feibel, R. L. Bernor, N. Fesseha, and K. Mowbray. "2.5-Million-Year-Old Stone Tools from Gona, Ethiopia." *Nature* 385, no. 6614 (1997): 333.

Shreeve, Jamie. "The Birth of Bipedalism." *National Geographic,* July 2010.

———. "Evolutionary Road." *National Geographic,* July 2010.

———. "Oldest Skeleton of Human Ancestor Found." *National Geographic,* October 1, 2009. https://www.nationalgeographic.com/science/2009/10/oldest-skeleton-human-ancestor-found-ardipithecus.

———. "Sunset on the Savanna." *Discover* (July 1996): 116.

Seidler, Horst. "Fossil Hunters in Dispute over Ethiopian Sites" (letter to editor). *Nature* 411, no. 6833 (2001): 15.

Seidler, Horst, Wolfram Bernhard, Maria Teschler-Nicola, Werner Platzer, Dieter Zur Nedden, Rainer Henn, Andreas Oberhauser, and Thorstein Sjovold. "Some Anthropological Aspects of the Prehistoric Tyrolean Ice Man." *Science* 258, no. 5081 (1992): 455.

Senut, Brigitte, Martin Pickford, Dominique Gommery, Pierre Mein, Kiptalam Cheboi, and Yves Coppens. "First Hominid from the Miocene

(Lukeino Formation, Kenya)." *Comptes Rendus de l'Académie des Sciences*, Series IIA-Earth and Planetary Science 332, no. 2 (2001): 137.

Sigmon, Becky A. "Physical Anthropology in Socialist Europe." *American Scientist* 81, no. 2 (1993): 130.

Simpson, George Gaylord. "The Nature and Origin of Supraspecific Taxa." *Cold Spring Harbor Symposia on Quantitative Biology* 24 (1959): 255.

———. "Some Principles of Historical Biology Bearing on Human Origins." In *Cold Spring Harbor Symposia on Quantitative Biology* 15, 55–66. Cold Spring Harbor Laboratory Press, 1950.

———. *Tempo and Mode in Evolution*. New York: Columbia University Press, 1944.

Simpson, Scott W., Naomi E. Levin, Jay Quade, Michael J. Rogers, and Sileshi Semaw. "*Ardipithecus ramidus postcrania* from the Gona Project Area, Afar Regional State, Ethiopia." *Journal of Human Evolution* 129 (2019): 1.

Sobotta, Johannes. *Atlas and Textbook of Human Anatomy*. Edited with additions by J. Playfair McMurrich. Philadelphia: W.B. Saunders Company, 1909.

Standring, Susan. "A Brief History of Topographical Anatomy." *Journal of Anatomy* 229, no. 1 (2016): 32.

Stanford, Craig. *The New Chimpanzee: A Twenty-First-Century Portrait of Our Closest Kin*. Cambridge, MA: Harvard University Press, 2018.

Stanford, Craig B. "Chimpanzees and the Behavior of *Ardipithecus ramidus*." *Annual Review of Anthropology* 41 (2012): 139.

Stanley, Henry Morton. *Through the Dark Continent*. New York: Harper and Brothers Publishers, 1878.

Stern, Jack T., Jr. "Climbing to the Top: A Personal Memoir of *Australopithecus afarensis*." *Evolutionary Anthropology* 9, no. 3 (2000): 113.

Stern, Jack T., and Randall L. Susman. "The Locomotor Anatomy of *Australopithecus afarensis*." *American Journal of Physical Anthropology* 60, no. 3 (1983): 279.

Stewart, T. Dale. "Adolph Hans Schultz., 1891–1976." In *Biographical Memoirs*, 325–49. Washington, D.C.: National Academy of Sciences, 1983.

Straus, William L. "The Riddle of Man's Ancestry." *The Quarterly Review of Biology* 24, no. 3 (1949): 200.

Stringer, Christopher, and Robin McKie. *African Exodus: The Origins of Modern Humanity*. New York: Henry Holt and Company, 1996.

Stringer, Chris. "The Origin and Evolution of *Homo sapiens*." *Philosophical Transactions of the Royal Society B: Biological Sciences* 371, no. 1698 (2016): 20150237.

Suh, H. Anna, ed. *Leonardo's Notebooks: Writing and Art of the Great Master.* New York: Black Dog and Leventhal Publishers, 2005.

Susman, Randall L., Jack T. Stern Jr, and William L. Jungers. "Arboreality and Bipedality in the Hadar Hominids." *Folia Primatologica* 43, no. 2–3 (1984): 113.

Suwa, Gen. "The Paleoanthropological Inventory of Ethiopia and the Discovery of Konso-Gardula, the Earliest Acheulean." *Nilo Ethiopian Studies Newsletter* (1993): 12.

Suwa, Gen, Berhane Asfaw, Reiko T. Kono, Daisuke Kubo, C. Owen Lovejoy, and Tim D. White. "The *Ardipithecus ramidus* Skull and Its Implications for Hominid Origins." *Science* 326, no. 5949 (2009): 68.

Suwa, Gen, Yonas Beyene, and Berhane Asfaw. "Konso-Gardula Research Project." *The University Museum The University of Tokyo Bulletin*, no. 48 (2015): 1.

Suwa, Gen, Reiko T. Kono, Shigehiro Katoh, Berhane Asfaw, and Yonas Beyene. "A New Species of Great Ape from the Late Miocene Epoch in Ethiopia." *Nature* 448, no. 7156 (2007): 921.

Suwa, Gen, Reiko T. Kono, Scott W. Simpson, Berhane Asfaw, C. Owen Lovejoy, and Tim D. White. "Paleobiological Implications of the *Ardipithecus ramidus* dentition." *Science* 326, no. 5949 (2009): 69.

Swisher, Carl C., Garniss H. Curtis, and Roger Lewin. *Java Man: How Two Geologists Changed Our Understanding of Human Evolution.* Chicago: University of Chicago Press, 2000.

Taieb, Maurice. *Sur la Terre des Premiers Hommes*. Paris: Robert Laffont, 1985.

Tattersall, Ian. "Once We Were Not Alone." *Scientific American*, January 2000.

———. *The Strange Case of the Rickety Cossack: and Other Cautionary Tales from Human Evolution*. New York: St. Martin's Press, 2015.

Taylor, Jeremy. *Not a Chimp: The Hunt to Find the Genes That Make Us Human*. Oxford: Oxford University Press, 2009.

Todd, T. Wingate. "The Physician as Anthropologist." *Science* 83, no. 2164 (1936): 588.

Traufetter, Gerald. "Kabale um die Knochen." *Der Spiegel* (Hamburg, Ger.), August 6, 2001.

Tubiana, Raoul, Jean-Michel Thomine, and Evelyn Mackin. *Examination of the Hand and Wrist*. London: Martin Dunitz, 1998.

Tuttle, Russell H. "Footprint Clues in Hominid Evolution and Forensics: Lessons and Limitations." *Ichnos* 15, no. 3–4 (2008): 158.

Tyson, Edward. *Orang-outang, sive Homo Sylvestris: or, The Anatomy of a Pygmie Compared with that of a Monkey, an Ape, and a Man [. . .]*. London: Thomas Bennet and Daniel Brown, 1699.

Ungar, Peter S., and Matt Sponheimer. "The Diets of Early Hominins." *Science* 334, no. 6053 (2011): 190.

Ungar, Peter S., John Sorrentino, and Jerome C. Rose. "Evolution of Human Teeth and Jaws: Implications for Dentistry and Orthodontics." *Evolutionary Anthropology* 21, no. 3 (2012): 94.

Villmoare, Brian, William H. Kimbel, Chalachew Seyoum, Christopher J. Campisano, Erin N. DiMaggio, John Rowan, David R. Braun, J. Ramón Arrowsmith, and Kaye E. Reed. "Early Homo at 2.8 Ma from Ledi-Geraru, Afar, Ethiopia." *Science* 347, no. 6228 (2015): 1352.

Vrba, Elisabeth S. "The Pulse That Produced Us." *Natural History* 102, no. 5 (1993): 47.

Walker, Alan, and Pat Shipman. *The Ape in the Tree: An Intellectual and Natural History of Proconsul*. Cambridge, MA: Belknap Press, 2005.

Walker, Alan, and Chris Stringer, eds. *The First Four Million Years of Human Evolution*. Special Issue of the *Philosophical Transactions of the Royal Society B: Biological Sciences* 365, no. 1556 (2010): 3263.

Wallace, A. R. *Darwinism: An Exposition of the Theory of Natural Selection with Some of its Applications*. London: Macmillan and Company, 1889.

Walter, Robert C. "Age of Lucy and the First Family: Single-Crystal 40Ar/39Ar Dating of the Denen Dora and Lower Kada Hadar Members of the Hadar Formation, Ethiopia." *Geology* 22, no. 1 (1994): 6.

Ward, Carol V., William H. Kimbel, and Donald C. Johanson. "Complete Fourth Metatarsal and Arches in the Foot of *Australopithecus afarensis*." *Science* 331, no. 6018 (2011): 750.

Ward, Carol V., Thierra K. Nalley, Fred Spoor, Paul Tafforeau, and Zeresenay Alemseged. "Thoracic Vertebral Count and Thoracolumbar Transition in *Australopithecus afarensis*." *Proceedings of the National Academy of Sciences* 114, no. 23 (2017): 6000.

Ward, Steve, Barbara Brown, Andrew Hill, Jay Kelley, and Will Downs. "Equatorius: A New Hominoid Genus from the Middle Miocene of Kenya." *Science* 285, no. 5432 (1999): 1382.

Washburn, Sherwood L. "Behaviour and the Origin of Man." *Proceedings of the Royal Anthropological Institute of Great Britain and Ireland* (1967): 21.

———. "The Evolution Game." *Journal of Human Evolution* 2, no. 6 (1973): 557.

———. "Evolution of a Teacher." *Annual Review of Anthropology* 12, no. 1 (1983): 1.

———. "Fifty Years of Studies on Human Evolution." *Bulletin of the Atomic Scientists* 38, no. 5 (1982): 37.

———. "The New Physical Anthropology." *Transactions of the New York Academy of Sciences*, 2nd series, 13, no. 7 (1951): 298.

Washburn, Sherwood L., and Ruth Moore. *Ape Into Man: A Study of Human Evolution*. Boston: Little, Brown and Company, 1974.

Weaver, William. "Flying First Class." *Architectural Digest*, March 1995.

Weber, Gerhard W. "Virtual Anthropology (VA): A Call for Glasnost in Paleoanthropology." *The Anatomical Record* 265, no. 4 (2001): 193.

Weiner, Tim. *Legacy of Ashes: The History of the CIA*. New York: Anchor Books, 2007.

Wendorf, Fred. Review of "Impure Science. Fraud, Compromise, and Political Influence in Scientific Research," by Robert Bell. *American Journal of Physical Anthropology* 92, no. 3 (1993): 401.

West, Kevin. "Pacific Heights." *W*. January 1, 2007.

White, Tim. "At Large in the Mountains." In *Curious Minds: How a Child Becomes a Scientist*, edited by John Brockman, 203–10. New York: Vintage Books, 2005.

———. "Early Hominids—Diversity or Distortion?" *Science* 299, no. 5615 (2003): 1994.

———. "Ethiopian Paleotourism: Integrating Science and Development." (Address to the Ethiopian National Symposium on Eco-tourism and Paleo-tourism in Ethiopia, with special reference to the Afar Region, 2004).

———. "On the Environment of Aramis" (Reply to Gani and Gani). *Current Anthropology* 57, no. 2 (2016): 220.

———. "Paleoanthropology: Five's a Crowd in Our Family Tree." *Current Biology* 23, no. 3 (2013): R112.

———. "Why Combining Science and Showmanship Risks the Future of Research." *Guardian*, November 26, 2015. https://www.theguardian.com/science/blog/2015/nov/26/why-combining-science-and-showmanship-risks-the-future-of-research.

White, Tim D. "African Omnivores: Global Climatic Change and Plio-Pleistocene Hominids and Suids." In *Paleoclimate and Evolution, with Emphasis on Human Origins*. Edited by Elisabeth S. Vrba et al., 369–78. New Haven, CT: Yale University Press, 1995.

———. "Cut Marks on the Bodo Cranium: A Case of Prehistoric Defleshing." *American Journal of Physical Anthropology* 69 (1986): 503.

———. "Feud Over Old Bones." Letter to the editor. *The Times* (London), April 2, 1995.

———. "Fossils for All." Letter to the editor. *Scientific American*, January 2010.

———. "Human Evolution: How Has Darwin Done?" In *Evolution Since Darwin: The First 150 Years*, edited by Michael A. Bell et al., 519–60. Sunderland, MA: Sinauer Associates, 2010.

———. "Human Origins and Evolution: Spring Harbor, Deja Vu." *Cold Spring Harbor Symposia on Quantitative Biology* 74. (2009): 335.

———. "Ladders, Bushes, Punctuations, and Clades: Hominid Paleobiology in the Late Twentieth Century." In *The Paleobiological Revolution: Essays on the Growth of Modern Paleontology*, editors David Sepkoski and Michael Ruse, 122–48. Chicago: University of Chicago Press, 2009.

———. "Monkey Business." Review of *The Monkey in the Mirror. Essays on the Science of What Makes us Human*, by Ian Tattersall. *BioEssays* 24, no. 8 (2002): 767.

———. "Once We Were Cannibals." *Scientific American*, August 2001.

———. "Paleontological Case Study: 'Ardi,' the *Ardipithecus ramidus* Skeleton from Ethiopia." In *Human Osteology*, 3rd ed., by Tim D. White, Michael T. Black, and Pieter A. Folkens, 543–58. Burlington, MA: Academic Press, 2012.

———. *Prehistoric Cannibalism at Mancos 5MTUMR-2346*. Princeton, NJ: Princeton University Press, 1992.

———. "A View on the Science: Physical Anthropology at the Millennium." *American Journal of Physical Anthropology* 113, no. 3 (2000): 287.

White, Tim D., Stanley H. Ambrose, Gen Suwa, Denise F. Su, David DeGusta, Raymond L. Bernor, Jean-Renaud Boisserie et al. "Macrovertebrate Paleontology and the Pliocene Habitat of *Ardipithecus ramidus*." *Science* 326, no. 5949 (2009): 67.

White, Tim D., Berhane Asfaw, Yonas Beyene, Yohannes Haile-Selassie, C. Owen Lovejoy, Gen Suwa, and Giday WoldeGabriel. "*Ardipithecus ramidus* and the Paleobiology of Early Hominids." *Science* 326, no. 5949 (2009): 64.

White, Tim D., Berhane Asfaw, David DeGusta, Henry Gilbert, Gary D. Richards, Gen Suwa, and F. Clark Howell. "Pleistocene *Homo sapiens* from Middle Awash, Ethiopia." *Nature* 423, no. 6941 (2003): 742.

White, Tim D., Michael T. Black, and Pieter A. Folkens. *Human Osteology*. 3rd ed. Burlington, MA: Academic Press, 2012.

White, Tim D., C. Owen Lovejoy, Berhane Asfaw, Joshua P. Carlson, and Gen Suwa. "Neither Chimpanzee Nor Human, *Ardipithecus* Reveals the Surprising Ancestry of Both." *Proceedings of the National Academy of Science* 112, no. 16 (2015): 4877.

White, Tim D., and Gen Suwa. "Hominid Footprints at Laetoli: Facts and Interpretations." *American Journal of Physical Anthropology* 72, no. 4 (1987): 485.

White, Tim D., Gen Suwa, and Berhane Asfaw. "*Australopithecus ramidus*, A New Species of Early Hominid from Aramis, Ethiopia." *Nature* 371, no. 6495 (1994): 306.

White, Tim D, Gen Suwa, and Berhane Asfaw. "Corrigendum: *Australopithecus ramidus*, A New Species of Early Homind from Aramis, Ethiopia." *Nature* 375, no. 6526 (1995): 88.

White, Tim D., Giday WoldeGabriel, Berhane Asfaw, Stan Ambrose, Yonas

Beyene, Raymond L. Bernor, Jean-Renaud Boisserie et al. "Asa Issie, Aramis and the Origin of *Australopithecus*." *Nature* 440, no. 7086 (2006): 883.

Whiten, Andrew, William C. McGrew, Leslie C. Aiello, Christophe Boesch, Robert Boyd, Richard W. Byrne, Robin I. M. Dunbar et al. "Studying Extant Species to Model Our Past." Letter to the editor. *Science* 327, no. 5964 (2010): 410.

Wiebel, Jacob. "Revolutionary Terror Campaigns in Addis Ababa, 1976–1978." Thesis for Doctor of Philosophy in History, University of Oxford, 2014.

Wilford, John Noble. "Ethiopia Bones Called Oldest Ancestor of Man." *New York Times*, June 11, 1982.

———. "New Fossils Reveal the First of Man's Walking Ancestors." *New York Times*, August 17, 1995.

———. "Scientists Challenge 'Breakthrough' on Fossil Skeleton." *New York Times*, May 27, 2010.

———. "Tempers Flare as Fossil Theft and Claim-Jumping Are Charged." *New York Times*, April 25, 1995.

Williams, Martin. *Nile Waters, Saharan Sands: Adventures of a Geomorphologist at Large*. Switzerland: Springer International Publishing, 2016.

WoldeGabriel, Giday, Stanley H. Ambrose, Doris Barboni, Raymonde Bonnefille, Laurent Bremond, Brian Currie, David DeGusta et al. "The Geological, Isotopic, Botanical, Invertebrate, and Lower Vertebrate Surroundings of *Ardipithecus ramidus*." *Science* 326, no. 5949 (2009): 65.

WoldeGabriel, Giday, Grant Heiken, Tim D. White, Berhane Asfaw, William K. Hart, and Paul R. Renne. "Volcanism, Tectonism, Sedimentation, and the Paleoanthropological Record in the Ethiopian Rift System." In *Volcanic Hazards and Disasters in Human Antiquity*, edited by Floyd W. McCoy; Grant Heiken, 83–99. Special Paper 345. Boulder, CO: Geological Society of America, 2000.

WoldeGabriel, Giday, Tim D. White, Gen Suwa, Paul Renne, Jean de Heinzelin, William K. Hart, and Grant Heiken. "Ecological and Temporal Placement of Early Pliocene Hominids at Aramis, Ethiopia." *Nature* 371, no. 6495 (1994): 330.

WoldeGabriel, Giday, Tim D. White, Gen Suwa, Sileshi Semaw, Yonas Beyene, Berhane Asfaw, and Robert Walter. "Kesem-Kebena: A Newly Discovered Paleoanthropological Research Area in Ethiopia." *Journal of Field Archaeology* 19, no. 4 (1992): 471.

Wolde-Georgis, Tsegay, Demlew Aweke, and Yibrah Hagos. "The Case of Ethiopia: Reducing the Impacts of Environmental Emergencies through Early Warning and Preparedness: The Case of the 1997–98 El Niño."

http://archive.unu.edu/env/govern/ElNIno/CountryReports/pdf/ethiopia.pdf.

Wood, Bernard. "The Oldest Hominid Yet." *Nature* 371, no. 6495 (1994): 280.

Wood, Bernard, and Terry Harrison. "The Evolutionary Context of the First Hominins." *Nature* 470, no. 7334 (2011): 347.

Wulff, Henrik R. "The Language of Medicine." *Journal of the Royal Society of Medicine* 97, no. 4 (2004): 187.

Yasin, Yasin Mohammed. "Political History of the Afar in Ethiopia and Eritrea." *Africa Spectrum* (2008): 39.

Yellen, John, Alison Brooks, David Helgren, Martha Tappen, Stanley Ambrose, Raymonde Bonnefille, James Feathers et al. "The Archaeology of Aduma Middle Stone Age Sites in the Awash Valley, Ethiopia." *PaleoAnthropology* 10, no. 25 (2005): e100.

Yitbarek, Salaam. "A Problem of Social Capital and Cultural Norms?" International Conference on African Development. Paper 100 (2007).

Zane, Damian. "Ethiopia's Pride in Herto Finds." BBC News, June 11, 2003. http://news.bbc.co.uk/2/hi/science/nature/2978800.stm.

Zena, Ashenafi Girma. "Archaeology, Politics and Nationalism in Nineteenth- and Early-Twentieth-Century Ethiopia: The Use of Archaeology to Consolidate Monarchical Power." *Azania: Archaeological Research in Africa* 53, no. 3 (2018): 398.

Zimmer, Carl. "The Human Family Tree Bristles with New Branches." *New York Times*, May 27, 2015.

Zuckerkandl, Emile, and Linus Pauling. "Evolutionary Divergence and Convergence in Proteins." In *Evolving Genes and Proteins*, edited by V. Bryson and H.J. Vogel, 97–166. New York: Academic Press, 1965.

MANUSCRIPTS

Asfaw, Berhane. Letter to Higher Education Commissioner, Addis Ababa, Ethiopia. October 11, 1982. Kalb papers in author's possession.

Clark, J. Desmond. Letter to John Yellen, August 31, 1978. Kalb papers.

———. Letter to John Yellen, December 29, 1981. Kalb papers.

Clark, J. Desmond, and F. Clark Howell. "Construction of a Storage/Laboratory Unit in Addis Ababa to house the collections made by United States based and financed expeditions in Ethiopia" (proposal to U.S. National Science Foundation). November 30, 1978. (Wendorf papers, box 18, folder 29).

Graburn, Nelson H. H. Memo, Berkeley Department of Anthropology, July 27, 1981. White papers in author's possession.

Kalb, John E. "Statement Regarding the Rift Valley Research Mission in Ethiopia." May 27, 1977. Kalb lawsuit exhibit.

Korn, David. "Archeological Expedition Stalled." Telegram from American Embassy Addis Ababa to Secretary of State Washington, D.C. September 30, 1982. Kalb papers.
Yellen, John. "Report on the Paris and Addis Ababa Conferences Concerning Future Research in the Ethiopian Rift Valley." National Science Foundation memo (1980). Kalb papers.

NATIONAL SCIENCE FOUNDATION PROPOSALS AND CASE FILES

NSF award 7908342 "Construction of a Storage/Laboratory Unit in Addis Ababa to House the Collections Made By United States Based and Financed Expeditions in Ethiopia."
NSF award SBR 8210897 "Paleoanthropological Research in the Middle Awash Valley, Ethiopia."
NSF award 9318698 "Pliocene Paleontology and Geology of the Middle Awash Valley, Ethiopia."
NSF award SBR 9632389 case file "Mio-Pliocene of the Middle Awash Valley, Ethiopia." (Panel summary and reviews provided by MARP.)
NSF award 9910344 "Field Research in the Middle Awash Valley, Ethiopia." (Panel reviews provided by MARP.)
NSF proposal 0218392 "Revealing Hominid Origins Initiative." (Unfunded proposal; panel summary and reviews provided by MARP.)
NSF award BCS 0321893 "Revealing Human Origins." (Panel reviews provided by MARP.)
NSF proposal 1138062 (Unfunded proposal; panel summary and reviews provided by MARP.)
NSF proposal 1241314 (Unfunded proposal; panel summary and reviews provided by MARP.)
NSF proposal EAR1332712 (Unfunded proposal; panel summary and reviews provided by MARP.)

FIELD NOTES AND RECORDS

MARP video archive 1981–2000. Private Collection.
White, Tim. Field notes and photograph collection. Personal Communications.

FILM AND VIDEO

Cartmill, Matthew, Jeremy DeSilva, William Kimbel, and William Jungers. "Proclaiming *Ardipithecus:* A Revolution in Our Understanding of Human Origins?" Boston University Dialogues in Biological Anthropology 8. Live webcast, December 12, 2013. Part 1 YouTube video 1:00:13 "BU Dialogues, Dec. 12, 2013—*Ardipithecus*." https://

www.youtube.com/watch?v=zwybCokNIJI. Part 2 video 1:27:02 "*Ardipithecus* discussion BU Dialogues Dec 12 2013." https://www.youtube.com/watch?v=OYQwsbhfhOw.

Hoffman, Milton B. *The First Family.* Cleveland, OH: WVIZ-TV, 1980. DVD 60 min.

Johanson, Donald, Peter Jones, Paula S. Apsell and Michael Gunton. *Nova.* "In Search of Human Origins." Parts 1–3. Boston: WGBH Educational Foundation, 1994. DVD.

Lichtman, Flora. "Desktop Diaries: Tim White." Science Friday. August 9, 2013. Video 5:48. https://www.sciencefriday.com/segments/desktop-diaries-tim-white.

Noxon, Nicolas. *The Man Hunters: An MGM Documentary.* Narrated by E.G. Marshall. Los Angeles: Metro-Goldwyn-Mayer, 1970. DVD.

Paul, Rod. *Discovering Ardi.* Produced by Primary Pictures for Discovery Channel. Chatsworth, CA: Image Entertainment, 2010. DVD.

Scholl, Tillmann. *Knochenkrieg* (Bonewar). Germany: Spiegel TV GmbH, 2002. DVD.

Smeltzer, David. *Lucy in Disguise.* Youngstown, OH: Smeltzer Films, 1981. DVD 58 min.

Spurlock, Linda. "Burying the Hatchet, the Body, and the Evidence." Filmed October 17, 2014 at Ursuline College, Cleveland, OH. TEDx video posted January 20, 2016. 17:21 https://www.youtube.com/watch?v=zEVjrZlDhPo.

Townsley, Graham, Jay O. Sanders, Robert Neufeld, WGBH, and National Geographic Studies. *Nova,* "Dawn of Humanity." Arlington, VA: Public Broadcasting Service, 2015. https://www.pbs.org/wgbh/nova/video/dawn-of-humanity.

Von Gutten, Matthias. *Coincidence in Paradise.* New York: First Run Icarus Films, 1991. Videocassette (VHS).

White, Tim, interview by Elizabeth Farnsworth. *MacNeil/Lehrer NewsHour,* PBS. September 22, 1994. 40:26–50:59. American Archive of Public Broadcasting. https://americanarchive.org/catalog/cpb-aacip_507-w08w951k0t.

White, Tim, interview by Harry Kreisler. "On the Trail of Our Human Ancestors: Timothy White." September 18, 2003. Institute of International Studies, the University of California at Berkeley. YouTube video 53:35 https://www.youtube.com/watch?v=z1ycbn0gYBY.

Wood, Bernard. "Darwin in the 21st Century: Nature, Humanity, and God." Lecture at November 1–3, 2009 conference, University of Notre Dame, Indiana. YouTube video posted April 5, 2012. 39:49. https://www.youtube.com/watch?v=7hvxeUbemuw.

Wood, Bernard, interview by Chiara Ceci, pikaia.eu, August 2010. YouTube video January 24, 2011. 5:53. https://www.youtube.com/watch?v=4mbHt_gHKOI.

ARCHIVES, COLLECTIONS, AND COURT FILES

C. Loring Brace Papers: 1954–2009. Bentley Historical Library, University of Michigan.

The Foreign Affairs Oral History Collection, the Association for Diplomatic Studies and Training. Frontline Diplomacy, Manuscript Division, Library of Congress, Washington, D.C.

Paul B. Henze Papers. Collection Number 2005C42. Hoover Institution Archives. Stanford University.

John Ervin Kalb v. the National Science Foundation. Civil Case Number 86-3557. U.S. District Court for the District of Columbia.

Jon Kalb Papers. Private collection (documents provided to author by Kalb). This collection includes a large number of documents obtained from the U.S. National Science Foundation by Kalb under the Freedom of Information Act.

Papers of Ernst Mayr, 1931–1993. Harvard University Archives.

Fred Wendorf Papers. Southern Methodist University Archives.

Tim White papers. Private collection (documents provided to author by White).

ORAL HISTORIES

The Association for Diplomatic Studies and Training (ADST). "Frederick L. Chapin" (interview by Arthur Tienken, 1988).

———. "Ambassador Arthur W. Hummel, Jr." (interview by Dorothy Robins Mowry, 1989).

———. "Ambassador David A. Korn" (interview by Charles Stuart Kennedy, 1990).

———. "Ambassador Tibor Peter Nagy, Jr." (interview by Charles Stuart Kennedy, 2010).

———. "Ambassador Owen W. Roberts" (interview by Charles Stuart Kennedy, 1991).

———. "David Hamilton Shinn" (interview by Charles Stuart Kennedy, 2010).

———. "Ambassador Arthur T. Tienken" (interview by Charles Stewart Kennedy). 1989.

Clark, J. Desmond. "An Archaeologist at Work in African Prehistory and Early Human Studies," an oral history conducted in 2000–2001 by Suzanne Riess, Regional Oral History Office, The Bancroft Library, University of California, Berkeley, 2002.

Isaac, Barbara. Oral history interview by Pamela J. Smith, University of Cambridge Department of Archaeology and Anthropology.

Credits

Page 15: Daniel Huffman

Page 28: Saxon Donnelly, photograph courtesy of UC Berkeley

Page 32: Daniel Huffman

Page 42: Maurice Taieb

Page 48: Kermit Pattison

Page 68: UC Berkeley

Page 100: Kermit Pattison

Page 102: Daniel Huffman

Page 106: Drawings by Luba Dmytryk Gudz originally appeared in *Lucy: The Beginnings of Humankind,* © Donald C. Johanson and Maitland A. Edey. Reprinted with permission of artist.

Page 109: Scott W. Simpson

Page 116: Schultz, "Die Körperproportionen der erwachsenen catarrhinen Primaten, mit spezieller Berücksichtigung der Menschenaffen." Anthropologischer Anzeiger H. 2/3 (1933). Used by permission.

Page 121: Illustrations copyright 2020 Stephen D. Nash/International Union for Conservation of Nature (IUCN) Species Survival Commission (SSC) Primate Specialist Group. Used with permission.

Page 158: Adapted from Fig. 2 of Schultz, "The Physical Distinctions of Man." *Proceedings of the American Philosophical Society* 94, no. 5 (1950). Reprinted with permission.

Page 202: From Fig. 5 of Schultz "Form und Funktion der Primatenhande," pp. 9–30. In B. Rentsch, ed., Handgebrauch und Verständigung bei Affen und Frühmenschen. Bern: Verlag Hans Huber, 1968. Reprinted with permission.

Page 205: Image adapted from Fig. 1 of White et al., "Neither Chimpanzee nor Human." *Proceedings of the National Academy of Sciences* 112, no. 16 (2015). Reprinted with permission.

Page 224: Daniel Huffman

Page 225: Kermit Pattison

Page 232: Illustration © and courtesy of Henry Gilbert, Human Evolution Research Center. Originally from White et al., "*Ardipithecus ramidus* and the Paleobiology of Early Hominids." *Science* 326, no 64 (2009).

Page 241: Daniel Huffman

Page 260: Illustration by Waterhouse Hawkins. From frontpiece of Thomas Henry Huxley, *Evidence as to Man's Place in Nature,* 1863.

Page 279: Image adapted from Fig. 1 White et al., "Neither Chimpanzee nor Human." *Proceedings of the National Academy of Sciences* 112, no. 16 (2015): 4877. Reprinted with permission.

Page 283: Image © and courtesy of Gen Suwa, University of Tokyo Museum. Adapted from Fig. 1 of Lovejoy et al., "Combining Prehension and Propulsion: The Foot of *Ardipithecus ramidus.*" *Science* 326, no. 72 (2009).

Page 293: From Robert Hartmann, *Der Gorilla; Zoologisch-Zootomische Untersuchungen,* 1880. Image courtesy of the Nash Collection of Primates in Art and Illustration, University of Wisconsin.

Page 302: Adapted from Fig. 1 of White et al., "Neither Chimpanzee nor Human." *Proceedings of the National Academy of Sciences* 112, no. 16 (2015). Reprinted with permission.

Page 308: Digital rendering © 2009 Gen Suwa, Adapted from Fig 2 of Suwa et al., "The *Ardipithecus ramidus* Skull and Its Implications for Hominid Origins." *Science* 326, 68 (2009).

Page 309: Image adapted from Fig. 1 of White et al., "Neither Chimpanzee nor Human." *Proceedings of the National Academy of Sciences* 112, no. 16 (2015): 4877. Reprinted with permission.

Page 311: Illustration by William Kimbel and used by permission. Chimpanzee and *afarensis* photographs by William Kimbel. Ardi scan courtesy of Gen Suwa and Tim White.

Page 319: From Fig. 4 of Tom Hogervorst and Evie E. Vereecke. "Evolution of the Human Hip. Part 1: The Osseous Framework." *Journal of Hip Preservation Surgery* 1, no. 2 (2014).

Page 321: Digital rendering © 2009 Owen Lovejoy. From Lovejoy et al., "The Pelvis and Femur of Ardipithecus Ramidus: The Emergence of Upright Walking." *Science* 326, no. 5949 (2009).

Page 329: Images from Figs 21–23 of Johannes Sobotta, *Atlas and Text-Book of Human Anatomy*, 1909.

Page 331: Adapted from Fig. 6 of Adolph Schultz, "The Physical Distinctions of Man." Reprinted with permission of the American Philosophical Society.

Page 351: © 2009 J. H. Matternes

Page 355: © 2009 J. H. Matternes

Page 394: Kermit Pattison

Page 416: Kermit Pattison

Index

Italic page references indicate images.